Mechanisms of Viral Toxicity in Animal Cells

Editor

Luis Carrasco, Ph.D.
Professor of Microbiology
Centro de Biología Molecular
Universidad Autónoma
Madrid, Spain

CRC Press, Inc.
Boca Raton, Florida

Library of Congress Cataloging-in-Publication Data

Mechanisms of viral toxicity in animal cells.

Includes bibliographies and index.
1. Host-virus relationships. 2. Viruses. 3. Cell death. 4. Lysogeny. I. Carrasco, Luis, 1949—
[DNLM: L. Cytotoxicity, Immunologic. 2. Vertebrate Viruses. QW 164 M486]
QR482.M43 1987 591.2'234 87-673
ISBN 0-8493-6742-5

Direct all inquiries to CRC Press, Inc., 2000 Corporate Blvd., N.W., Boca Raton, Florida, 33431.

International Standard Book Number 0-8493-6742-5

Library of Congress Card Number 87-673
Printed in the United States

FOREWORD

The idea that virus development destroys infected cells can be traced back to the discovery that phages lyse bacteria. Animal viruses are able to interfere with their hosts in a number of different ways. Upon infection, they sometimes stimulate cells to divide, inducing malignancy and producing a tumor. In other instances, they kill cells. The harm caused to the host varies from a simple cold to a life-threatening disease, such as poliomyelitis, yellow fever, or AIDS. Some viruses feel more comfortable in a latent state and the host harbors them for years, or even for a lifetime. Thus, herpesviruses once installed in their host, settle into a latent state, producing recurrent infections at more or less regular intervals.

Although we now know the steps of viral replication for a number of viruses reasonably well in molecular terms, we understand little about the mechanisms used by viruses to cause injury to infected cells. For picornaviruses, which are one of the best known group of animal viruses at the molecular level, various problems such as which viral component interferes with host translation, how the cell cytoskeleton and membrane permeability are altered, and so on, still remain unanswered. The new sophisticated methods of manipulation of the genetic information carried by viruses will facilitate further advances in our knowledge of the mechanisms of virus-induced cell toxicity in the forseeable future. Such advances will, in turn, strengthen our ability to combat animal virus diseases.

The present book is intended to give an account of the state of the art on how animal viruses induce cytotoxic effects in cells. The volume starts with a review on the mechanisms used by animal viruses to get into their host. This is a very debatable area of research and of great interest at present. It is followed by reviews on how animal viruses interfere with the synthesis of host nucleic acids and proteins. Finally, the effects of interferon on virus-induced cell toxicity exemplify how interference with cell killing can be achieved in some instances and also how cell killing ensues even though viral progeny formation is blocked. Future research in other areas such as the basic mechanisms of virus development, studies on viral cytotoxic agents, the structure and function of cell membranes, cloning of selected pieces of viral genomes, and so on, will benefit further development of our knowledge of virus-induced cell damage. We hope that this book will encourage more people to direct their efforts to ascertain the detailed molecular events by which animal viruses kill their host cells. I am indebted to all contributors who kindly did their best to review each chapter of this volume in a critical fashion.

THE EDITOR

Luis Carrasco, Ph.D., is a Full Professor of Microbiology, Universidad Autónoma de Madrid. He is also Director of the Department of Microbiology and Chief of the group of the Animal Virus Infection Unit at the Centro de Biología Molecular Consejo Superior de Investigaciones Científicas, Universidad Autónoma de Madrid. (CSIC-UAM).

Professor Carrasco received his M.S. and Ph.D. degrees from Universidad Complutense de Madrid in 1971 and 1974, respectively. After receiving postdoctoral training at the Imperial Cancer Research Fund at London, he joined the Centro de Biología Molecular (CSIC) at the Universidad Autónoma de Madrid. He has been the recipient of several short-term fellowships to work in several institutions in Europe and the U.S.A.

His current major research interests include the molecular mechanism of virus-induced cell killing and the discovery and study of the mode of action of new antiviral agents. He has authored 100 research papers in these areas.

CONTRIBUTORS

Henry R. Bose, Jr., Ph.D.
Professor of Microbiology
Department of Microbiology
University of Texas
Austin, Texas

Dennis T. Brown, Ph.D.
Professor of Microbiology and Director
Cell Research Institute
University of Texas
Austin, Texas

Jose Luis Castrillo, Ph.D.
Postdoctoral Fellow
Department of Microbiology
Centro de Biología Molecular
Universidad Autónoma
Madrid, Spain

Kevin M. Coombs, Ph.D.
Postdoctoral Research Associate
Department of Microbiology
Cell Research Institute
University of Texas
Austin, Texas

Mariano Esteban, Ph.D.
Professor
Department of Biochemistry
Downstate Medical Center
Brooklyn, New York

Carlos Fernández-Tomás, Ph.D.
Full Professor
Centro de Investigacion y de Estudios Avanzados del IPN
Mexico D.F., Mexico

Robert F. Garry, Ph.D.
Assistant Professor
Department of Microbiology & Immunology
Tulane School of Medicine
New Orleans, Louisiana

Alberto Muñoz, Ph.D.
Colaborador Cientifico
Centro de Biología Molecular
Universidad Autónoma
Madrid, Spain

Eduardo Paez, Ph.D.
Clinical Instructor
Department of Biochemistry
Downstate Medical Center
Brooklyn, New York

Emin T. Ulug, Ph.D.
Robert A. Welch Postdoctoral Fellow
Department of Microbiology
University of Texas
Austin, Texas

TABLE OF CONTENTS

Chapter 1

MECHANISMS OF VIRUS-INDUCED CELL TOXICITY: A GENERAL OVERVIEW

L. Carrasco

Since the beginning of this century, viruses have been known to cause disease in man and animals. The initial studies in 1898 of Löffler and Frosch on the etiological agent of foot and mouth disease in cattle, and those of Rose and Reed in 1901 on yellow fever in man which showed that these illnesses were produced by filterable organisms, initiated the science of Animal Virology. Soon, viruses were also described as the causative agents of leukemias and solid tumors in birds. Early attempts to cultivate animal viruses in tissue culture date from 1913 when Steinhardt, Israel, and Lambert achieved the multiplication of vaccinia virus in the corneal epithelium of guinea pig. The chorioallantoic membrane of chick embryos was used by a number of investigators to study the growth of several filterable microorganisms. Although viruses were detectable by their destructive effects on these systems, direct examination of viral particles was not achieved until 1939, when the tobacco mosaic virion was visualized by electron microscopy.

Viruses produced clearly detectable cellular and histological lesions in those in vitro systems, leading to the idea that the traumatic lesions caused by virus multiplication in the organism were amenable to study in more simplified situations. Since the middle of this century, animal cell culture has been gradually introduced as a routine technique in many laboratories, aiding the growth and molecular analysis of viruses. Soon it became obvious that intracellular virus growth led to cell toxicity. During the last 40 years, several virus/cell systems have been used to ascertain the molecular events that viruses put in motion to cause harm in their host cells. Although many systems are now available for study and have been the object of intensive research, in my opinion many fundamental questions in the field still remain unanswered. At least two of these unresolved questions that apply to any cytolytic virus are (1) Which viral component is involved in cell death? and (2) What is the molecular mechanism of action of this viral component? The answer to these questions is the current challenge faced by scientists working in this field. An exact knowledge of how viruses kill their host cells is needed if we hope to understand how viruses produce trauma and cause disease in infected animals.

Analyses of the modifications induced in cell morphology and cell metabolism during virus growth present a complicated picture. Two excellent reviews have been written on this subject.[1,2] This complication arises in the first place due to the fact that a given cytolytic virus interferes with an array of biochemical events, including inhibition of cellular macromolecular synthesis, interference with the synthesis of ATP, increase in the synthesis of lipids, modifications in membrane permeability, and disruption of the cell cytoskeleton, etc. At least some of these disturbances may constitute the molecular basis of the changes observed in cell morphology and are depicted in Figure 1. Furthermore, the kind of morphological alterations induced by each virus and the kinetics of interference with these cellular processes differ for each virus/cell system considered. Therefore, it has been suggested that the underlying molecular events responsible for toxicity may be different for each animal virus. This assumption has hampered our knowledge of the mechanisms of virus-induced cell toxicity. Consistent with this idea of diversity is the supposition that each cell process modified by viral infection might be caused by a different virus product. A good example of this situation would be the picornaviruses such as poliovirus and mengovirus, which can interfere with host translation by different mechanisms.[3] Moreover, we could assume that

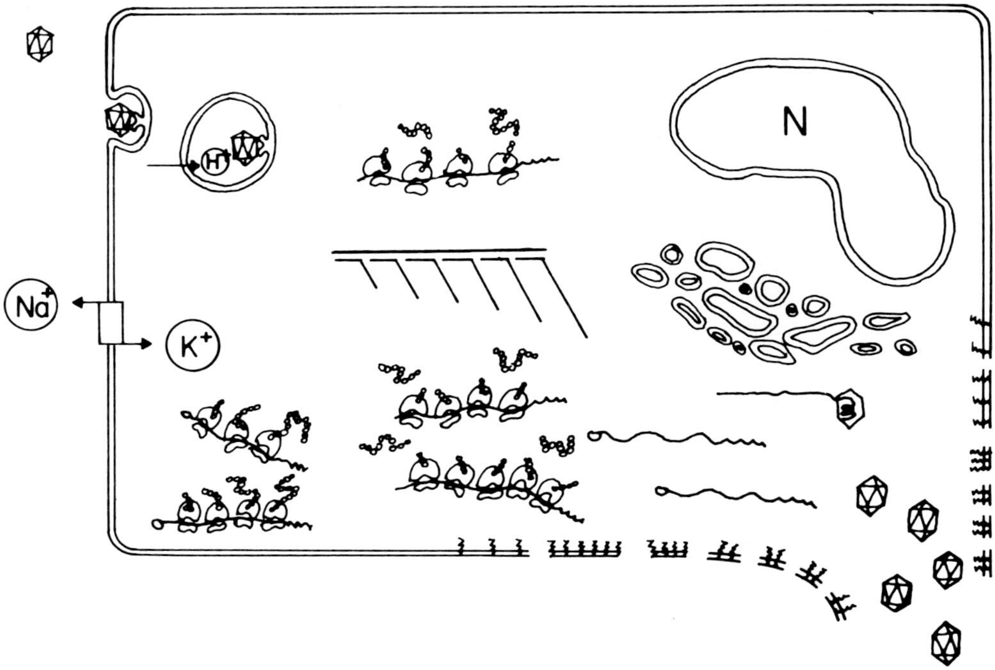

FIGURE 1. Schematic representation of the cellular modifications that take place during picornavirus growth. After entry and decapsidation of the virion particle, the genome is translated, thus giving rise to the polyprotein which is cleaved to form the mature viral proteins. At least one of those proteins forms part of the replicase that will synthesize a number of viral mRNAs. As infection progresses, a number of morphological and biochemical alterations occurs in the infected cell. Thus, cellular transcription and translation become blocked and ATP synthesis declines, while the synthesis of phospholipids increases and the gradients of ions maintained by the membrane disappear. The morphological alterations include the distortion of the nucleus, a great proliferation in the internal membranes, cell rounding, and finally, lysis of the cell membrane.

poliovirus inhibits host protein synthesis by a given viral component still unidentified, but that another viral product will be responsible for other cellular alterations, such as inhibition of transcription, stimulation of lipid synthesis, etc. On the other hand, a more simple explanation could be that most cytolytic viruses use a basic mechanism to kill cells, using the cell membrane as a target.[4,5] This would, in turn, affect a number of cell functions in a pleiotropic manner (including alterations in cell morphology) that will finally lead to cell death.

At least two ideas can be put forward to account for cell death during virus growth. One is that the synthesis and accumulation of the progeny virus lyses and destroys the cell. Another possibility is that a specific virus product is involved in cell killing. This second idea implies that some viruses code for a toxin-like agent involved in cell damage. Therefore, even if progeny virus formation is blocked, cells will die as long as the virus toxic product is made. The concept that viruses possess these killer factors is rather old and is supported by several lines of evidence. Thus, many virion preparations and even some isolated viral components, i.e., the fiber capsid protein from adenoviruses[6] or the glycoprotein from vesicular stomatitis virus,[7] have a cytotoxic effect. In addition, a number of observations on abortive viral infections clearly shows that cell killing occurs when mature virions are not produced. Furthermore, many virus inhibitors do not protect cells from death, though the formation of new virions is greatly diminished.[8] However, conclusive evidence implicating the expression of a single viral component responsible for cell toxicity is scanty.

Regarding the mechanism of action of those putative virus components involved in cell killing, there are some lines of research at present that may guide us to conclusive results in the foreseeable future. Continuing with the idea that viruses code for toxin-like proteins, I have advocated the thesis that viral components located on cell membranes could alter their function, leading to interference with several cell processes and, finally, to lysis and destruction of the cell.[4,5] Some lines of research support this hypothesis, but more evidence to confirm this possibility is still necessary. An attractive aspect of this idea resides in its simplicity; thus, a single virus component would be responsible for interference with host cell metabolism. Support for this hypothesis comes from two independent findings: (1) that viruses damage the host cell membrane during their replication,[5] and (2) that a number of molecules exists in nature (such as ionophores and some toxin proteins) that have the membrane as their single cell target and in this way influence a number of cell functions in parallel.[9,10] The mechanisms of action of the different membrane-active agents differ when they are analyzed in detail, as might also occur with the different virus components involved in membrane modification and cell killing.

A number of laboratories are directing their efforts to ascertain the mechanism used by animal viruses to modify cellular membranes and to identify the viral components responsible for those modifications. The results obtained will help to clarify the relationship between virus-induced membrane damage and the cytotoxic effects caused by virus replication.

ACKNOWLEDGMENTS

FIS and Comisión Asesora para el Desarrollo de la Investigación Científica y Técnica are acknowledged for financial support.

REFERENCES

1. **Bablanian, R.,** Structural and functional alterations in cultured cells infected with cytocidal viruses, *Prog. Med. Virol.*, 19, 40, 1975.
2. **Wagner, R. R.,** Cytopathic effects of viruses: a general survey, in *Comprehensive Virology*, Vol. 19, Fraenkel-Conrat, H. and Wagner, R. R., Eds., Plenum Press, New York, 1984, 1.
3. **Ehrenfeld, E.,** Picornavirus inhibition of host cell protein synthesis, in *Comprehensive Virology*, Vol. 19, Fraenkel-Conrat, H. and Wagner, R. R., Eds., Plenum Press, New York, 1984, 177.
4. **Carrasco, L.,** The inhibition of cell functions after viral infection. A proposed general mechanism, *FEBS Lett.*, 76, 11, 1977.
5. **Carrasco, L. and Lacal, J. C.,** Permeabilization of cells during animal virus infection, *Pharmacol. Ther.*, 23, 109, 1983.
6. **Levine, A. Y. and Ginsberg, H. A.,** Mechanism by which fiber antigen inhibits multiplication of type 5 adenovirus, *J. Virol.*, 1, 747, 1967.
7. **McSharry, J. J. and Choppin, P. W.,** Biological properties of the VSV glycoprotein. I. Effects of the isolated glycoprotein on host macromolecular synthesis, *Virology*, 84, 172, 1978.
8. **DeClercq, E. and Walker, R. T., Eds.,** *Targets for the Design of Antiviral Agents*, Plenum Press, New York, 1984.
9. **Pressmann, B. C.,** Biological applications of ionophores, *Annu. Rev. Biochem.*, 45, 501, 1976.
10. **Cuatrecasas, P., Ed.,** The specificity and action of animal, bacterial and plant toxins, in *Receptors and Recognition* (Series B), Vol. 1, Chapman & Hall, London, 1977.

Chapter 2

THE PENETRATION OF ANIMAL CELLS BY VIRUSES

K. Coombs and D. T. Brown

TABLE OF CONTENTS

I. INTRODUCTION

The earliest events in virus infection have been arbitrarily divided into the three stages: attachment, penetration, and uncoating. For the sake of clarity, all three will be dealt with separately in this review, although the reader is reminded that while division into these stages implies that they occur in discrete steps, it is becoming increasingly clear that, at least for some viruses, some of these steps occur simultaneously.

The passage of a virus particle from the surrounding milieu to its point of replication within a target cell is one of the most difficult events to study in the life cycle of a virus. The early events of attachment, penetration, and uncoating of animal viruses have received considerable attention over the past 20 years. At this time, however, relatively little of these processes is understood as compared to those events concerning the replication of the virus in an established infection. Although the early events in virus replication are incompletely understood, a number of excellent literature reviews has appeared that describes in great detail experimentation in this important area of investigation.[1-12] The purpose of this chapter is to describe the current state of our understanding of those events that lead to the establishment of a virus infection. The reader is referred to the above-mentioned reviews for the experimental detail and historical development leading to the hypotheses outlined below.

A great deal of controversy has been generated by interpretations of the investigations designed to elucidate early events in virus infection. These studies have all suffered from common problems related to the methodology employed. The infection of a living cell is initiated by contact with a single infectious virion. While this statement is so self-evident that it is trite, it forms the caveat that has rendered studies of virus attachment, penetration, and uncoating suspect and controversial. In no animal virus system has it yet been possible to demonstrate that all virus particles are infectious. In the very best of circumstances, the particle to plaque-forming unit (PFU) ratio is ten to one, and generally it is much worse, being on the order of hundreds or thousands to one. In many virus systems, no reliable assay is available to determine the infectivity of a virus particle. Establishing the relative infectivity of a preparation of viruses requires the use of a biological assay. Plaque assays, focus formation assays, and infectious center assays are relatively sensitive indicators of the concentration of infectious virus particles. Hemagglutination titrations and infectious dose determinations in whole animals are quite inaccurate in establishing the relative amount of infectious virus. The processes by which populations of virus are purified to homogeneity so that infectivity assays may be carried out are probably responsible for the generation of much of the damaged noninfectious virions. At the same time, relatively sensitive assays such as the plaque assay may not represent a true measure of the infectious potential of a virus population because of the physiological state of the indicator cells themselves. The same preparation of virus particles frequently yields higher titers when assayed on one cell type relative to another, and on subconfluent as opposed to confluent monolayers of the same cell type. The problems related to the percentage of infectious particles in a virus population are compounded by the limitations in technology employed to examine their interactions with a cell population. Experimental examination of virus/host cell interactions has employed two primary techniques, electron microscope observation and biochemical analysis of virus particles that have been mixed with a population of susceptible cells.

A typical animal cell is about 50 μm in diameter. These cells may be cut into 500:1000-Å-thick ultrathin sections for electron microscopy. Thus, to ensure that 10 particles will be visible in a particular section, 5000 virions must be added for each cell in the mixture. Biochemical assays in which radioactively labeled virus is examined after exposure to a preparation of cells are similarly demanding of large amounts of starting material. For example, if a virus genome of 4 kb (kilobases) is 100% uridine, and if this virus can be labeled totally with the highest specific activity tritiated uridine that is available (about 50

C/mmol), this would result in a population of virus particles which have about 10^{-4} decays/min/virion. If all of this radioactivity can be efficiently counted, one must adjust the input of virus particles per cell to obtain enough radioactivity in the experiment for subsequent analysis. If this required amount is 20,000 counts/min, 10^6 cells must be infected with about 500 particles/cell, or 5×10^8 cells must be infected with 1 particle/cell, in order to obtain enough material for subsequent enzymatic or physical assays on the status of the incoming virion. The presence of other bases in the viral genome besides uridine, and the inability of scintillation spectrometers to efficiently count all the input radioactivity necessitates increasing the MOI, or number of cells, by a factor of at least four. In addition, the presence of a significant host pool of unlabeled nucleic acid precursors that compete with input label results in a further reduction in the specific activity of the virus population and requires another increase in either MOI or number of cells employed.

The demands for the large quantities of material needed in these experiments might not be important if it could be established that all virus particles interacting with a single cell were processed through an identical pathway, one which would lead to infection by a single virus particle. It seems certain at this time that several host cell components are essential to the process of replication of a number of virus particles. In some instances, it appears that these components are present in limited amounts. With the exception of cell surface receptors, little information is available on the presence, nature, or quantity of cell components that participate in the processes of virus penetration and uncoating. Thus, if it were possible to obtain virus populations which were 100% infectious (by whatever assay), one might easily exceed the capacity of the cell to process this virus by a route leading to infection. Rossenwirth and Eggers[13] have shown that although tissue-cultured green monkey kidney cells are capable of internalizing thousands of particles of echovirus 12 (a picornavirus), only about 100 particles are uncoated. When only 10% or less of the virus population is infectious (as discussed above), the situation is clouded by the prospect that noninfectious virus may compete with infectious virus for limited cell components.

The dilemma facing the investigator is, therefore, two-pronged. First, in the best of circumstances, one is required to draw conclusions about the process of infection of a cell by a virus in which 90% of the starting virus population may not be capable of successfully infecting the cell. Second, the insensitivity of the procedures employed require that large quantities of virus particles be applied to cells, and these quantities may exceed the amounts that can enter the cell by a pathway leading to a productive infection. The second problem may be circumvented by using low-MOI infections. However, the number of cells required to generate statistically significant data in enzymatic or radioactive input studies, such as that described above, is prohibitively large.

Given the limitations in experimental approach that are outlined above, it is not surprising that research designed to elucidate the processes of virus attachment, penetration, and uncoating has generated controversy surrounding models relating to the process by which virus particles transverse the cell membrane to gain access to the cell interior. The unfavorable particle to PFU ratio, plus the fact that many infectious virus particles may react with host cells through nonproductive pathways, have rendered suspect observations made by traditional procedures of biochemistry and electron microscopy. In recent years, a more productive approach has been to examine the properties of virus and cellular components and to surmise how these properties might facilitate the transport of a virion from an extracellular environment to its site of replication within the host cell.

II. INTERACTION OF VIRUS PARTICLES WITH CELL SURFACES

A virus particle is a gene delivery system, and as such, it is required to perform a number of important functions. It is assembled from its various components into a highly organized, three-dimensional geometric structure of high stability. This assembly may take place at one or more locations within the host cell. During and after its release from the confines of the host cell, this structure is required to protect the virus genome from the extracellular environment. The virus particle must ultimately recognize and attach to a susceptible host cell and perform a series of intricate maneuvers which will deliver the genome to the interior of the cell. During this process, the otherwise stable virus particle must partially or completely disaggregate to expose the virus genome to the host cell biochemical machinery. Therefore, virus particles must be metastable structures that are specifically aggregated to disassemble when a certain set of conditions are met. The signals that initiate this disaggregation are most likely provided as the virus traverses the barrier of the cell surface membrane to achieve access to the cell cytoplasm. Therefore, the irreversible sequence of events leading to the uncoating of a virus particle is probably initiated at some stage in the attachment of the virus to the cell surface.

The surface of a virus particle is composed of a lipid-containing membrane and/or protein that presents a charged and highly hydrated surface to the environment. During the early events in the interaction of the virion with the host cell, the repulsive forces and the hydrophobic barriers existing on the surface of the virus and host cell must be overcome. It seems logical that the recognition of particular cell surface components by specialized virus proteins act, in large part, to overcome these barriers. It also seems highly probable that changes in the medium that surround the virion and host cell surface may also negate the repulsive forces, resulting in a random or nonspecific interaction between virus and cell. Thus, distinguishing between specific and nonspecific associations of viruses with cell surfaces has proven a most difficult task.

Many experimental methods have been developed to examine the virus/receptor interaction(s). Experiments have been designed to quantitate virus receptors by attempting to saturate these sites with virus particles. In some instances, specificity of virus/cell interaction has been defined as ''tight'' vs. ''loose'' association of the virion with the cell surface. Competition studies have been employed to determine if two related or unrelated viruses utilize the same cell surface receptor. From these experiments, the number of virus receptors on a cell has been variously estimated to be between 10^4 and 10^6/cell.[14-18] Viruses such as adenovirus and cardiovirus utilize the same receptors, and these receptors seem to be different from those utilized by poliovirus.[14,19-21] Flu, paramyxo, and polyoma viruses utilize neuraminic acid-containing structures as receptors.[22-24] The complete chemical identification and characterization of particular virus receptors on cell surfaces have proven to be difficult. Putative receptors have been isolated in experiments in which a virus structural protein has been demonstrated to bind to a particular host cell protein.[25-27] In many cases, interpretation of these kinds of experiments has been clouded by subsequent observations that cells which did not express the particular protein in question were still infected by the virus under study.[28] The use of cell systems, which for genetic reasons fails to express a particular protein and simultaneously loses the ability to be infected by a virus, might demonstrate more convincingly host proteins that serve as specific virus receptors. It has been frequently assumed that cell surface components that serve as receptors for a particular virus or class of viruses would be highly conserved and ubiquitous components of plasma membranes. In some instances, this may prove to be correct, however in cases of particular viruses that successfully infect and grow in cells of diverse phylogeny (e.g., the alphatogaviruses which successfully infect cells of both vertebrate and invertebrate origin), the constraints on the constancy of conformation, composition, and sequence of a particular surface protein cannot be rigidly enforced. A unique approach for the identification of cellular proteins that serve as receptors has developed as a result of the ability to produce monoclonal antibodies specific in their binding for virus protein domains that are responsible for attachment to host cell receptors. Assuming that the virus attachment protein/cell receptor protein interaction is a stereo-topological match (lock and key), monoclonal antibodies that block virus attachment are utilized to produce anti-idiotype antibodies which are then used to immune precipitate the putative host cell receptor protein.

III. VIRUS PENETRATION

A specific interaction between a virus particle and the cell surface (attachment) presumably sets the stage for the translocation of the virus from the cell surface to its site of replication in the cell interior. The process of uncoating the virus genome may be initiated as this translocation event occurs. For many viruses, two major approaches for gaining access to the cell interior have been proposed. According to one hypothesis, the internalization of virus particles occurs through the utilization of normal cellular processes designed to internalize hormones, metabolites, and macromolecules.[9,29-35] According to the other hypothesis, viruses are capable of passing from the cell surface to the cell cytoplasm without relying on any cellular event for internalization.[36-43] Because the process of internalization of virus particles by endocytosis is readily detected in the electron microscope, it has received considerable attention over the years. Indeed, cells are capable of internalizing a wide variety of particulate material of biological and nonbiological origin by endocytosis. The only constraint seems to be the size of the particulate material itself. Therefore, it is not surprising that when large numbers of virus particles of any type are added to cells, these particles are found to accumulate in internal vesicles. Virus particles will accumulate, with time, in lysosomal vesicles, leading to hypothesis that the battery of degradative enzymes present in these compartments might play an important role in the uncoating process of the virus particle.

A major revival of interest in the role of endocytosis in virus infection has developed in recent years. The single most important factor in refocusing attention on the endosome as a participant in the process of virus penetration of cells came with the discovery that many of those membrane-containing viruses that were thought incapable of inducing membrane fusion could be demonstrated to do so if the virus/membrane complex was exposed briefly to acidic pH.[9,31,44] This observation, when combined with the knowledge that lysosomal compartments of cells are acidic, along with the observation that infectivity of some viruses (i.e., nonenveloped reoviruses) was enhanced after brief in vitro exposure to many of the enzymes commonly found in lysosomes,[45] refocused attention on this organelle as a potential mediator of virus penetration. The presumed detrimental effects of the lysosomal degradative enzymes on the structure and infectivity of many viruses have led to the suggestion that the arrival of a virus within a lysosome sets the stage for a competition between two events. These events constitute a "race" in which the virus particle is forced to fuse with the lysosomal membrane before being degraded by the battery of enzymes located within this organelle. This dilemma was eliminated by the discovery that prelysosomal vesicles also provided suitably low pH environments.[46] The hypothesis that intracellular vesicles with low pH were essential for penetration of enveloped viruses was further supported by the demonstration by Helenius and co-workers[9] that various compounds (e.g., chloroquine, ammonium chloride, and amantadine) that could be shown to raise the pH of intracellular compartments[47] also prevented virus replication when present at very early times during infection.[9] The ability of some of these compounds to inhibit the replication of influenza virus in tissue cultured cells was first demonstrated a quarter of a century ago.[48,49]

Much research has been conducted to examine the events surrounding the low pH-mediated virus/cell membrane fusion event. It has been demonstrated that brief low-pH treatment induces irreversible conformational changes in the virus surface proteins of a number of enveloped viruses. In the case of some viruses such as influenza virus, the molecular rearrangements accompanying these conformational changes have been carefully described.[50-55] The conformational changes induced by exposure to low pH are believed to expose a hydrophobic domain of the virus surface protein that is capable of initiating the fusion of the virus membrane with a host cell membrane. Studies with model membranes (liposomes) of varying chemical composition suggest that the target membrane for virus-mediated fusion needs only to contain cholesterol to participate in the fusion event.[30,56] The composition of the membrane, with respect to other lipids and phospholipids, was thought to play no role in this process. Furthermore, it has been suggested that the pH parameters to which the virus/membrane complex must be exposed are determined solely by the virus. Evidence to support this proposition includes the observation that low pH-induced fusion of cells can be demonstrated after expression of only the viral glycoprotein(s)[57-59] and the selection of virus mutants that demonstrate different acid-pH optima for fusion.[60] The collective data outlined above have been assembled into a popular model that proposes enveloped viruses exploit a normal cellular function to gain entry to the cell cytoplasm, and that cell surface proteins used by the viruses as receptors are the same as those employed by the cell as receptors for essential ligands such as growth hormones, low density lipids, etc. These proteins combine with a ligand to form a stable complex that is incorporated into a coated pit at the cell surface. The coated pit encloses the receptor/ligand complex as it forms a closed vesicle from the cell plasma membrane. As the vesicle (endosome) is established, a proton pump increases the hydrogen ion concentration in the vesicle interior. Many ligand/receptor complexes are unstable at acid pH, and the low pH condition of the endosome separates the ligand from the receptor. A subsequent partitioning of the endosome allows the receptor to be returned to the cell surface while the vesicle that contains the ligands may fuse with a lysosome. It has been suggested that membrane-containing viruses attach to cell receptors instead of the normal ligand of the protein. Formation of the virus/receptor complex leads to the formation of an endosome that contains the virus particle. The subsequent decrease in the endosomal pH triggers the fusion of the virus membrane with the endosomal membrane, releasing the virus genome into the cytoplasm of the cell. The elegant simplicity of this model, combined with the supporting evidence outlined above, has resulted in such a wide acceptance that as a result, schematic illustrations of the events described above now appear in major university-level textbooks.[61] Furthermore, the caveat that acid compartments are essential for penetration of cells via the endosomal route has led investigators to conclude that if infection of a cell by a particular virus was blocked by an agent capable of raising the internal pH of the endosome, then its route of entry into the cell must follow this pathway.

Although the model described above provides an attractive mechanism for the passage of a virus particle into the cell interior, a number of experimental observations has recently appeared that questions the essential role of the acidic compartment in the fusion/penetration event as it may apply to particular virus/cell combinations. It has been shown that cytochalasin (an inhibitor of cell surface motility) prevents the uptake of Sindbis virus into intracellular vesicles, but has no effect on the process of infection of cells by this virus.[41] It is also somewhat paradoxical that when virus internalization is examined in the electron microscope, large numbers of particles that are internalized by nondrug-treated cells are transported into lysosomes. Since the acid condition that is believed to be required for fusion of the virus membrane with cellular membranes is met in prelysosomal vesicles, one might expect the particles to have fused at this early point in endocytosis and not to have appeared in lysosomal vesicles. (The fact that they do further illustrates the nonselectivity of endocytosis as a process which internalizes any particulate material of suitable size.) The situation is further clouded by the observations that the lysosomotropic weak bases chloroquine and ammonium chloride block replication of alphavirus RNA[43] and prevent maturation of virus proteins into mature virions[41] when applied to cells in concentrations typically used to raise the pH of intracellular compartments. Therefore, these drugs may effect many discrete stages in the life cycle of a virus. More importantly, chloroquine and ammonium chloride were found to not block the expression of alphavirus-induced homologous interference (the ability of a virus particle to prevent the replication of subsequently infecting virus of the same type) when added to cells prior to and during the process of infection.[43] It has been demonstrated that homologous interference is the result of the expression of a virus gene. Thus, translation of the alphavirus genome must take place in the presence of this drug, an event which requires that the genome first reaches the cell cytoplasm. Furthermore, a variety of lysosomotropic weak bases was shown to not significantly reduce the formation of infectious centers of Sindbis virus when applied to cells during the period of infection.[43]

Further evidence that these drugs may affect some early event after transport of the viral genome into the cell cytoplasm was provided by experiments in which the induction of interferon by vesicular stomatitis virus (VSV) was examined. Interferon induction by VSV results from the production of double-stranded RNA molecules in the cytoplasm of the infected cell after the entry of the parental negative strand into the cell cytoplasm. When cells were infected with wild-type VSV in the presence of ammonium chloride, progeny virus was not produced and interferon induction was greatly suppressed, implying that either the virus genome did not reach the cell cytoplasm (penetration block) or that if it did reach the cell cytoplasm, a double-stranded molecule was not formed (RNA synthesis block). When cells were infected in the presence of ammonium chloride with a VSV defective interfering particle that has the ability to form a double-stranded molecule in the absence of RNA synthesis by folding back upon itself (snap-back defective interfering particle), interferon induction took place.[62] This experiment suggested that VSV RNA reached the cell cytoplasm in the presence of ammonium chloride, but did not undergo replication to allow the formation of a double-stranded interferon-inducing RNA. This experiment agrees with the alphavirus experiment presented above; alphavirus RNA reached the cell cytoplasm and was translated in the presence of drug, but replication did not take place.

The effects of amantadine on early events in virus replication were first demonstrated by Neumayer and co-workers[63] and Hoffman et al.[64] whose basic experimental procedures for investigating the effects of the lysosomotropic weak bases were subsequently used by most later workers. The effects of amantadine on the pH of lysosomes was not known at the time, so Kato and Eggers[65] concluded that it blocked virus uncoating. This conclusion is consistent with the more recent observations that virus mutations which convey resistance to concentrations of amantadine typically used to block infection have been mapped in the virus matrix protein[66] rather than in the envelope protein(s), as would be expected if the drug affected events involved in membrane fusion. These data imply that the inhibitory effects of amantadine occur after the virus has penetrated into the cell cytoplasm. In addition, Richman and co-workers[67] have demonstrated that when cell monolayers are treated with amantadine at concentrations sufficient to block infection with influenza (and subsequently washed to remove the drug), most of the drug remains cell-associated (and presumably in acidic compartments). However, the washed cells are readily infected by flu virus, implying that the drug needs only to be present in the extracellular milieu in order to express is antiviral effects.

Paradoxically, the various lysosomotropic weak basic compounds have also been demonstrated to block infection by viruses such as herpes virus[68] and parainfluenza viruses (i.e., Sendai virus),[69] which fuse with cell surfaces under neutral pH conditions. These viruses presumably have no need to be provided with the internal acidic environments that these drugs alter.

The supposition that an acid-induced conformational change in a virus protein is responsible for the fusion of the virus membrane with a host cell membrane has also been recently challenged. Most dramatically, Haywood and Boyer[70] have demonstrated that wild-type influenza virus can attach to, and fuse with, artificial membranes at pH 7.5 without prior acid treatment. Furthermore, these investigators demonstrated that the chemical composition of the liposome determined whether or not exposure to acid conditions was necessary for influenza virus to fuse with these artificial membranes. These observations are in striking contrast to those of Kielian et al.,[60] who concluded that characteristics of the virus alone determined the degree of acidity required for the fusion event to take place. Edwards and Brown[71] also found that the same preparation of Sindbis virus demonstrated different pH requirements to initiate cell fusion, depending upon the cell type and the media in which the cells were grown. The expression of Sindbis virus-induced fusion in vertebrate BHK-21 cells which had been grown in Eagle's medium required brief exposure of infected cells to pH 5.3. In contrast, the expression of fusion in the invertebrate cells of *Aedes albopictus* (an alternate host) which has been grown in the same medium as the BHK-21 cells required exposure to pH 4.6. *A. albopictus* cells that had been passaged in insect cell culture medium could not be fused after exposure of the virus/cell complex to any pH measured between 4.0 and 8.0. Even though these cells demonstrated different pH requirements for virus-mediated fusion, they were all infected equally efficiently with Sindbis virus. This experiment appears to separate the process of infection of cells by Sindbis virus from the process of low pH-mediated cell fusion. It is also important to note that although fusion of infected cell monolayers was mediated by a variety of virus particles after exposure to low pH, the fusion event itself did not occur until the monolayers were returned to near-neutral pH conditions.[72] Cell monolayers that are maintained under acidic pH conditions for hours do not show detectable levels of fusion.[73] This observation seriously questions the role that the acid-containing endosome may play in virus infection because the pH of these intracellular compartments does not fluctuate from acid to neutral.

These data (particularly the data of Haywood and Boyer[70]) strongly suggest that viruses such as flu are capable of fusing membranes to which they efficiently attach, and that the function of the low pH environment may solely be to overcome the repulsive charges associated with the membranes of particular types of cells that prevent the appropriately efficient contact of virus particles with the cell surface. It is not yet known if the conformational change demonstrated in virus membrane proteins (i.e., flu) actually occurs at low pH, or if the stage is simply set for this change and the change itself occurs after restoration to neutral pH conditions. It is possible that experimental investigation of the events concerning low pH-induced virus-mediated fusion may simply demonstrate the ability of a virus membrane protein to efficiently produce membrane fusion under artificial conditions. The conformational change(s) in the membrane proteins that is required for fusion of the virus envelope with the host cell membrane may be produced under neutral pH conditions after a critical event in the binding of the virus to the cell surface has been met. In support of this view, Houk and co-workers[74] have recently produced electron micrographs that clearly show fusion of the alphavirus Western equine encephalitis with the surface of cells in the gut of mosquitos which have ingested a blood meal. The pH of the insect gut is normally alkaline, and insect cell membranes contain no cholesterol. The contents of the gut could be only slightly buffered toward neutral pH by the presence of serum in the blood meal itself. Thus, a member of the group of viruses that were initially used to produce the model that acidic compartments are essential for infection by membrane-containing viruses infects cells by directly fusing with the cell surface in a neutral-to-alkaline pH environment.

It is also possible that the attachment of a virus particle to a host cell membrane produces a local perturbation in the hydrogen ion pumps, which may result in formation of the necessary transient acidic microenvironment immediately surrounding the virus proteins that are attached to the appropriate host receptors(s). As the excess of hydrogen ions in this microenvironment diffuses, the microenvironment is returned to near-neutral pH conditions that would allow fusion of the two membranes. Such a scenario would be expected to be more easily carried out at the cell surface rather than within an endosome where the enclosed nature and small volume of the vesicle would interfere with the reestablishment of the near-neutral pH conditions that are required for fusion to occur.

A number of enveloped viruses, (i.e., herpes virus) must be transported into the nucleus in order to be replicated. These viruses are faced with the problem that they must traverse two membranes (the cellular membrane and the nuclear membrane) in order to arrive at their site of replication. The extra barrier requires an additional mode of penetration because in fusing with the first membrane, the virus particles lose their fusion-specific organelles. The method(s) by which such de-enveloped nucleocapsids ultimately gain access to the nucleus may be very similar to some of those methods used by nonenveloped viruses to gain access to their host cells (see below).

Although considerable debate exists regarding the cellular location and mechanism by which enveloped viruses achieve entry into the cell, there is unanimity in accepting that this process involves fusion of the virus membrane with some cellular membrane. The situation with nonenveloped viruses is somewhat more confused, as these agents are faced with the same barriers with which enveloped viruses are required to deal. They must somehow traverse a membrane bilayer without the benefit of a fusion organelle. In spite of the obvious differences between the enveloped and noneveloped viruses, the controversy regarding the process of penetration of cells by these viruses is very similar to that existing with enveloped virus particles. The two primary possible mechanisms by which these virions can obtain entry into the cell cytoplasm seem to be endocytosis, as in the case of membrane-containing viruses, or direct penetration. Much of the support for one route or the other is derived from morphological studies. A process of direct penetration of cell membranes by polio and adenoviruses has been suggested by ultrastructural studies involving ultrathin sectioning and freeze-fracture techniques.[36,37,40] Likewise, internalization of nonenveloped viruses by endocytosis and their transport to lysosomal compartments have been suggested by electron microscopy[2,75-77] and biochemically.[78-80] It is difficult to imagine how direct penetration can be used by viruses such as adenoviruses or polioviruses. It is possible that specific interactions of these viruses with cell surface proteins result in the development of a disorganized region of the membrane bilayer through which the viruses can pass. The observation that endosomal compartments are readily acidified has led to speculation that exposure to an acidic environment results in conformational changes in the virus that activate enzymes capable of dissolving the endosomal membrane, effectively placing the virus in the cell cytoplasm.[81] Indeed, conformational changes have been reported to occur in nonenveloped virus particles as they penetrate host cells. Most of these changes seem to involve the loss of particular virus structural components, accompanied by changes in the overall morphology and physical properties of the virion.[14,38,82-84] However, it has been suggested that in the case of some picornaviruses, the particles reach the cell interior in an unaltered form.[20,85] Adenoviruses find their site in replication in the nucleus, and, whether the virus enters the cell cytoplasm by direct penetration of the cell surface or enzymatically dissolves cell membranes which engulf it, it must be transported to the nuclear site of replication after entering the cell cytoplasm. Evidence suggests that adenoviruses accomplish this by interacting with cell cytoskeletal components that presumably direct the partially uncoated particle to the nuclear membrane.[86-88] Biochemical and morphological evidence suggests that final entry of the virus into the nucleoplasm is achieved by a process in which the virus associates with a nucleopore and extrudes its genome through this structure by an injection-like mechanism.[36,89] A similar event has been suggested to occur in the case of certain baculoviruses,[90] although entry of intact baculovirus nucleocapsids into the nucleus has also been reported.[91] Likewise, parvoviruses such as adeno-associated viruses seem to achieve access to the nucleus by a route that takes them through the cell cytoplasm. Evidence in this system suggests that this transport event occurs without involvement of the cell cytoskeletal proteins.[92,93]

Other nonenveloped viruses that have the nucleus as their site of replication may be transported to the nucleus without being exposed to the environment of the cell cytoplasm. Nonenveloped papovaviruses seem to acquire an envelope as they are taken into the cell by a process of endocytosis that incorporates a single virion into a single vesicle. This vesicle is transported across the cell cytoplasm to the nuclear membrane where a fusion event between the vesicle and the nucleomembrane occurs, releasing the subviral structure into the nucleoplasm.[77,94,95] It is unclear how these events take place; however, the endocytotic vesicle that envelopes the virus particle may be modified by incorporation of proteins from the virus itself, and these proteins may specifically direct the vesicle to be transported to the nucleomembrane. Evidence in other systems has been obtained that demonstrates that membrane proteins can direct the intracellular transport of protein-containing vesicles to particular target organelles.[96-99] These observations suggest that the papovaviruses may, in fact, be enveloped viruses that acquire the envelope necessary for the final penetration event (fusion with the nuclear membrane) during the process of cellular penetration itself.

IV. VIRUS UNCOATING

Once it successfully penetrates to its site of replication within a host cell, the invading virus must disassemble (uncoat) to such an extent that it is capable of setting in motion a sequence of events that will eventually lead to the production of progeny virions. This sequence of events is initiated by virally encoded products, which implies that at least part of the viral genome be expressed. The expression of the viral genome that requires conversion of the host's macromolecular precursors into virus macromolecules is dependent upon accessibility of those precursors and the presence of various host factors. These dependencies require that the incoming virus be disassembled to such an extent that the genome becomes accessible.

Uncoating is a necessary event for viruses whose genomes are protected from the external environment by a lipid membrane and/or protein shell. It converts the genome from a highly secluded and protected state into one which can direct its transcription (or translation if it is an ssRNA virus whose genome serves directly as message). The protective lipid membrane of enveloped viruses is lost as the virus penetrates the cell (see above), so that penetration and uncoating are not completely synonymous for many enveloped viruses because the genome may still lie within a protective protein-containing nucleocapsid that must also be partially or completely unraveled in order for viral-encoded functions to commence. As described above, some nonenveloped viruses (i.e., picornaviruses) may enter the cell cytoplasm in an essentially intact form. These require uncoating in order for replication to be initiated. Other nonenveloped viruses appear to penetrate cells by a process that results in conformational changes and/or loss of protein components. These may require further uncoating before they can be replicated.

The degree to which a virus must be uncoated to allow replication depends upon the particular virus. For example, incoming VSV genomes retain most, if not all, of their N protein.[100,101] Indeed, this protein appears to be required, along with other virus and host proteins, to direct accurate transcription of the genome and regulate the switch between transcription and translation of the transcribed genome.[102-105] Reovirus subviral cores remain essentially intact to transcribe progeny dsRNA. At the other extreme are viruses that are disaggregated to such an extent that they are morphologically unrecognizable (e.g., the aphaviruses and influenza viruses). Replicating influenza virus is known to contain some of the viral protein within an active ribonucleoprotein complex.[106] However, it is currently not known whether any of the alphavirus proteins remain associated with the genome to direct transcription and translation. Experimental evidence indicates that adenovirus must be uncoated almost completely to allow accurate replication.[107]

The conditions that promote in vivo uncoating are poorly understood. As indicated above, the virus lipid membrane barrier is removed from enveloped viruses as the particles fuse with cell membranes to enter the cytoplasm. Likewise (as discussed earlier), a number of nonenveloped viruses has been shown to undergo conformational changes and lose some of their protein as they penetrate into host cells. However, in both cases a protein shell remains that may need to be topologically altered further to allow viral-encoded replication. A variety of in vitro treatments has been found to induce conformational changes in viruses and subviral components, including alterations in pH,[108] ionic strength, concentrations of various cations,[109,110] and exposure to a variety of proteases.[111] While it is likely that an entering virus may encounter one or more of these conditions, it is unclear what effects these alterations have on in vivo uncoating. The lysosome appears to play a major role in the uncoating process of reoviruses.[112,113] Ribosomal subunits have been implicated as being partly responsible for the uncoating of some alphaviruses[114,115] and picornaviruses[116] because of evidence that virus components may associate with them during assembly. Likewise, proteolytic cleavage events that occur late in the pathway of assembly of some viruses have been suggested as representing a priming reaction for subsequent uncoating of those viruses.[116] Much valuable information about this important aspect of entry may be gained from studies of virus mutants that are blocked in uncoating, such as temperature-sensitive mutants of adenovirus that have been recently isolated and partially characterized.[117]

V. CONCLUSIONS

Experiments conducted during the past few years have reawakened interest in the early events of virus infection. Although we have learned a great deal about the contributing cellular events and maneuvers through which an invading virus must enter a host cell to initiate infection, much remains to be determined about these important processes. The cellular requirements for efficient attachment of a number of viruses are still poorly understood. The presence of cholesterol in the target membrane of vertebrate cells appears to be a requirement for the attachment of enveloped viruses to these cells, but the lack of this sterol in the membranes of other hosts (some insect cells) which may be infected as efficiently as the vertebrate cells, raises doubts as to the rigor of this requirement. In addition, the results of a number of investigators have demonstrated that acidic conditions may not be absolutely required for enveloped virus penetration, raising questions about the role that endocytosis and acidic compartments may play in infection. Much also remains to be determined about the process(es) by which nonenveloped viruses enter cells. Finally, the requirements for uncoating of the genomes of most viruses so that replication may be initiated are very poorly understood, with respect to both the functions provided by the host cell and the degree to which triggering events during virion assembly prime eventual disassembly.

ACKNOWLEDGMENTS

The authors were supported by Public Health Service grants AI19545 and AI14710 from the National Institutes of Health, and through generally appropriated funds from the State of Texas to the Cell Research Institute.

Kevin Coombs is the recipient of a predoctoral fellowship through NIH grant number T32-CA09182.

REFERENCES

1. **Burrows, R.,** Early stages of virus infection: studies *in vivo* and *in vitro, Symp. Soc. Gen. Microbiol.,* 22, 303, 1972.
2. **Dales, S.,** Early events in cell-animal virus interactions, *Bacteriol. Rev.,* 37, 103, 1973.

3. **Lonberg-Holm, K. and Philipson, L.**, Early interactions between animal viruses and cells, in *Monographs in Virology*, Vol. 9, Melnick, J. L., Ed., S. Karger, Basel, 1974, 1.
4. **Lonberg-Holm, K. and Philipson, L.**, Molecular aspects of virus receptors and cell surfaces, in *Cell Membranes and Viral Envelopes*, Blough, H. A. and Tiffany, J. M., Eds., Academic Press, London, 1980, 789.
5. **Meager, A. and Hughes, R. C.**, Virus receptors, in *Receptors and Recognition*, Vol. 4, Cuatrecasas, P. and Greaves, M. F., Eds., Chapman & Hall, London, 1977, 141.
6. **Bächi, T., Deas, J. E., and Howe, C.**, Virus-erythrocyte interactions, in *Cell Surface Reviews*, Vol. 2, Poste, G. and Nicholson, G. L., Eds., North Holland, Amsterdam, 1977, 83.
7. **Kohn, A.**, Early interactions of viruses with cellular membranes, *Adv. Virus Res.*, 24, 223, 1979.
8. **Howe, C., Coward, J. E., and Fenger, T. W.**, Viral invasion: morphological, biochemical, and biophysical aspects, in *Comprehensive Virology*, Vol. 16, Fraenkel-Conrat, H. and Wagner, R. R., Eds., Plenum Press, New York, 1980, 1.
9. **Helenius, A., Kartenbeck, J., Simons, K., and Fries, E.**, On the entry of Semliki Forest virus into BHK-21 cells, *J. Cell Biol.*, 84, 404, 1980.
10. **Dimmock, N. J.**, Initial stages in infection with animal viruses, *J. Gen. Virol.*, 59, 1, 1982.
11. **Bukrinskaya, A. G.**, Penetration of viral genetic material into host cells, *Adv. Virus Res.*, 27, 141, 1982.
12. **White, J., Kielian, M., and Helenius, A.**, Membrane fusion proteins of enveloped animal viruses, *Q. Rev. Biophys.*, 16, 151, 1983.
13. **Rossenwirth, B. and Eggers, H. J.**, Early processes of echovirus 12-infection: elution, penetration, and uncoating under the influence of rhodanine, *Virology*, 97, 241, 1979.
14. **Philipson, L., Lonberg-Holm, K., and Pettersson, U.**, Virus-receptor interaction in an adenovirus system, *J. Virol.*, 2, 1064, 1968.
15. **Birdwell, C. R. and Strauss, J. H.**, Distribution of the receptor sites for Sindbis virus on the surface of chicken and BHK cells, *J. Virol.*, 14, 672, 1974.
16. **Linser, P. L., Bruning, H., and Armentrout, R. W.**, Specific binding sites for a parvovirus, minute virus of mice, on cultured cells, *J. Virol.*, 24, 211, 1977.
17. **Fries, E. and Helenius, A.**, Binding of SFV and its spike glycoproteins to cells, *Eur. J. Biochem.*, 97, 213, 1979.
18. **Smith, A. L. and Tignor, G. H.**, Host cell receptors for two strains of Sindbis virus, *Arch. Virol.*, 66, 11, 1980.
19. **Lonberg-Holm, K., Crowell, R. L., and Philipson, L.**, Unrelated animal viruses share receptors, *Nature (London)*, 259, 679, 1976.
20. **Baxt, B. and Bachrach, H. L.**, Early interactions of foot-and-mouth disease virus with cultured cells, *Virology*, 104, 42, 1980.
21. **McClintock, P. R., Billups, L. C., and Notkins, A. L.**, Receptors for encephalomyocarditis virus on murine and human cells, *Virology*, 106, 261, 1980.
22. **Tillack, T. W., Scott, R. E., and Marchesi, V. T.**, The structure of erythrocyte membranes studied by freeze-etching. II. Localization of receptors for phytohemagglutinin and influenza virus to the intramembranes particles, *J. Exp. Med.*, 135, 1209, 1972.
23. **Wu, P. S., Ledeen, R. W., Udem, S., and Isaacson, Y. A.**, Nature of the Sendai virus receptor: glycoprotein versus ganglioside, *J. Virol.*, 33, 304, 1980.
24. **Fried, H., Cahan, L. D., and Paulson, J. C.**, Polyoma virus recognizes specific sialyoligosaccharide receptors on host cells, *Virology*, 109, 188, 1981.
25. **Robinson, P. J., Hunsmann, G., Schneider, J., and Schirrmacher, V.**, Possible cell surface receptor for Friend murine leukemia virus isolated with viral envelope glycoprotein complexes, *J. Virol.*, 36, 291, 1980.
26. **Svensson, U., Persson, R., and Everitt, E.**, Virus-receptor interaction in the adenovirus system. I. Identification of the virion attachment proteins of the HeLa cell plasma membrane, *J. Virol.*, 38, 70, 1981.
27. **Maassen, J. A. and Terhorst, C.**, Identification of a cell-surface protein involved in the binding site of Sindbis virus on human lymphoblastic cell lines using a heterobifunctional cross-linker, *Eur. J. Biochem.*, 115, 153, 1981.
28. **Oldstone, M. B. A., Tishon, A., Dutko, F., Kennedy, S. I. T., Holland, J. J., and Lambert, P. W.**, Does the major histocompatibility complex serve as a specific receptor for Semliki Forest virus?, *J. Virol.*, 34, 256, 1980.
29. **Maeda, T. and Ohnishi, S.**, Activation of influenza virus by acidic media causes hemolysis and fusion of erythrocytes, *FEBS Lett.*, 122, 283, 1980.
30. **White. J. and Helenius, A.**, pH-Dependent fusion between Semliki Forest virus membrane and liposomes, *Proc. Natl. Acad. Sci. U.S.A.*, 77, 3273, 1980.
31. **Lenard, J. and Miller, D. K.**, pH-Dependent hemolysis by influenza, Semliki Forest virus and Sendai virus, *Virology*, 110, 479, 1981.

32. **Talbot, P. J. and Vance, D. E.,** Biochemical studies on the entry of Sindbis virus into BHK-21 cells and the influence of NH_4Cl, *Virology,* 118, 451, 1982.
33. **Matlin, K., Reggio, H., Helenius, A., and Simons, K.,** The infective entry of influenza virus into MDCK-cells, *J. Cell Biol.,* 91, 601, 1981.
34. **Matlin, K., Reggio, H., Helenius, A., and Simons, K.,** The pathway of vesicular stomatitis entry leading to infection, *J. Mol. Biol.,* 156, 609, 1982.
35. **Redmond, S., Peters, G.,and Dickson, C.,** Mouse mammary tumor virus can mediate cell fusion at reduced pH, *Virology,* 133, 393, 1984.
36. **Morgan, C., Rosenkranz, H. S., and Mednis, B.,** Structure and development of virus as observed in the electron microscope. X. Entry and uncoating of adenovirus, *J. Virol.,* 4,, 777, 1969.
37. **Dunnebacke, T. H., Levinthal, J. D., and Williams, R. C.,** Entry and release of poliovirus as observed by electron microscopy of cultured cells, *J. Virol.,* 4, 505, 1969.
38. **Lonberg-Holm, K. and Philipson, L.,** Early events of virus-cell interaction in an adenovirus system, *J. Virol.,* 4, 323, 1969.
39. **Heine, J. W. and Schnaitman, C. A.,** Entry of vesicular stomatitis virus into L cells, *J. Virol.,* 8, 786, 1971.
40. **Brown, D. T. and Burlingham, B. T.,** Penetration of host cell membranes by Adenovirus 2, *J. Virol.,* 12, 386, 1973.
41. **Coombs, K., Mann, E., Edwards, J., and Brown, D. T.,** Effects of choroquine and cytochalasin B on the infection of cells by Sindbis virus and vesicular stomatitis virus, *J. Virol.,* 37, 1060, 1981.
42. **Edwards, J., Mann, E., and Brown, D. T.,** Conformational changes in Sindbis virus envelope proteins accompanying exposure to low pH, *J. Virol.,* 45, 1090, 1983.
43. **Cassell, S., Edwards, J., and Brown, D. T.,** Effects of lysosomotropic weak bases on infection in BHK-21 cells by Sindbis virus, *J. Virol.,* 52, 857, 1984.
44. **White, J., Matlin, K., and Helenius, A.,** Cell fusion by SFV, influenza, and VS viruses, *J. Cell Biol.,* 89, 674, 1981.
45. **Spendlove, R. S., McClain, M. E., and Lenette, E. H.,** Enhancement of reovirus infectivity by extracellular removal or alteration of the virus capsid by proteolytic enzymes, *J. Gen. Virol.,* 8, 83, 1970.
46. **Marsh, M., Bolzau, E., and Helenius, A.,** Penetration of Semliki Forest virus from acidic prelysosomal vacuoles, *Cell,* 32, 931, 1983.
47. **Ohkuma, S. and Poole, B.,** Fluorescent probe measurement of the intralysosomal pH in living cells and the perturbation of pH by various agents, *Proc. Natl. Acad. Sci. U.S.A.,* 75, 3327, 1978.
48. **Jensen, E. M., Force, E. E., and Unger, J. B.,** Inhibitory effect of ammonium ions on influenza virus in tissue culture, *Proc. Soc. Exp. Biol. Med.,* 107, 447, 1961.
49. **Eaton, M. D. and Scala, A. R.,** Action of ammonium ion on an early phase of viral infection, *Proc. Soc. Exp. Biol. Med.,* 108, 766, 1961.
50. **Skehel, J., Bayley, P., Brown, E., Martin, S., Waterfield, M., White, J., Wilson, I., and Wiley, D.,** Changes in the conformation of influenza virus hemagglutinin at the pH optimum of virus-mediated membrane fusion, *Proc. Natl. Acad. Sci. U.S.A.,* 79, 968, 1982.
51. **Yewdell, J. W., Gerhard, W., and Bächi, T.,** Monoclonal anti-hemagglutinin antibodies detect irreversible antigenic alterations that coincide with the acid activation of influenza virus A/PR/834-mediated hemolysis, *J. Virol.,* 48, 239, 1983.
52. **Webster, R. G., Brown, L. E., and Jackson, D. C.,** Changes in the antigenicity of the hemagglutinin molecule of H3 influenza virus at acidic pH, *Virology,* 126, 587, 1983.
53. **Daniels, R. S., Douglas, A. R., Skehel, J. J., and Wiley, D. C.,** Analysis of the antigenicity of influenza hemagglutinin at the pH optimum for virus-mediated membrane fusion, *J. Gen. Virol.,* 64, 1657, 1983.
54. **Daniels, R. S., Downie, J. C., Hay, A. J., Knossow, M., Skehel, J. J., Wang, M. L., and Wiley, D. C.,** Fusion mutants of the influenza virus hemagglutinin glycoprotein, *Cell,* 40, 431, 1985.
55. **Doms, R. W., Helenius, A., and White, J.,** Membrane fusion activity of the influenza hemagglutinin: the low pH-induced conformational change, *J. Biol. Chem.,* 260, 2973, 1985.
56. **Kielian, M. and Helenius, A.,** Role of cholesterol in fusion of Semliki Forest virus with membranes, *J. Virol.,* 52, 281, 1984.
57. **Marsh, M., Bolzau, E., White, J., and Helenius, A.,** Interaction of Semliki Forest virus spike glycoprotein rosettes and vesicles with cultured cells, *J. Cell Biol.,* 96, 455, 1983.
58. **Kondor-Koch, C., Burke, B., and Garoff, H.,** Expression of Semliki Forest virus proteins from cloned cDNA. I. The fusion activity of the spike glycoprotein, *J. Cell. Biol.,* 97, 644, 1983.
59. **Florkiewicz, R. Z. and Rose, J. K.,** A cell line expressing vesicular stomatitis virus glycoprotein fuses at low pH, *Science,* 225, 721, 1984.
60. **Kielian, M., Keranen, S., Kääriäinen, L., and Helenius, A.,** Membrane fusion mutants of Semliki Forest virus, *J. Cell Biol.,* 98, 139, 1984.
61. **Alberts, B., Bray, D., Lewis, J., Raff, M., Roberts, K., and Watson, J. D.,** *Molecular Biology of The Cell,* Garland Publishing, New York, 1983, 324.

62. **Svitlik, C. and Marcus, P. I.,** Interferon induction by viruses. XI. Early events in the induction process, *J. Interferon Res.*, 4, 585, 1984.
63. **Neumayer, E. M., Haff, R. F., and Hoffman, C. E.,** Antiviral activity of amantadine hydrochloride in tissue culture and *in ovo*, *Proc. Soc. Exp. Biol. Med.*, 119, 393, 1965.
64. **Hoffman, C. E., Neumayer, E. M., Haff, R. F., and Goldsby, R. A.,** Mode of action of the antiviral activity of amantadine in tissue culture, *J. Bacteriol.*, 90, 623, 1965.
65. **Kato, N. and Eggers, H. J.,** Inhibition of uncoating of fowl plaque virus by 1-adamantanamine hydrochloride, *Virology*, 37, 632, 1969.
66. **Hay, A. J., Kennedy, N. C. T., Skehel, J. J., and Appleyard, G.,** The matrix protein gene determines amantadine sensitivity of influenza viruses, *J. Gen. Virol.*, 42, 189, 1979.
67. **Richman, D. D., Yazaki, P., and Hostetler, K. Y.,** The intracellular distribution and antiviral activity of amantadine, *Virology*, 112, 81, 1981.
68. **Schambach, B., Muller, S., Richter, I. E., and Falke, D.,** Fusion-From-Within and Fusion-From Without by HSV, presented at 9th Int. Herpes Workshop, Seattle, Washington, 1984.
69. **Skehel, J., Hay, A. J., and Armstrong, J. A.,** On the mechanism of inhibition of influenza virus replication by amantadine hydrochloride, *J. Gen. Virol.*, 38, 97, 1977.
70. **Haywood, A. M. and Boyer, B. P.,** Fusion of influenza virus membranes with liposomes at pH 7.5, *Proc. Natl. Acad. Sci. U.S.A.*, 82, 4611, 1985.
71. **Edwards, J. and Brown, D. T.,** Sindbis virus induced fusion of tissue cultured *Aedes albopictus* (mosquito) cells, *Virus Res.*, 1, 705, 1984.
72. **Mann, E., Edwards, J., and Brown, D. T.,** Polycaryocyte formation mediated by Sindbis virus glycoproteins, *J. Virol.*, 45, 1083, 1983.
73. **Edwards, J. and Brown, D. T.,** Sindbis virus mediated cell fusion from without is a two step event, *J. Gen. Virol.*, 67, 377, 1986.
74. **Houk, E. J., Kramer, L. D., Hardy, J. L., and Chiles, R. E.,** Western equine encephalomyelitis virus: *in vivo* infection and morphogenesis in mosquito mesenteronal epithelial cells, *Virus Res.*, 2, 123, 1985.
75. **Dales, S.,** Penetration of animal viruses into cells, *Prog. Med. Virol.*, 7, 1, 1965.
76. **Silverstein, S. C. and Dales, S.,** The penetration of reovirus RNA and initiation of its genomic function in L-strain fibroblasts, *J. Cell. Biol.*, 36, 197, 1968.
77. **Mackay, R. L. and Consigli, R. A.,** Early events in polyoma virus infection: attachment, penetration, and nuclear entry, *J. Virol.*, 19, 620, 1976.
78. **Lonberg-Holm, K.,** The effects of concanavalin A on the early events of infection by rhinovirus type 2 and poliovirus type 2, *J. Virol.*, 28, 313, 1975.
79. **Madshus, I. H., Olsnes, S., and Sandvig, K.,** Mechanism of entry into the cytosol of poliovirus type 1: requirement for low pH, *J. Cell Biol.*, 98, 1194, 1984.
80. **Seth, P., Willingham, M. C., and Pastan, I.,** Adenovirus-dependent release of Cr-51 from KB cell at an acidic pH, *J. Biol. Chem.*, 259, 4350, 1984.
81. **Svensson, U. and Persson, R.,** Entry of Adenovirus 2 into HeLa cells, *J. Virol.*, 51, 687, 1984.
82. **Sussenbach, J. S.,** Early events in the infection process of adenovirus type 5 in HeLa cells, *Virology*, 33, 567, 1967.
83. **Gutteman, N. and Baltimore, D.,** A plasma membrane component able to bind and alter virions of poliovirus type 1: studies on cell-free alteration using a simplified assay, *Virology*, 82, 25, 1977.
84. **DeSena, J. and Torian, B.,** Studies on the *in vitro* uncoating of poliovirus. III. Roles of membrane-modifying and -stabilizing factors in the generation of subviral particles, *Virology*, 104, 149, 1980.
85. **Cavanagh, D., Rowlands, D. J., and Brown, F.,** Early events in the interaction between foot-and-mouth disease virus and primary pig cells, *J. Gen. Virol.*, 41, 255, 1978.
86. **Dales, S. and Chardonnet, Y.,** Early events in the interaction of adenovirus with HeLa cells, IV. Association with microtubules and the nuclear pore complex during vectorial movement of the inoculum, *Virology*, 56, 465, 1973.
87. **Luftig, R. B. and Weihing, R. R.,** Adenovirus binds to rat brain microtubules *in vitro*, *J. Virol.*, 16, 696, 1975.
88. **Weatherbee, J. A., Luftig, R. B., and Weihing, R. R.,** Binding of adenovirus to microtubules. II. Depletion of high-molecular weight microtubule-associated protein content reduces specificity of *in vitro* binding, *J. Virol.*, 21, 732, 1977.
89. **Chardonnet, Y. and Dales, S.,** Early events in the interaction of adenovirus with HeLa cells. I. Penetration of type 5 and intracellular release of the DNA genome, *Virology*, 40, 462, 1970.
90. **Summers, M. D.,** Electron microscopic observations on granulosis virus entry, uncoating, and replication processes during infection of midgut cells or *Trichoplusia ni*, *J. Ultrastruct. Res.*, 35, 606, 1971.
91. **Granados, R. R.,** Early events in the infection of *Heliothis zea* midgut cells by a baculovirus, *Virology*, 90, 170, 1978.
92. **Rose, J. A. and Koczot, F.,** Adenovirus associated virus multiplication. VII. Helper requirement for viral deoxyribose nucleic acid and ribonucleic acid synthesis, *J. Virol.*, 10, 1, 1972.

93. **Linser, P. L., Bruning, H., and Armentrout, R. W.,** Uptake of minute virus of mice into cultured rodent cells, *J. Virol.,* 31, 537, 1979.
94. **Bolen, J. B. and Consigli, R. A.,** Differential absorption of polyoma virions and capsids to mouse kidney cells and guinea pig red blood cells, *J. Virol.,* 32, 679, 1979.
95. **Chlumecka, V., D'Obrenan, P., and Colter, J. S.,** Early events in the infection of permissive cells with polyoma virus: comparison of chymotrypsin-treated and untreated virus, *Virology,* 94, 219, 1979.
96. **Rodriguez-Boulan, E. and Sabatini, D. D.,** Asymmetric budding of viruses in epithelial monolayers: a model system for study of epithelial polarity, *Proc. Natl. Acad. Sci. U.S.A.,* 75, 5071, 1978.
97. **Roth, M. G., Fitzpatrick, J. P., and Compans, R. W.,** Polarity of maturation in MDCK cells: lack of a requirement for glycosylation of viral glycoproteins, *Proc. Natl. Acad. Sci. U.S.A.,* 76, 6430, 1979.
98. **Rodriguez-Boulan, E.,** Membrane biogenesis, enveloped RNA viruses, and epithelial polarity, in *Modern Cell Biology,* Satir, B. H., Ed., Alan R. Liss, New York, 1983, 119.
99. **Matlin, K. and Simons, K.,** Sorting of an apical plasma-membrane glycoprotein occurs before it reaches the cell surface in cultured epithelial-cells, *J. Cell. Biol.,* 99, 2131, 1985.
100. **Emerson, S. U. and Wagner, R. R.,** Dissociation and reconstitution of the transcriptase and template activities of vesicular stomatitis B and T virions, *J. Virol.,* 10, 297, 1972.
101. **Banerjee, A. K., Abraham, G., and Colonno, R. J.,** Vesicular stomatitis virus: mode of transcription, *J. Gen. Virol.,* 34, 1, 1977.
102. **Simonsen, C. C., Hill, V. M., and Summers, D. F.,** Further characterization of the replicative complex of vesicular stomatitis virus, *J. Virol.,* 31, 494, 1979.
103. **Blumberg, B. M., Giorgi, C., Rose, K., and Kolakofsky, D.,** Preparation and analysis of the nucleocapsid proteins of vesicular stomatitis virus and Sendai virus, and analysis of the Sendai virus leader-NP gene region. *J. Gen. Virol.,* 65, 769, 1984.
104. **Arnheiter, H., Davis, N. L., Wertz, G., Schubert, M., and Lazzarini, P. A.,** Role of the nucleocapsid protein in regulating vesicular stomatitis virus RNA synthesis, *Cell,* 41, 259, 1985.
105. **Emerson, S. U.,** Rhabdoviruses, in *Virology,* Fields, B. N., Ed., Raven Press, New York, 1985, 1119,
106. **Rees, P. J. and Dimmock, N. J.,** Kinetics of the synthesis of influenza virus ribonucleoprotein structures, *J. Gen. Virol.,* 53, 125, 1982.
107. **Vayda, M. E., Leong, K., and Flint, S. J.,** Transcription of adenovirus cores *in vitro, Virology,* 139, 152, 1984.
108. **Keeling, J. and Matthews, R. E. F.,** Mechanism for release of RNA from turnip yellow mosaic virus at high pH, *Virology,* 119, 214, 1982.
109. **Borsa, J., Sargent, M. D., Copps, T. P., Long, D. G., and Chapman, J. D.,** Specific monovalent cation effects on modification of reovirus infectivity by chymotrypsin digestion *in vitro, J. Virol.,* 11, 1017, 1973.
110. **Wengler, G., Wengler, G., Boege, U., and Wahn, K.,** Establishment and analysis of a system which allows assembly and disassembly of alphavirus core-like particles under physiological conditions *in vitro, Virology,* 132, 401, 1984.
111. **Joklik, W. K.,** Studies on the effect of chymotrypsin on reovirions, *Virology,* 49, 700, 1972.
112. **Silverstein, S. C., Astell, C., Levin, D. H., Schonberg, M., and Acs, G.,** The mechanisms of reovirus uncoating and gene activation *in vivo, Virology,* 47, 797, 1972.
113. **Silverstein, S. C., Astell, C., Levin, D. H., Schonberg, M., and Acs, G.,** The role of lysosomes in the uncoating and activation of the reovirus genome, *Adv. Biosci.,* 11, 3, 1974.
114. **Ulmanen, I., Soderlund, H., and Kääriäinen, L.,** Role of protein synthesis in the assembly of Semliki Forest virus nucleocapsid, *Virology,* 99, 265, 1979.
115. **Wengler, G. and Wengler, G.,** Identification of a transfer of viral core protein to cellular ribosome during the early stages of alphavirus infection, *Virology,* 134, 435, 1984.
116. **Wilson, T. M. A.,** Nucleocapsid disassembly and early gene expression by positive-strand RNA viruses, *J. Gen. Virol.,* 66, 1201, 1985.
117. **Hannan, C., Raptis, L. H., Dery, C. V., and Weber, J.,** Biological and structural studies with an Adenovirus type-2 temperature sensitive mutant defective for uncoating, *Intervirology,* 19, 213, 1983.

Chapter 3

VIRUS-DIRECTED SUPPRESSION OF HOST TRANSCRIPTION

C. Fernańdez-Tomás

TABLE OF CONTENTS

I. INTRODUCTION

The inhibition of host genetic expression is a common but not invariable consequence of virus infection. It is apparent that different cellular genes are affected by viruses in different ways. Evidence published suggests the existence of different classes of cellular genes, distinguished by the effect of viral infection on their expression. On this basis, host cellular genes could be generally classified as induced, suppressed, and nonmodified genes. Some of these genes could be activated and suppressed during infection, with a similar genetic expression described for some early DNA viral genes. This is the case in several virus-infected cells where induction and repression of heat-shock, β-tubulin, and dihydrofolate-reductase genes (among others) are detected during different stages of infection (two-step kinetics expression). In these cases, one plausible explanation for suppression is that it results from a reversal of the activation mechanism, for instance, a decline in the levels of cellular or viral factors required for the activation mechanism, as has been demonstrated by several laboratories.

Since the inhibition of host cell mRNA production by animal viruses is exercised by different strategies and at different levels, to understand the molecular mechanism involved in virus-directed suppression it is necessary to analyze the nuclear and cytoplasmic mRNA from a single gene. This is a difficult task because of the low levels of the products from any individual gene. This could be partially overcome, analyzing a particular gene amenable to study and subjected to control in a defined manner. The use of these systems to characterize viral effects on host cellular gene expression offers advantages principally due to the availability of clones containing gene sequences that can be easily detected. Several examples of these cDNA clones have been reported already, such as the human heat-shock (HS) mRNA, β-tubulin mRNA, dihydrofolate-reductase (DHFR) mRNA, etc. Also, the availability of cell lines containing several copies of a structural gene per cell which produce mRNA and protein to a level higher than that found in other cells, has overcome the limitation of low levels of gene products. Such clones and cell lines provide ideal host systems in which to study the effect of virus infection on the expression of a single gene. Another approach often employed has been the analysis of the expression of DNA virus genomes as polyoma virus (Py), simian virus 40 (SV40) and herpes simplex virus (HSV) under coinfection with other viruses, usually RNA viruses (rhabdovirus, picornavirus, etc.). The use of these DNA viral systems as models to study eukaryotic gene regulation offers advantages over that of cellular genes, principally due to the availability of well-defined genomes and gene products, and the possibility of obtaining mutants defective in functions involved in regulatory events.

Besides the quantitation of a specific gene product by hybridization of total mRNA, the transcriptional activity of a gene could be measured by using in vitro transcription in isolated nuclei which has been demonstrated to faithfully elongate transcripts initiated at the time of harvest but to be defective in the initiation of new transcripts.[1] Therefore, the amount of label incorporated into a specific transcript in vitro accurately reflects the amount of transcriptional activity of that gene in vivo at the time of nuclei harvest. Such an assay provides a reasonable measurement of the primary transcription rates of a gene and allows a good correlation with the stability of mRNA, which could be characterized by the steady-state of host mRNA. Given a constant relative synthesis of specific mRNA, the relative steady-state nuclear levels of this RNA should reflect its relative stability and rate of transport to the cytoplasm. If, for example, a decrease in the cytoplasmic appearance of a particular RNA is due solely to a block in RNA transport, one might expect an accumulation of the RNA in the nucleus. However, if the decreased cytoplasmic levels reflect a decreased nuclear stability of this RNA, one would expect low nuclear levels of the specific RNA at this time. Measurement of the relative concentration of steady-state specific nuclear RNA should help distinguish between such alternatives.

As will be described, all these approaches have been very useful in the understanding of the effect of virus on the mRNA production in infected cells.

II. MODIFICATION IN THE EXPRESSION OF SOME EUKARYOTIC GENES DURING INFECTION

In this section, the characteristics of some eukaryotic genes that have been employed to study viral effects on host cell mRNA production will be summarized. An important consideration is that the eukaryotic genes studied have been assayed, more for the accessibility to cDNA clones and specific cell lines than for the eventual metabolic implications during viral infection, maybe with the exception of cytoskeleton (see below) and stress proteins.

The eukaryotic genes that will be described were classified depending on the effect of viral infection on their expression (see Table 1): (1) induced and repressed genes (two-step kinetics), and (2) repressed genes.

A. Induced and Repressed Eukaryotic Genes (Two-Step Kinetics of Expression)

Although the main purpose of this review is suppression of host cellular mRNA production during virus infection, it is important to call attention to the fact that some of these repressed genes are activated before shut-off takes place. One example of this transient stimulation has been described in picornavirus-infected cells. Studying the number of chromatin-bound host RNA polymerase II in polio-infected HeLa cells, Flores-Otero et al.[2] have reported that very early in infection (see Figure 1A) there is a transient increase in the number of molecules of the enzyme engaged in chain elongation. This is coincident with a transient stimulation of RNA synthesis (see Figure 1B). Besides this general pattern of induced-repressed host cellular RNA synthesis, some reports of two-step kinetics of expression have been reported for specific genes. A brief description of some of these eukaryotic genes and their expression during virus infection is included.

1. β-Tubulin Gene

Microtubules are filamentous structures present in virtually all eukaryotic cells and are principal components of the cytoskeleton. The major components of the microtubules are the α- and β-tubulin. Heterodimers of these proteins form subunits that are polymerized into microtubules. In mammals the genes that encode α- and β-tubulin proteins are members of two distinct families that contain 15 to 20 members and include pseudogenes as well as functionally expressed sequences. The ongoing task of identifying all of the functional human

Table 1
EUKARYOTIC GENES MODIFIED DURING VIRUS INFECTION

	Expression	Infecting virus	Ref.
Host genes[a]			
DHFR	Two-step kinetics	Adeno-2	15
β-tubulin	Two step kinetics	Adeno-5	4
HS	Two-step kinetics	Adeno-5	9, 14
		HSV-1	10, 11
		NDV	189
		SV40, Py	12
Class I major histocompatibility antigen	Repressed	Adeno-12	190
β-Actin	Repressed	Vaccinia	34
		Influenza	32, 33
		HSV	29
β-Globin	Repressed	HSV	25, 28, 29
GPD	Repressed	Influenza, HSV	32
Viral genes			
SV40	Two-step kinetics	Poliovirus	191
		Adeno	91
	Repressed	VSV	20, 21
Py	Repressed	HSV	114, 115
HSV	Repressed	VSV	137
Adeno	Repressed	HSV	113, 116
ALV	Repressed	Influenza	33

[a] Included are those genes more extensively studied.

tubulin genes has resulted in the complete characterization of three β-tubulin genes. Because tubulins are ubiquitous in eukarotic cells and because of their role in diverse cellular functions, an important question concerns the pattern of tubulin gene regulation. Studies of the gene expression of this protein have indicated that (1) the level of unpolymerized tubulin itself regulates the level of tubulin mRNA, (2) tubulin genes are differentially regulated during development, and (3) some kind of regulation exists that depends on tissue specificity.[3]

A good example of the modifications of α-tubulin gene expression during virus infection is that of Stein and Ziff,[4] which showed that the transcription of host β-tubulin nuclear RNA is stimulated and then repressed during the early phase of adenovirus 5 (adeno-5) infection. As can be seen in Figure 2A, the rate of transcription of β-tubulin mRNA followed a kinetics that parallels transcription of the adeno early genes themselves, as has also been described for HS genes (see below).

2. HS-70 kdalton Genes

In response to an increase in temperature, cells from a wide range of organisms rapidly synthesize a small set of proteins, the heat shock proteins (HSP) (molecular weights 100, 70, and 37 kdalton), while repressing the synthesis of most other proteins. A similar response can be induced at constant temperature by other stressful agents including amino acid analogues, chelatin drugs, heavy metal ions, arsenite, and ionophores. The most highly conserved HS protein is a 70-kdalton polypeptide whose expression could be controlled by heat shock in cells where cloned DNA copies of the gene were introduced, suggesting a considerable conservation of the regulatory signals.

The expression and characterization of the HS gene products were initially described in

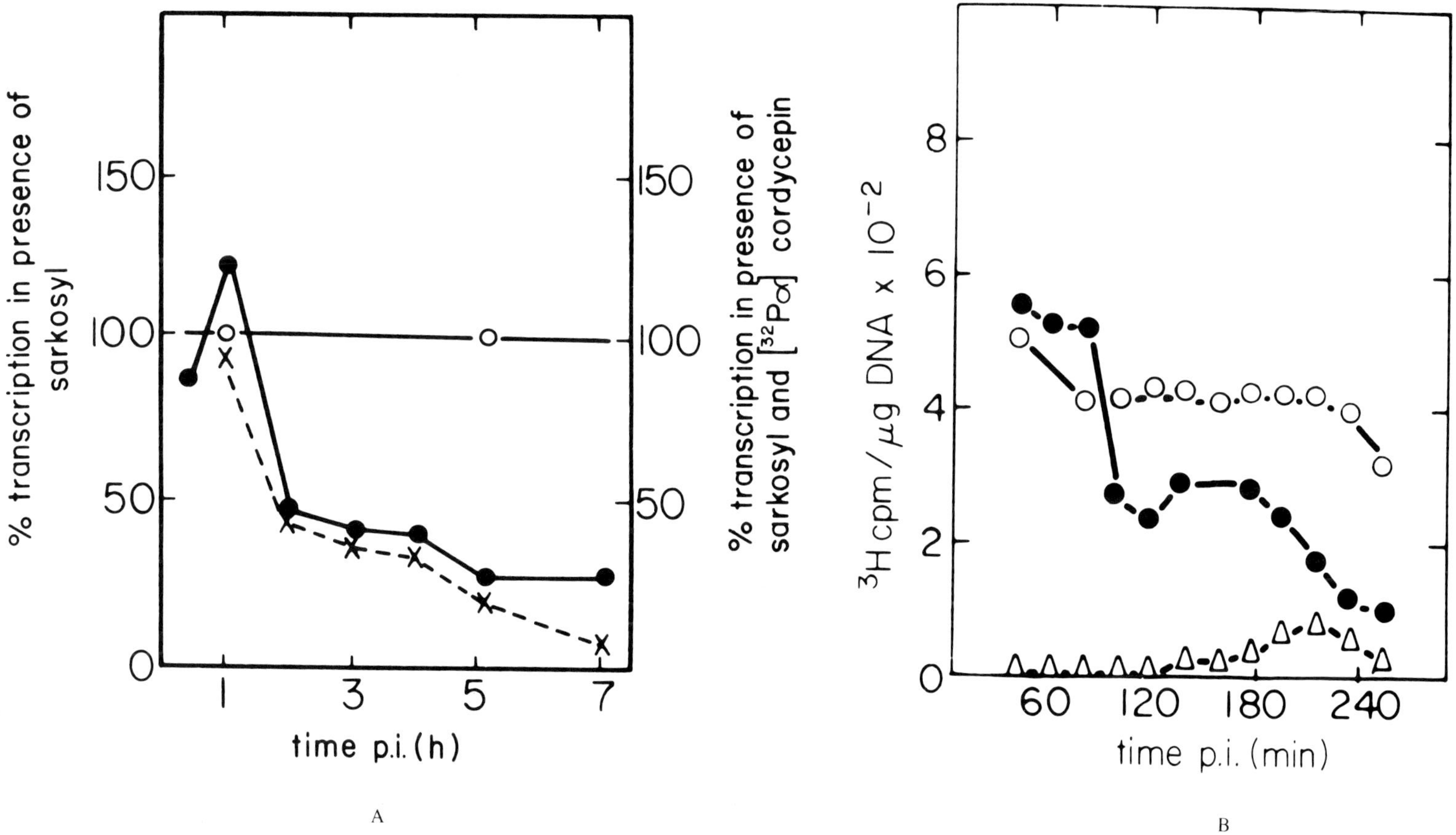

FIGURE 1. Poliovirus suppression of host transcription in HeLa cells.[2] (A) Percentage of control nuclear transcription activity (○), measured as Sarkosyl-stimulated endogenous RNA polymerase activity by (^{3}H)UMP (●) or cordycepin (α-^{32}P)triphosphate (X) incorporated in polio-infected nuclei. (B) RNA synthesis in HeLa cells after poliovirus infection, determined by (^{3}H)uridine incorporation; (●) poliovirus-infected cells; (Δ) poliovirus-infected cells in the presence of actinomycin D; (○) mock-infected cells.

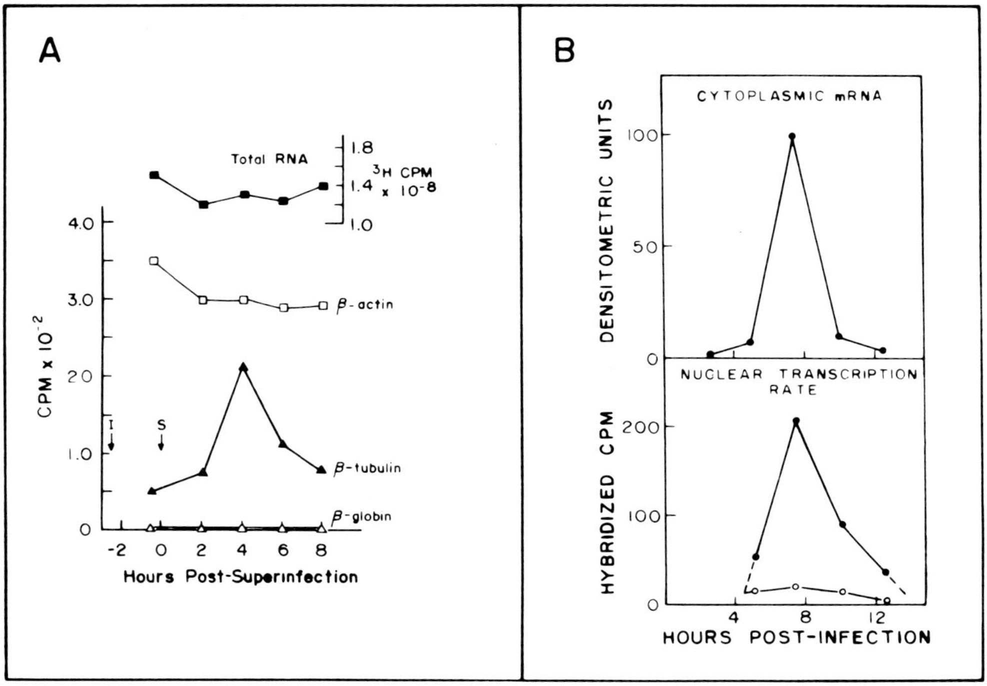

FIGURE 2. Transient stimulation of host cellular genes in adenovirus-infected cells. (A) HeLa cells were infected with a deletion E1a mutant (d1313) and then superinfected with a helper virus (d11327). (^{3}H)Uridine labeled-nuclear RNA was hybridized to the described specific probes.[4] (B) Cytoplasmic and (^{32}P)UMP labeled-nuclear RNA from adenovirus-infected cells were hybridized to HS-specific cDNA probes.[9] (●) Infection with wild-type virus; (○) infection with the E1a deletion mutant d1312.[9]

Drosophila, and the intensive study of these genes has led to the cloning of the copies of all seven of the HS genes and to sequencing of their promotor regions.[5,6] Pelham[7] described two functional elements in HS-70 gene, the TATA box and an upstream element, somewhat variable in position that are required in vivo for efficient use of the TATA box. It seems that the upstream element is analogous to that described for other eukaryotic promoters (see Section III), but is only functional in heat-shocked cells. Also Ayme et al.[8] described a sequence adjacent to the TATA box responsible for heat-shock activation.

Because of the interest in these stress genes, several laboratories have studied their expression under virus infection. An increased synthesis of the 70 kdalton protein has been reported in cells infected with several virus, as adeno-5,[9] herpes simplex virus type 1 (HSV-1),[10] HSV-2,[11] simian virus 40 (SV40), and polyoma (Py) virus.[12]

Kao and Nevins[9] studied the amount of HS-70 kdalton cytoplasmic poly(A)-RNA and the amount of primary transcription product in adeno-infected HeLa cells. As can be seen in Figure 2B, the abundance of the heat-shock mRNA initially increased dramatically as a result of adenovirus infection. Measuring transcription directly by allowing isolated nuclei to label nascent RNA in vitro, an early 20-fold increase of the HS gene was detected. Although it remains to be determined whether the HS induction in these cases is due to the action of a specific viral product, a tight control of expression of the HS gene after infection seems to exist. As shown in Figure 2B shortly after the peak transcription rate is achieved, transcription of the gene drops to the same level as that before infection.

This induced-repressed expression of HS genes is in contrast with the expression of the majority of cellular genes that continue to be transcribed during adenovirus infection at a normal rate.[13] In fact, the HS genes appear to be regulated in much the same manner as the early adenovirus genes, probably by viral products (see Section IV). Since the increased synthesis of the HS proteins is only one of many changes observed in infected cells, it is difficult to speculate about their possible role, except for the recent report by Kao et al.,[14] which hypothesized that an adeno-Ela-like activity was responsible for the cell cycle control of the HSP70 expression.

3. Dihydrofolate Reductase (DHFR) Gene

DHFR catalyzes the reduction of dihydrofolate to tetrahydrofolate, an essential cofactor for *de novo* synthesis of glycine, purines, and thymidine. The folate analogue methotrexate (Mtx) inhibits DHFR activity and is lethal to cells that do not have exogenous sources of glycine, purines (hypoxanthine), and thymidine. Resistance to Mtx in mammalian cells can be acquired by amplification of the DHFR genes and the consequent increased levels of DHFR synthesis. Mtx-resistant cells that express an altered form of DHFR exhibit decreased affinity for Mtx.[15] A cDNA copy of this altered gene has been isolated, and can serve as a dominant selectable marker in wild-type culture cells.[15]

The analysis of the mode of expression of the DHFR gene in mammalian cells has revealed an unexpected complexity due to the existence of multiple forms of DHFR mRNAs. These multiple mRNAs differ mainly in the length of the 3′-untranslated tail,[16] which can be up to severalfold longer than the reading frame. The evidence is compatible with the idea that the same primary transcript could give rise to different forms of mRNA by processing and polyadenylation at different sites of the molecule. Some other kind of regulation in DHFR genes has been recently detected by Masters and Attardi,[17] who reported other minor 5′ ends in DHFR mRNA mapping in nucleotide positions −449 to −448 upstream of the DHFR gene; these quite different mRNAs account for approximately 1% of the DHFR-specific polysomal poly(A)-containing RNA.[17]

Yoder et al.[15,18] examined the effect of adenovirus infection on the expression of DHFR gene in a line of human cells (S3-500) resistant to Mtx that contains amplified and protein DHFR-specific RNA. Threefold induction of enzyme synthesis was accompanied by concomitant increases in the level of steady-state DHFR cytoplasmic mRNA at 19 hr p.i. (postinfection), and in the rate of appearance of mRNA in the cytoplasm. Late in infection, a decline in DHFR protein and mRNA (below that observed in uninfected cells) was detected.

4. SV40 Late Genes

The genome of SV40 encodes two classes of viral RNA, early and late, which are transcribed during lytic infection. RNA is synthesized from the early SV40 genes throughout the lytic cycle, whereas late genes are temporally regulated in such a way that activation does not occur until the onset of viral DNA replication. The SV40 late promoter does not contain a consensus TATA sequence (see Section III), and its optimal expression requires the presence of the 21 and 72 bp (base pair) repeats and an intact origin of replication.[19]

The genome of SV40 can be regarded as an extrachromosomal transcription unit whose expression is accomplished with enzymes provided by the host cell. Because of its genetic simplicity, SV40 has served as a model for many studies on the molecular details of DNA or RNA synthesis under coinfection with VSV[20,21] and adenovirus.[22] Diaz-Moreno et al.[191] studied the effect of poliovirus infection on SV40 late transcription. As can be seen in Figure 3, the SV40 transcription complex (TC) isolated from poliovirus-coinfected CV-1 cells shows a transient stimulation of in vitro RNA chain elongation very early after coinfection. This is coincident with the temporal stimulation of host cell RNA synthesis detected in polio-infected HeLa cells (see Figure 1).

B. Repressed Eukaryotic Genes

The suppression of eukaryotic genes during viral infection has been well-documented.

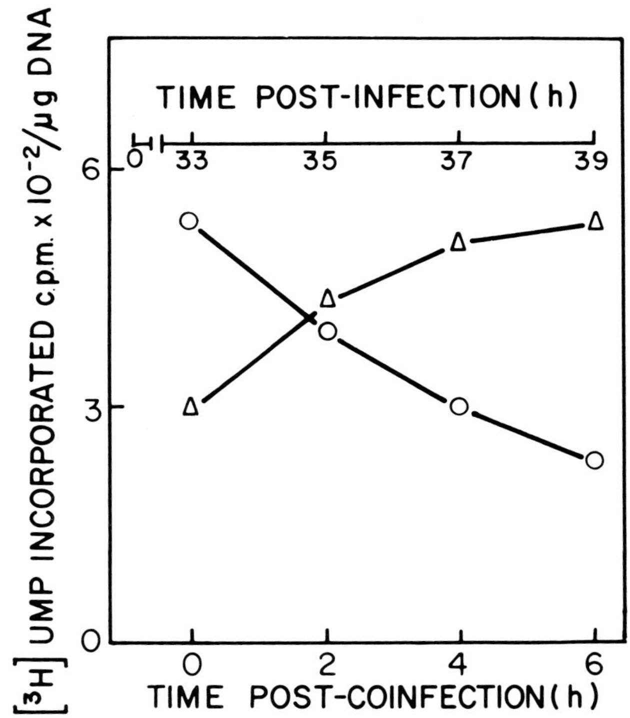

FIGURE 3. Activity in vitro of the SV40 transcriptional complex isolated from SV40-infected (△) and poliovirus-coinfected CV-1 cells (○).[191] Thirty-two hr after SV40 infection cells were superinfected with poliovirus and at different times after coinfection, the transcriptional activity of the SV40-TC was measured as (^{3}H)UMP incorporated during 30 min in vitro transcription.

Although most of the virus-infected cell systems studied have shown that RNA polymerase II transcripts are primarily affected, host mRNA degradation has also been reported. VSV infection of mouse cells (MPV-11)[23] (see Figure 4) induces an immediate and rapid inhibition of host cell RNA synthesis with no alteration of mRNA stability. While in HSV-1-infected cells, this effect is characterized by a progressive and eventual loss of RNA polymerase II activity;[24] host mRNA degradation has also been reported.[25] Since the suppressor effect has been studied in the expression of particular eukaryotic genes, a brief analysis of those repressed specific genes is described. A more detailed discussion on viral suppressor strategies is included in Section IV.

1. Globin Genes

Hemoglobin is composed of heme residues and two chains of α-like globin polypeptides. Each type of chain is encoded by a family of genes under differential but coordinate control during development. Blotting studies on genomic DNA have shown that the activity of a particular globin gene may be affected by deletions occurring some distance from it, which supports the idea that the human α- and β-like genes are ordered in two separate clusters. This has been confirmed in chicken, rabbit, and mouse β-like genes.[26]

The knowledge of globin genes has advanced during the last 5 years, but there are still some questions which remain to be answered, for instance, the mechanism used for the regulation during development. It seems likely that sequences immediately upstream from

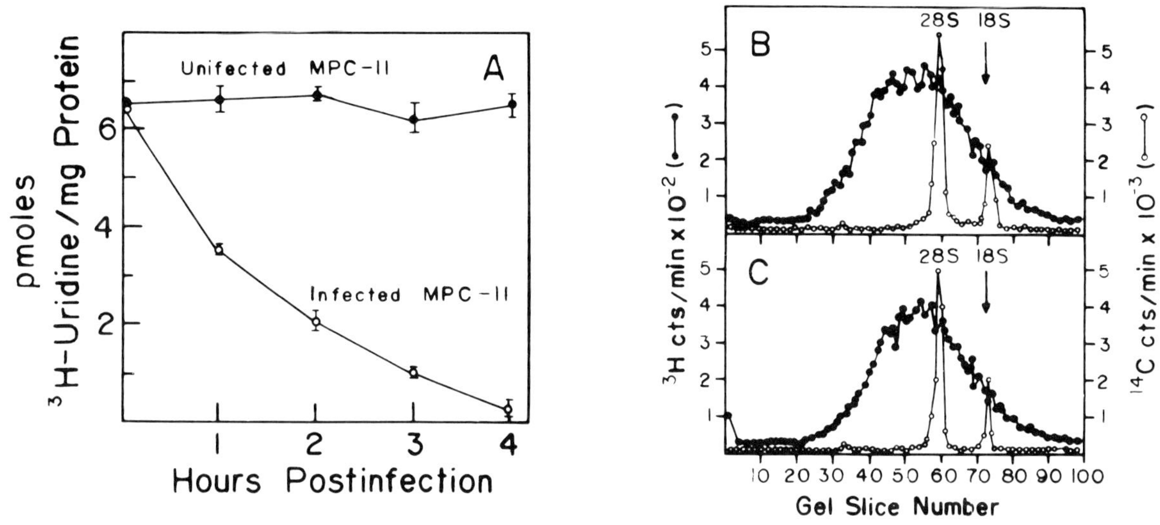

FIGURE 4. Effect of VSV infection on the synthesis and stability of MPC-11 cellular RNA.[33] (A) Acid-insoluble (^{3}H)uridine in mock and VSV-infected cells. (B) and (C) Electropherograms of host RNA from mock and VSV-infected MPC-11 cells, respectively, showing the polydisperse nature of the nuclear (^{3}H)uridine labeled-RNA. (^{14}C)Uridine labeled-28 and 18*S* ribosomal RNA were included as a marker.

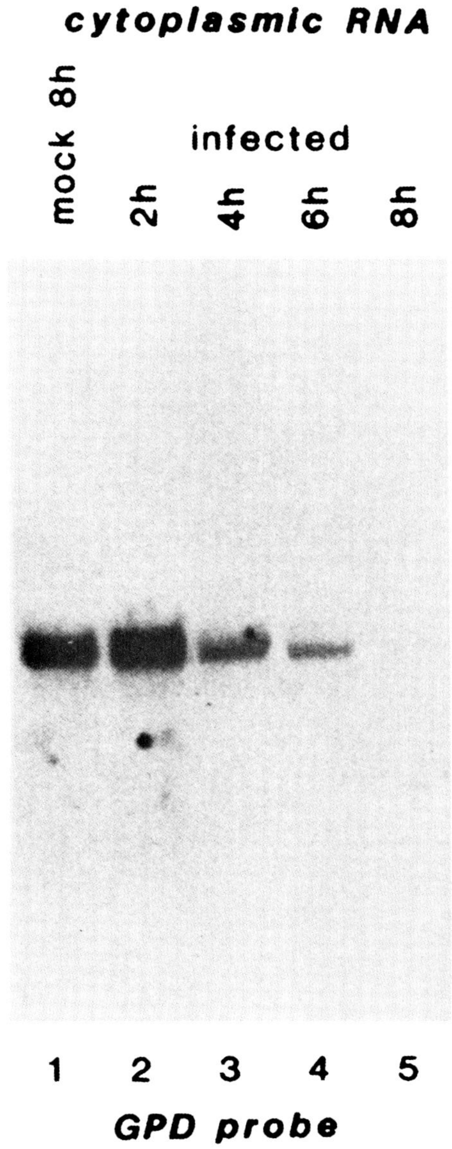

A

FIGURE 5. Stability of GPD gene transcripts after FPV (A) and HSV (B) infection.[32] Cytoplasmic RNA was extracted at various times after infection and, after transfer by blotting procedures, the sample was hybridized to (^{32}P)labeled plasmid DNA containing a copy of GPD gene.

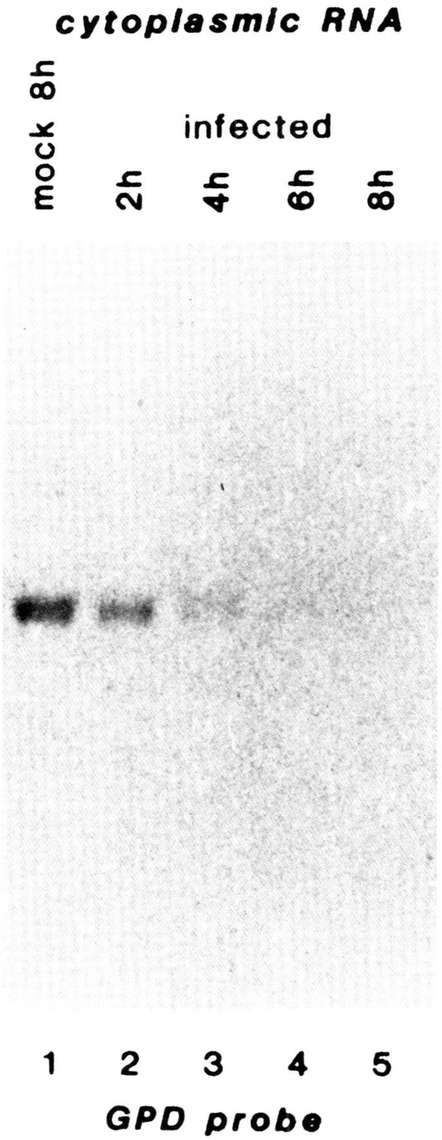

FIGURE 5B

the cap site around the putative promoter sites are not responsible for this regulation since a comparison of different β-globin genes (expressed at different stages of development) showed small differences over the first 200 nucleotides upstream of the cap site. Perhaps the more striking feature of β-globin promoters is the 5′ termini RNA located at 235 nucleotides upstream of the β-globin mRNA cap site. These upstream RNA sequences are covalently linked to mRNA sequences derived from the structural gene. In this respect, an interesting observation was made by Carlson and Ross,[27] showing that the 5′ end of these low abundance RNAs is transcribed in vitro by RNA polymerase III. These observations suggest that transcription of these RNAs is initiated by RNA polymerase III but that they extend into the mRNA-coding region, usually transcribed by RNA polymerase II.

Several laboratories have studied the expression of globin genes during HSV-1 infection. Nishioka and Silverstein,[25,28] using friend erythroleukemia (FL) cells, have shown that soon after infection with HSV-1, there is a dramatic reduction in *de novo* synthesis of globin. Recently, Mayman and Nishioka[29] demonstrated that the steady-state of globin mRNA decreased progressively at early stages after HSV-1 infection.

2. Glyceraldehyde-3-Phosphate Dehydrogenase (GPD)

GPD is a key enzyme in the control of glycolysis, consisting of an assembly of four identical subunits. The entire mRNA is 1269 nucleotides long and contains 71 and 196 bases of 5′ and 3′ noncoding regions, respectively. It is encoded by a single functional gene in man, mouse, and chicken genomes. It has been suggested that rat tissue-specific differences in the enzyme content are regulated at the posttranscriptional level and that an unusually high number of GPD-related sequences were present in the rat genome.[30] Recently Marty et al.[30] constructed and sequenced a full-length rat muscle GPD cDNA clone, and used it to corroborate that GPD mRNA contains strikingly conserved structural features, thus supporting the idea that the production of these mature mRNAs might be regulated at the posttranscriptional level.

Hanauer and Mandel[31] isolated cDNA clones corresponding to a 1210 bp sequence, which covers all of the protein-coding region, most of the 5′ noncoding region, and part of the 3′ noncoding region. This type of study has allowed the characterization of some features in the organization of this gene, such as the localization of the canonical TATA region, and 9 bp homologous sequences at −58 and −93 and at −434 and −355 positions. Unfortunately, the function of these inverted sequences remains unknown.

Inglis[32] reported experiments designed to test the stability of GPD-mRNA during infection of chicken embryo fibroblast (CEF) with the avian influenza fowl plague virus (FPV), and during HSV-1 infection of BHK cells. As shown in Figure 5, the analysis of cytoplasmic RNA, by hybridization with (^{32}P)-labeled probes of GPD gene, shows a decline in specific mRNA during infection with FPV (Figure 5A) and HSV-1 (Figure 5B).

3. Avian Leukosis Virus (ALV)

ALV RNA is transcribed, as integrated proviral DNA, by the host cell polymerase II and is subsequently processed like host cell transcripts. The location of the promoter for retroviral RNA synthesis was unclear until recently. Restriction enzyme analysis has revealed several structural features of the integrated viral DNA that play important roles in viral RNA synthesis (see Section III). One important feature of the viral DNA is a sequence of approximately 350 nucleotides, repeated at each end (LTR), which consists of three defined segments and appears to contain the viral promoter.

As has been mentioned, the use of viral systems to study gene regulation offers several advantages. With this approach, Katze and Krug[33] studied the effect of influenza virus coinfection on the rate of transcription of the ALV-specific RNA. Proviral DNA was integrated into CEF cells as a result of prior infection by ALV. At various times after influenza virus coinfection, cells were supplemented with (^{3}H)uridine and the amount of specific cytoplasmic and nuclear ALV RNA was determined by hybridization to filters containing the full-length DNA clone of virus. As shown in Figure 6A, ALV RNA synthesis was dramatically inhibited and degradation of nuclear ALV RNA was also observed.

4. β-Actin Genes

The actins play a major role in cytoskeletal structure of most eukaryotic cells. Although much information is available describing the biochemistry and structure of the actin proteins, questions such as gene number, genome structure, nucleotide sequence of actin genes, and possible regulation of actin gene expression have been described only recently. In mammals, at least six different actin proteins have been identified, indicating the presence of six actin

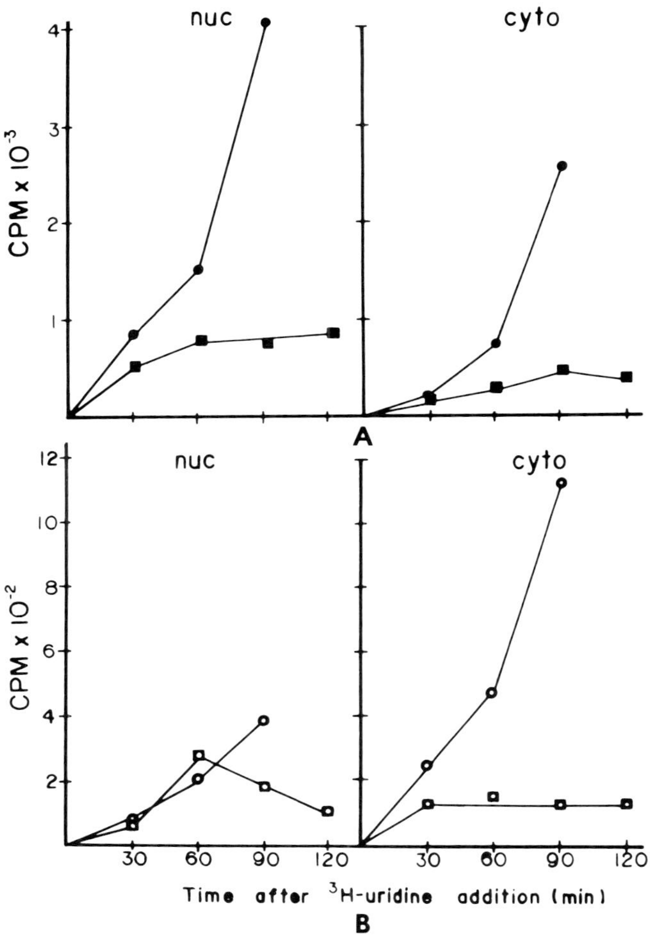

FIGURE 6. Kinetics of labeling of ALV (A) and β-actin (B) nuclear and cytoplasmic transcripts after influenza virus infection of CEF cells.[33] (^{3}H)Labeled nuclear and cytoplasmic RNA, obtained from mock-infected (●) or influenza-infected (■) cells were hybridized to specific DNA clone.

genes. The two cytoplasmic forms, β- and α-actins, are major components of the cytoskeleton system.

Several laboratories have studied the expression of actin mRNA during virus infection. Katze and Krug[33] reported that within 2 hr after influenza virus infection, the rate of transcription of β-actin decreased 40 to 60% and that the transcripts that continued to be synthesized did not appear in the cytoplasm as mature mRNA (see Figure 6B). Mayman and Nishioka[29] demonstrated that the steady-state levels of actin mRNA decreased progressively at early times after infection of friend erythroleukemia (FL) cells with HSV-1. Inglis[32] analyzed the cytoplasmic β-actin RNA isolated at different times after FPV and HSV-1 (an avian influenza virus) infection of chicken embryo fibroblast (CEF). Analysis by blot hybridization with plasmids containing sequences of the β-actin gene has shown, in both cases, a reduction in the amounts of cellular β-actin mRNA in the cytoplasm (Figure 7). The effect of vaccinia virus infection in mouse L-cells gene expression was also studied, using the highly conserved cDNA sequences that encoded chicken β-actin.[34] Hybridization analysis of cytoplasmic poly(A)$^+$ RNA demonstrated that infection induced a rapid and progressive degradation of β-actin mRNA.

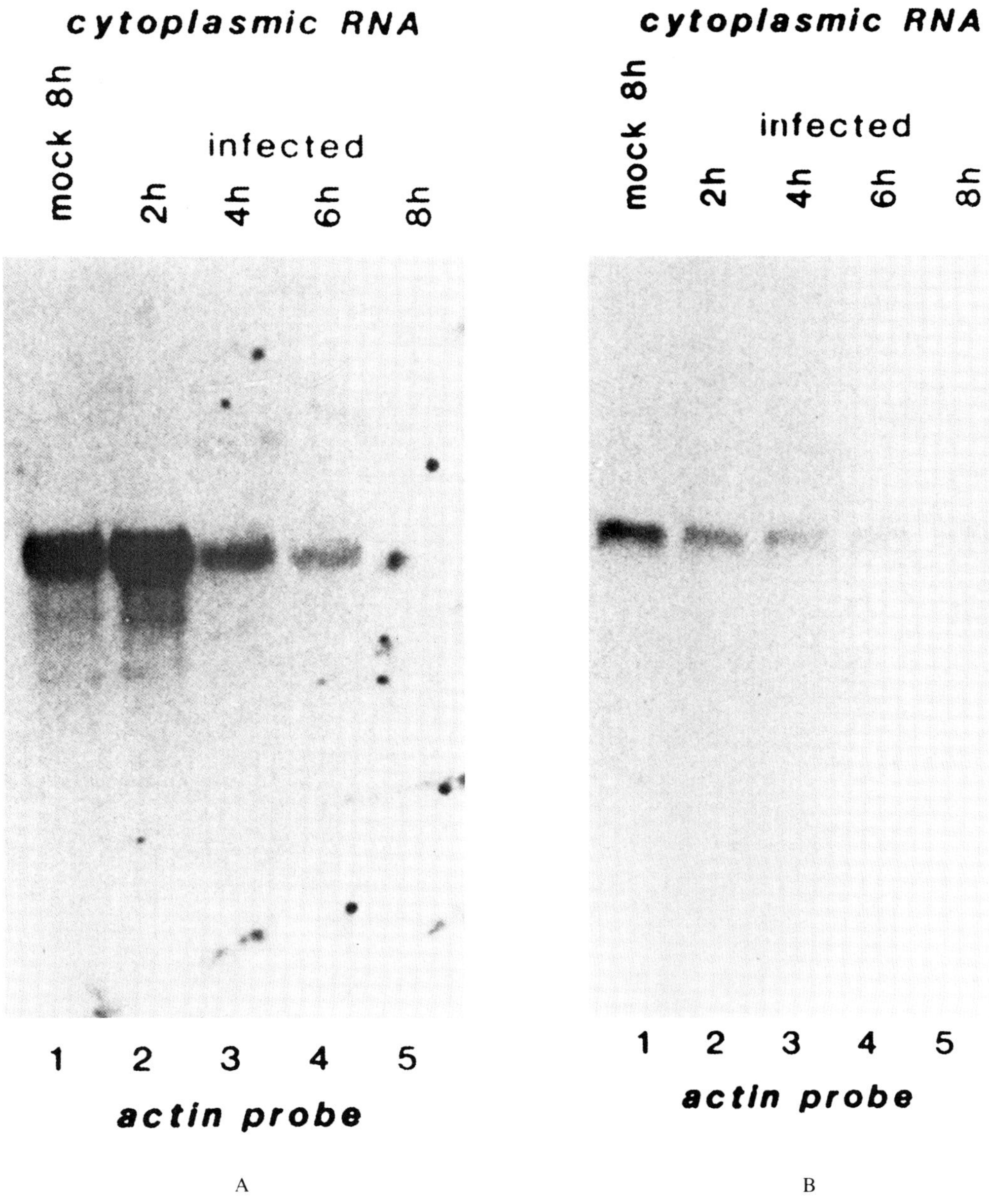

FIGURE 7. Stability of β-actin gene transcripts after FPV (A) and HSV-1 (B) infection.[32] The experiment was the same as that shown in Figure 5, except that the DNA probe contained the sequence of β-actin.

Some different results have been published recently by Khalili and Weinmann,[35] who reported the stabilization of actin mRNA during adenovirus infection. Employing homologous actin probes, these authors studied the transcription, accumulation, and turnover of human-specific mRNA in adeno-infected HeLa cells. Their observations suggest that although the rate of transport of actin-specific mRNA to the cytoplasm is reduced, an increase in mRNA half-life results in normal levels of cytoplasmic actin mRNA.

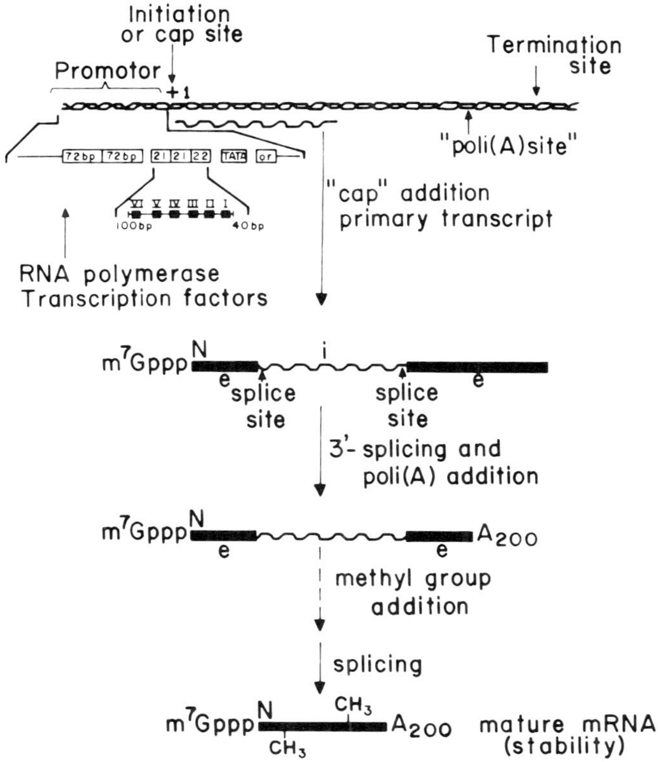

FIGURE 8. Steps in the regulation of mRNA production in eukaryotic cells.[53,76]

III. AN OVERVIEW OF GENE EXPRESSION MEDIATED BY RNA POLYMERASE II

Since, as has been described, during virus infection host cell mRNA is affected at different levels, a brief review of the different elements in the control of eukaryotic mRNA production is included (see Figure 8). This section is divided into (1) regulation at the transcriptional level, (2) control during mRNA processing, and (3) stability of mature cytoplasmic mRNA.

A. Regulation at the Transcriptional Level

The analysis of the signals for transcription was facilitated by progress in the manipulation of recombinant DNA, and by the development of in vitro systems that allow the assay of the effects of template-directed alterations. Deletion analysis of specific promoter elements allowed the characterization of some regions required for both in vitro and in vivo genetic expression. Point mutations in the promoter region have also demonstrated that specific sequences play an important role in determining the level of transcription in a variety of eukaryotic promoters.

1. Initiation (The Promoter)

Recent progress in transcriptional initiation regulation has identified several classes of eukaryotic transcription control regions that seem to be important for the regulation of the transcriptional process.[36] Although the mechanism by which these DNA elements regulate transcription is not exactly known, the 5′-upstream sequences of certain eukaryotic genes are known to have unusual structures in vivo.[37-41]

Perhaps the system most extensively studied is the SV40 promoter region. The major regulatory sequence in this virus, which controls initiation of early (E-SV40) and late (L-SV40) transcription, is located within about 400 bp of DNA (see Figure 8). This is also true for the promoter region of adenovirus, polyoma virus, Rous sarcoma virus, HSV, and BK virus,[42] and for some regulatory elements in eukaryotic cells, i.e., the thymidine kinase gene,[39] the immunoglobulin heavy chain gene,[43] the insulin and chymotrypsin genes,[44] the HS gene,[45] and the α1-globin genes.[41]

At least three specially distinct elements comprise the eukaryotic promoters, although some differences between them have been reported. The elements include the 17 bp A-T-rich sequence, the G-C-rich 21 bp repeats, and the 72 bp repeat enhancer (see Figure 8).

The A-T-rich region (TATA box), which shows a strongly conserved distance of 30 bp from the nucleotide coding for the first base of the mature mRNA ("cap site"), positions the site of RNA initiation.[46] In promoters where the TATA box is absent, cap sites are detected even though TATA-like sequences are not found upstream from these cap sites. Such is the case with several viral genes, including the L-SV40 promoter, adenovirus E2a and IVa2, and the hepatitis B virus surface antigen gene, which do not have any TATA-like sequence upstream of the transcription initiation site. The localization of a substitute start signal has been appropriated to an 11-nucleotide control element located 21 to 30 bp upstream of the L-SV40 and hepatitis B virus surface antigen promoters.[47,48]

The more complicated 21 bp repeat region of SV40 contains G-C-rich motifs (see Figure 8) which (acting in a nonpolar fashion) are essential for efficient transcription of SV40, and stimulate the heterologous adeno-2 late promoter. As has been suggested by Vigneron et al.,[49] these regions may facilitate recognition of the TATA box. Four G-C-rich motifs (III to VI), which comprise the binding site for the transcription factor SPI,[50] are important for E-SV40 and L-SV40 promoters. The requirement for I and II appears to be unique for the E-SV40 promoter. This is in agreement with other reports,[51] suggesting that the SV40 G-C-rich constitute independent transcriptional elements of both promoters. Although less characterized, transcriptional elements at nucleotide position (np) 72 to 114 have also been reported.[52,53]

Since the discovery of transcription enhancer sequences in the genome of the DNA viruses, SV40 and polyoma,[46] these elements have been shown to play an important role in the control of both viral and cellular gene expression. The enhancer, localized upstream of the G-C rich 21 bp-repeats (see Figure 8), act in "cis" to increase the amount of RNA produced from linked genes. These elements from different eukaryotic promoters shared some common properties.[43,54-56] They stimulate transcription in an orientation-independent manner from the same starting sites on the natural promoters as well as on substitute promoter elements. They are most efficient if located directly upstream from the promoter element, but they still stimulate transcription when they are either immediately downstream from the promoter element or further upstream. Enhancers activate proximal promoter sequence in preference to more distal ones (the "distance effect"). Although the mechanism by which enhancers facilitate transcription is largely unknown, some evidence supports its role as the polymerase "entry site",[57] since SV40 enhancer increases the number of associated RNA polymerase II molecules.[58,59]

Another well-studied system, the long terminal repeat (LTR) of exogenous avian retrovirus, deserves a brief mention. This control region is highly efficient and capable of directing the transcription of very high levels of virus-specific RNA and acts in an unusually wide range of cells. LTRs cloned from different strains of avian retrovirus can have quite different but related sequence arrangements.[60] The enhancer elements of Maloney murine leukemia virus are localized in a tandemly repeating region (75 bp long) that lies 5′ to the TATA box in LTR. It has been shown that these tandem repeats are required for LTR promoter activity and biological activity.

Besides these promoter sequences, several proteins are involved in regulation of transcription initiation: the RNA polymerase and the so-called transcription factors. In eukaryotic cells, purified RNA polymerase II is known to be composed of three different subspecies named IIO, IIA, and IIB, with the last two constituting the bulk of most purified enzymes from different sources. However, several laboratories[61-65] have found, by immunodetection using anti-RNA polymerase antibodies, that IIO is present in whole cells in higher proportions than already reported. The finding of IIO associated to cellular chromatin suggests a role for this subspecies in the in vivo initiation event.[63-65] Finally, the multiplicity of protein factors other than polymerase involved in transcription initiation has been partially defined by chromatographic fractionation of cellular extracts,[50,66-70] and their activity in vitro under the direction of well-known eukaryotic promoters and purified RNA polymerase has been reported.[71-73]

2. mRNA Termination

Another transcriptional control is at the level of RNA transcription termination, which could be defined as the cessation of RNA chain elongation at a specific site, followed by the release of the new RNA molecule. In eukaryotic cells, the RNA polymerase III termination site has been related to a TTTT cluster flanked by GC residues required for its precise termination;[74] this has also been suggested for the type I enzyme.[75] In the case of RNA polymerase II, the 3′ end of the mRNA is produced by an endonucleolytic cleavage of a large precursor transcript, followed by poly(A) addition.[76] That the site for the addition of poly(A) does not represent the only signal for the termination site is supported by the evidence in the production of the nonpolyadenylated histone mRNA, whose 3′ end is also a product of RNA cleavage.[77]

Studies with the β-globin (major) gene have shown that transcription terminates within a region 600 to 1500 nucleotides downstream of the poly(A) site,[78-80] suggesting that no discrete sites of termination exist. The putative termination region (denominated the "F" region)[79] contains a short stem-loop structure in the first 200 bp followed by a stretch of T residues, and three polyadenylation signals (AATAAA) in a span of 300 nucleotides in the beginning of the F region. Falck-Pedersen et al.[80] speculated about two possibilities that might explain how termination elements affect mRNA production: a cis-acting element could influence the activity of promoters localized downstream from the termination site or, alternatively, a functioning terminator might interfere with the stimulating effect of distant enhancers.

3. mRNA Processing

The sequence of events that converts a primary RNA transcript into a mature cytoplasmic mRNA begins very soon after initiation of an RNA chain with the addition of a m7Gp residue (cap) to the first nucleotide (see Figure 8). Most of these cap sites detected both in vivo and in vitro are related to U or A residues in the RNA.[81]

The posttranscriptional cleavage and poly(A) addition usually precede the final splicing steps in mRNA processing, although splicing can occur in the absence of poly(A) addition.[82] If poly(A) synthesis is inhibited, mRNA is formed and can, at least in part, enter polysomes, but it does not accumulate as stable mRNA. From this evidence it was postulated that the role for the 5′ cap structure in the cytoplasmic stabilization of mRNA[83,84] was also true for poly(A).[85] Although the majority of the mature eukaryotic mRNAs contains the 3′ poly(A) segment and the 5′ cap structure, there are exceptions, such as the polio and reovirus mRNA at their 5′ end,[86,87] and the histone mRNA at their 3′ end.[88] Finally, although it is known that methylation of mRNA (1 out of 400 Ap residues) occurs before splicing,[76] the role of methylation in mRNA production remains unknown.

4. Stability of Mature Cytoplasmic mRNA

At the cytoplasmic level, the fate of the mature mRNA molecules differs, as has been demonstrated by the wide range in the half-lives of different mRNAs in the same cell,[76] and of the same mRNA under different conditions, as is the case of the casein,[89] vitellogenin, and the very low density lipoprotein[90] mRNAs, under hormonal stimulation.

IV. VIRAL STRATEGIES

Different viruses exert different strategies to suppress host mRNA production. Although no general rules can be established at this moment, it looks as though the primary transcription is almost always affected. Some viruses affect the quantity of mature cytoplasmic mRNA at more than one step in mRNA production.

A. DNA Viruses

Productive infection of eukaryotic cells by almost all DNA viruses results in an ordered series of events culminating in the death of the cell and the release of progeny virus. DNA virus infection alters host cell metabolism to such an extent that gene products become the dominant species during the late stages of infection. The virus-induced host RNA modifications apparently require viral genes to be expressed early in infection. This is the case with a variety of DNA viruses, although the suppressor mechanism is only partially understood. To date, the only viral DNA genes identified that function as regulators of transcription in mammalian cells are the adenovirus Ela gene and the HSV immediate early gene.

1. Adenovirus

With the type C adenoviruses (i.e., Adeno-2 and Adeno-5) the early viral gene expression program depends on products encoded by the viral immediate-early Ela transcription unit. This unit acts in "trans" to induce transcription from at least four other viral early promoters.[91,92] The Ela encodes two early proteins: 289 and 243 amino acid long (from the 13S and 12S mRNA, respectively);[93-95] the large polypeptide has been implicated in promoter activation, while both products have a role in cell transformation.[96,97] The mechanism for induction appears to be transcriptional[91] although there may also be posttranscriptional effects.[98] The possibility that this adenovirus control mechanism could also operate to induce or repress genetic expression in the host cell infected-cell has received support. Recently, several groups have described induction and repression of host and viral mRNA through the action of Ela gene product, although the regulatory mechanism of the Ela effect has not been clearly defined. This is the case of the early SV40 genes,[99] the β-tubulin gene,[4] the 70 kdalton HS gene,[9,100] the human histone genes,[98] and the human β-globin gene.[101] The activation of some of these particular genes has been discussed in Section II.

Adenovirus can repress primary transcription of mRNA and ribosomal RNA,[102] but it has also been documented that newly-synthesized cell mRNAs fail to reach the cytoplasmic, which must be due to a failure in their transport mechanism.

a. Suppression of Host Transcriptional Activity

As described in Figure 2, Kao and Nevins[9] reported an adenovirus-induced drop in the production of HS mRNA, preceded early in infection by an increase in cytoplasmic HS mRNA in cells infected with wild-type virus. This effect was not detected in HeLa cells superinfected with an Ela deletion mutant (d1312). In another set of experiments, the authors isolated the nuclei at different times after infection and incubated in vitro in the presence of (^{32}P)UTP. As with cytoplasmic mRNA, the nuclear labeled-(^{32}P)RNA were hybridized by Northern blotting to a HS cDNA clone, which encoded the major mRNA species of 2.7 kilobases in HS HeLa cells. After the maximal induction of transcription rate, with wild-type virus but not with d1312 mutant, this fell back to preinduced levels (see Figure 2B). This result is in sharp contrast to the activity of most cellular genes that are neither increased in transcriptional activity nor decreased as a result of adenovirus infection (see the synthesis of β-actin and β-globin in Figure 2A and Babich et al.[13]). The data demonstrate that the activity of the HS gene is tightly controlled during adenovirus infection, and suggest that HS gene expression is mediated by the action of Ela. Because of the finding that in the adenovirus-transformed 293 cell line, HS mRNA is expressed at very high levels, Kao et al.[14] investigated the related expression of adeno-Ela and HSP70 gene in synchronized 293 cells. The level of HSP70 mRNA in the blocked cells was quite low, and upon release in the *S* phase the level rose to a peak value between 9 and 15 hr postrelease; the same occurred for the Ela mRNA, but the activation occurred at an earlier time (6 to 9 hr postrelease). These data are consistent with the hypothesis that Ela or a cellular Ela-like activity is responsible for the regulation of HSP70 mRNA in adeno-infected cells.[103]

Stein and Ziff[4] (see Figure 2A), using hybridization assays of total RNA, and Velcich and Ziff,[99] using a transient expression assay, reported that the rate of transcription of cellular β-tubulin and early SV40 promoter was transiently induced during the early phase of adenovirus infection. On the other hand, Braithwaite et al.[104] suggested a role for the Ela gene in the regulation of the thymidine kinase gene. Thus, during adenovirus infection, these host cellular genes respond with kinetics that parallel the activation and subsequent repression of transcription of the adenovirus early viral gene.

Velcich and Ziff[99] postulate that the two proteins (289 and 243 aa) encoded in Ela gene have the ability, on their own, to repress the transcription of the early SV40 RNA. They suggest that Ela proteins may repress through an effect upon the SV40 enhancer, or upon promoter elements more closely linked to the RNA initiation site, including the 21 bp repeat and the TATA box. The experiments described by Green et al.[101] give some idea of the process. A single base change at position −87 in the β-globin promoter reduced the level of transcription when expressed under the influence of an enhancer, but the stimulation by Ela was maintained. By contrast, a nucleotide change in the TATA sequence strongly reduced activity in both cases; thus the transcriptional event in TATA sequences is important. Ela products could interact directly with DNA elements or, indirectly, modify cellular factors required for eukaryotic transcription.[96] This approach is interesting, since the "trans" effect of the Ela gene is opposite to the "cis" mechanism of enhancers, which activates promoters by a mechanism independent of position.[105] This suggests that the Ela proteins and the SV40 enhancer may be viral counterparts of the transcription mechanism that regulate eukaryotic gene expression.

Besides Ela, but less well-characterized, adenovirus Elb and E4 genes have been also implicated in host mRNA production during adenovirus late infection.[106]

b. HnRNA and mRNA Stability

Modifications of host mRNA transcription are not the only viral effects in adenovirus-infected cells. Yoder et al.[15] reported that synthesis of DHFR is induced early after adenovirus infection and inhibited late in infection, although the relative synthesis of DHFR-specific mRNA remains constant throughout infection. Recently Yoder and Bergett[18] measured the steady-state nuclear levels of DHFR-specific RNA. Infected and mock-infected Mtx-resistant HeLa cells (see Section II) were labeled for 2 hr with (^{3}H)uridine. At 16 and 30 hr p.i., nuclear RNA was obtained and subjected to hybridization to specific DHFR DNA probes, immobilized on nitrocellulose filters. As can be seen in Figure 9, late in infection, relative steady-state nuclear levels of DHFR-specific RNA initially increased and subsequently decreased. While at 15 hr p.i., a 2.6-fold increase over mock-infected cells was detected; at 30 hr p.i., the relative levels of DHFR-specific nuclear RNA dropped by 1.4-fold compared to that detected in uninfected cells. On the other hand, nuclear levels of DHFR RNA did not decline to a level low enough to account for the magnitude of the decline in cytoplasmic DHFR-specific RNA. These data, together with a constant incorporation of (^{32}P)UTP to

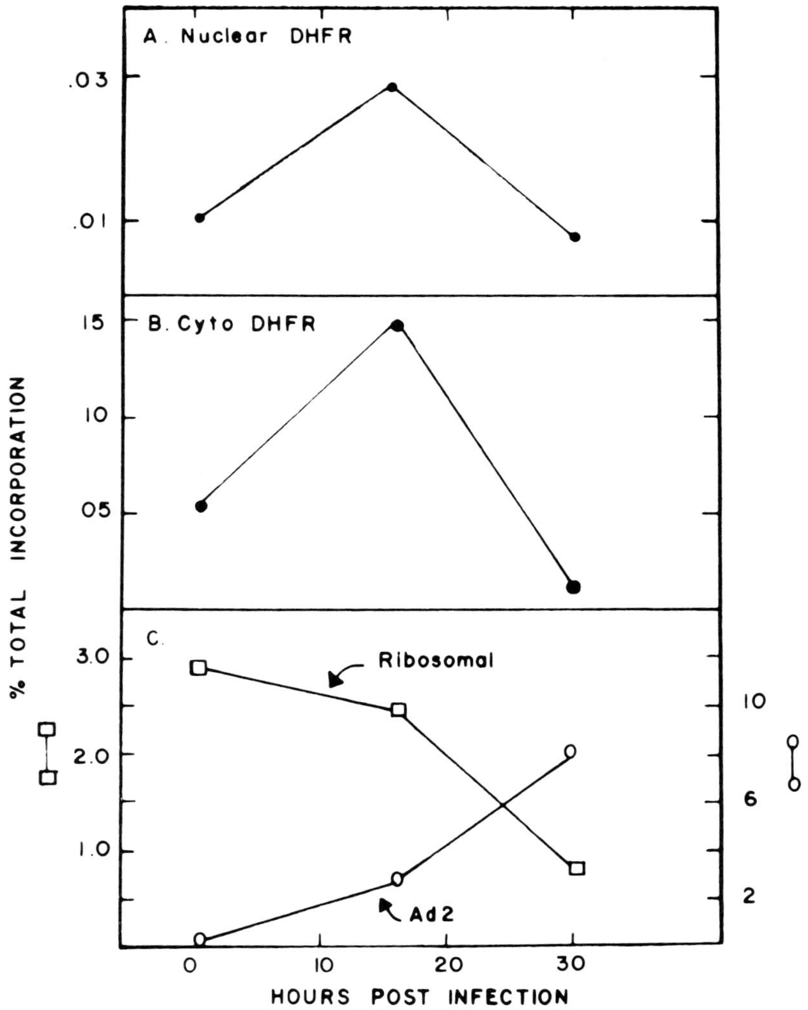

FIGURE 9. Levels of DHFR-specific RNA during adenovirus infection.[18] At different times after infection, cells were labeled with (^{3}H)uridine, and nuclear and cytoplasmic RNA were obtained and hybridized to cloned DHFR (A and B) and Adeno-2 or ribosomal DNA (C).

DHFR RNA in isolated nuclei in vitro founded by the authors, suggest an alteration in the steady-state levels and a blockage in the nuclear transport of host DFHR RNA late in adenovirus infection.

A similar effect has been reported for actin mRNA in adenovirus-infected cells. Khalili and Weinmann,[35] carrying out a comparative analysis of the expression of the actin genes in infected and mock-infected HeLa cells, showed that while the rate of actin gene transcription remained constant, the newly synthesized actin mRNA entered the cytoplasm at a reduced rate. Concomitantly, an increase in the half-lives of the cytoplasmic results in accumulation of mature actin mRNA in the cytoplasm of adenovirus-infected cells.

These results, together with others, support the posttranscriptional modification of host cellular mRNA during adenovirus infection. Thus, posttranscriptional control may be more prevalent than was previously thought. Although the molecular mechanism of such modifications is unclear at this time, it seems that adenovirus-mediated suppression does not alter polyadenylation of host transcripts.[22]

Finally, although an Ela-mediated increase in the cellular topoisomerase I in adeno-infected HeLa cells has been reported,[107] a putative effect on the structure or function of host cellular mRNA during adenovirus infection remains to be determined.

2. Herpes Simplex Virus (HSV)

HSV are large viruses with the potential to encode many polypeptides. Virus penetration into the cell usually results in a lytic productive infection, which is associated with an early shut-off of host cell-directed macromolecular synthesis, although the stimulation of several eukaryotic promoters has been reported.[92,108]

Early studies demonstrated that HSV-1 mediates the inhibition of cellular DNA and protein synthesis. A decrease in host RNA production also occurs, although the degree of its inhibition is not as severe as that observed with DNA and protein synthesis.[29,109]

The mechanism of host macromolecular synthesis shut-off mediated by HSV remains obscure, although a virion component[110] and an early gene product may be involved.[111] Recently La Thange et al.[11] reported that HSV-2 infection of BHK-21 cells induced a cellular stress response in the form of the increased synthesis of a 57 kdalton protein that could be involved in the shut-off of host cell translation. This 57 kdalton polypeptide also accumulates during HS of BHK-21 uninfected cells.

a. Change in the Structure and Function of Cellular Chromatin

The rate of synthesis of cellular DNA declines during the first hour after infection with herpes virus.[112] In order to determine whether the capacity of the virus to suppress host DNA synthesis was sensitive to UV light, Fenwick and Walker[110] infected Vero cells with irradiate virus, and the vDNA was separated and characterized by CsCl gradients. While host DNA synthesis was inhibited by 75% with nontreated virus, irradiated virus caused an even greater suppression. These experiments suggest that cellular DNA inhibition is caused by components of the infecting particle other than the viral DNA.

b. Suppression of Host Transcriptional Activity

Data from Stenberg and Pizer[113] indicate that very soon after infection, an HSV immediate early gene function is expressed that inhibits nuclear transcription. These investigators found that HSV suppressed nuclear adeno-RNA by 70 to 75% in 293 to 31 adenovirus-transformed cells. An examination of the accumulation of adeno-RNA in the cytoplasm showed that it was inhibited by 91% at 2 hr p.i. These results indicate that the HSV-induced suppression of adeno-RNA primary transcription causes the inhibition of cytoplasmic adeno-RNA content.

Pizer and Beard[114] studied the effect of HSV-1 infection on the synthesis and amount of Py specific RNA. By 5 hr after coinfection, the amount of Py-RNA was 20% of that found in control cells, as determined by hybridization of radioactive total RNA by Py-DNA immobilized on filters. This experiment shows that, as with the adeno-RNA, the synthesis of Py-RNA also declines rapidly after HSV-1 infection.

c. HnRNA and mRNA Stability

After the earliest reports on the degradation of host mRNA by Nishioka and Silverstein,[25] evidence has emerged that supports a differential stability of host cell mRNA during HSV-1 infection. A decrease in the steady-state level of several eukaryotic mRNAs during HSV-1 infection has been reported, e.g., the Py-mRNA,[114,115] the adeno-mRNA,[116] the globin mRNA,[25] the β-actin mRNA,[29,32] and the GPD mRNA.[32]

One exception to the general pattern of mRNA degradation directed by HSV has been described. Mayman and Nishioka[20] quantified histone H3, globin, and actin mRNAs in friend erythroleukemia (FL) cells during infection with HSV-1. The steady-state level of histone H3 mRNA, determined by molecular hybridization of total cellular RNA to a mouse

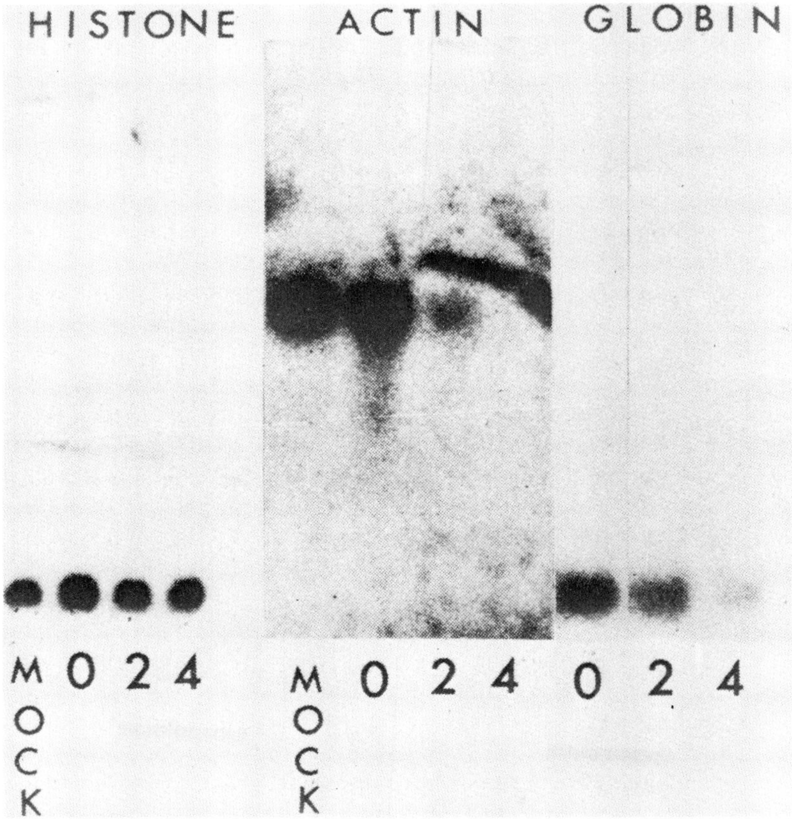

FIGURE 10. Steady-state levels of different host mRNA in FL cells infected with HSV-1.[29] Total RNAs from infected cells isolated at different times after infection were separated in formaldehyde-agarose gel and, after transfer to nitrocellulose paper, the RNA was hybridized to (^{32}P)labeled plasmid DNA containing histone, actin, and globin sequences.

clone of histone cDNA probe, remains unchanged for the first 4 hr p.i. (Figure 10), but the ratios of β-actin and globin mRNA to total RNA markedly decrease by 4 hr p.i., which is coincident with other reports.[25,32] In order to discriminate between the postulate stabilization of preexisting histone H3 mRNA or an increased rate of the histone mRNA transcription, the nuclei from HSV-1 infected FL cells were purified and incubated in the presence of (^{32}P)UTP and the total (^{32}P) RNA-hybridized to histone HS cDNA immobilized on a membrane filter. Although the procedure is semiquantitative, the experiment demonstrated that the synthesis of histone mRNA was about 60% of the control, coincident with the in vivo RNA synthesis assayed by (^{3}H)uridine incorporation. These experiments showed that the nonpolyadenylated histone H3 mRNA, while being inhibited at the transcriptional level as in the case in adenovirus-infected cells,[98] is refractory to HSV-1 induced mRNA degradation. The investigators proposed that host mRNA degradation could be initiated by virus-induced deanilation of mRNA. Additional evidence on the proposed primary effect of HSV-mediated deanilation of mRNA came from Nakai et al.[115] who, studying the effect of HSV-1 infection on Py-transformed BHK cells, reported a reduction in the cytoplasmic levels of Py-mRNA together with an increased amount of nonadenylated mRNA.

3. Simian Virus 40 (SV40) and Polyoma Virus (Py)

Unlike other DNA and RNA viruses, the lytic infection of mouse and monkey cell cultures

Table 2
STIMULATION OF PROTEIN SYNTHESIS BY VIRUS INFECTION AND THERMAL TREATMENT

	CV-1		Mouse kidney	
Protein	**Inf/mock**[a]	**Hs/mock**[a]	**Inf/mock**	**Hs/mock**
Total proteins	1.5	1.1	1.7	1.1
92,000	2.8	2.1	1.8	2.3
72,000	4.1	2.1	2.0	1.7
70,000	—	6.5	—	>6
Actin	1.1	1.0	1.2	1.1

Note: See Khandjian, E. W. and Turler, H., *Mol. Cell Biol.*, 1, 807, 1983 for details.

[a] Densitometer tracings of autoradiographs gave the ratio of the peak area of individual proteins or the area of total proteins from virus-infected (Inf) or thermally treated (Hs) cultures, to those from uninfected cultures kept at 37°C (mock).

with Py or SV40, respectively, induces an increased synthesis of the majority of cellular proteins.[19] The expression of the early viral gene (19S mRNA)[117] is rapidly followed by a mitogenic reaction of the host cell. This reaction includes virus-induced simulation of overall cellular RNA synthesis and an increase in total (mainly ribosomal RNA) activation of the host DNA-synthesis machinery leading to an increase in cytoplasmic and nuclear protein. In permissive cells (lytic infection), *S* cell phase is paralleled by replication of viral DNA and production of progeny virus.

Khandjian and Turler[12] studied the stimulation of host cell proteins during infection of CV-1 cells by SV40. Since it was reported by the same group that two proteins, possibly related to HS proteins (92 and 72 kdalton), were particularly stimulated during SV40 and Py infections, they compared the SV40 and polyoma-induced proteins with the terminally-induced proteins of CV-1 and mouse kidney cells, respectively. As can be seen in Table 2, the thermal treatment on both cell lines lead to an increase in (^{35}S)methionine-labeling of the 92 and 72 kdalton proteins and the appearance of a new labeled protein with a molecular weight of 70 kdalton. The analysis and quantification of these proteins after 15 hr of virus infection showed higher increases in 92, 72, and 70 kdalton, although in the case of SV40-infected CV-1 cells, the 70 kdalton protein was overlapped with a 72 kdalton virus band. The 92 and 72 kdalton proteins stimulated by virus infection comigrated in two-dimensional gel electrophoresis with the proteins induced by thermal treatment of uninfected cells. These results suggest that these proteins belong to the group of cellular stress proteins, and that virus infection induces stress proteins under ''physiological'' conditions, i.e., at 37°C and in the absence of toxic chemicals.[118] It would be interesting to see if papoviruses have an indirect role in modulation of host gene expression mediated by activation of stress proteins. The trans-acting activity of the large T antigens of SV40 and polyoma has been documented.[119-121] It is well-known that T antigens stimulate transcription from the late SV40 promoter and, recently, stimulation of adenovirus E2 and E3 promoters has been reported. The regulatory role of some other chromosomal factors in SV40 late transcription has been described.[121]

4. Vaccinia Virus

Vaccinia virus, a member of the poxvirus family, contains a large double-stranded DNA genome of 180 kilobase pairs. Early genes are expressed after penetration into the host cell and, after viral DNA replication, late genes encoding predominantly structural proteins are expressed.[122]

In contrast to other animal DNA viruses, vaccinia virus replicates in the cytoplasm and utilizes its own transcription system. Some characteristics important in this virus include a multisubunit RNA polymerase and proteins involved in modification of RNA, contained in the virus particles; the viral mRNAs are not spliced but have the 5′ cap and the 3′ poly(A) tail; vaccinia DNA is transcribed in vitro only by cell-free extracts from infected cells; the 5′ flanking regions of four early genes lack the regulatory elements of most of the eukaryotic promoters transcribed by RNA polymerase II (see Section III); expression of foreign DNA introduced into the viral genome is dependent on the use of vaccinia virus promoter; all necessary signals for regulation of late gene expression reside within 100 bp of the 5′ flanking sequence.[123]

a. Suppression of Host Transcriptional Activity

The inhibition of host cell RNA synthesis following poxvirus infection has received little attention. While some authors[124,125] reported a 3 hr lag prior to the onset of inhibition of cell RNA synthesis after vaccinia virus infection, Kit and Dubbs[126] suggest that the suppression occurs immediately after infection. In agreement with the latter report, Pedley and Cooper[24] showed that vaccinia virus induced an immediate and rapid inhibition of HeLa cell RNA synthesis. This early inhibition, determined by pulse-labeling studies with (^{3}H)uridine, showed that it is independent of the multiplicity of infection, but dependent on viral protein synthesis. Although by analysis of sedimentation profiles no selective inhibition of synthesis of different RNAs was found, only 42% of the activity of the host nuclear RNA polymerase II remained by 2 hr p.i. Finally, at 2 hr p.i., a new type of RNA polymerase was detected. This was demonstrated by several criteria such as the elution pattern in DEAE-Sephadex® chromatography, the ammonium sulfate optimum, and the sensitivity to α-amanitine.[24]

b. HnRNA and mRNA Stability

Since a large increase in the rate of degradation of host mRNA has been reported in other DNA virus-cell system (adeno and herpes virus), Pedley and Cooper[24] also studied the stability of mRNA in vaccinia-infected HeLa cells, prelabeling cells with (^{3}H)uridine 18 hr before infection and determining the acid-insoluble radioactivity in isolated nuclei. The results showed that the rate of host cellular RNA degradation, in contrast to other DNA viruses, was slightly greater in mock-infected cells.

B. RNA Viruses

Recently, Kaariainen and Ranki[127] have published a detailed review on the inhibition of cell functions by RNA viruses.

1. Reovirus

Reovirus is a double stranded RNA virus that alters cellular metabolism and cytoarchitecture (see below). This virus contains ten segments of dsRNA surrounded by a double capsid shell of proteins. The dsRNA segments are named according to size classes; there are three large segments (L1 to L3), three medium segments (M1 to M3), and four small segments (S1 to S4). An excellent review on the structure and function of the reovirus genome has been published by Joklik.[128]

a. Changes in the Structure or Function of Host Chromatin

Reovirus type 3 inhibits cellular DNA replication in transformed cells, although it has little or no effect on DNA replication in normal cells.[129] This inhibition has been found to be selective for the replicative function of cellular DNA and specific for the initiation step of DNA replication.

Hand and Kasupski[130] have shown that the amount of DNA synthesis during a ^{3}H-thymidine pulse of reovirus-infected L-929 was reduced by 51%. Studying the density of DNA that was sequentially labeled with analogues of thymidine, Hand and Kasupski[130] demonstrated that DNA fork movement was not modified in infected cells. These data provide direct evidence that the inhibitory effect is accompanied by a reduction in active chromatin initiation sites. Another interesting finding was the early specific inhibition of histone synthesis that precedes suppression of host DNA replication,[130] although unfortunately the levels of histone mRNA were not determined. In the same direction Hand and Tamm,[131] using DNA fiber autoradiography, have shown that the frequency of initiation events in localized regions of host DNA is inhibited in reovirus-infected L-929 cells.

On the other hand, Chaly et al.[132] showed by ultrastructural analysis that there was no recondensation of chromatin after the normal decompaction that follows DNA synthesis. It was suggested that a virus product may damage chromatin, preventing recondensation. Following a similar approach, Roner and Cox,[133] studying the host DNA sedimentation profiles during reovirus type 3 infection of HeLa cells, suggested that reovirus prevents the initiation of host cellular DNA synthesis by acting, directly or indirectly, in a manner which prevents normal levels of DNA supercoiling.

In an attempt to characterize the reovirus suppressor activities, several groups[134,135] have shown that the S1 double-stranded RNA segment is responsible for the capacity of reovirus type 3 to inhibit cellular DNA synthesis in mouse L cells. Sharpe and Fields[134] used a genetic approach with recombinant viruses containing dsRNA segments from serotypes 1 and 3. Recombinants containing a type 3 S1 dsRNA segment inhibited L cell DNA synthesis, whereas recombinant having a type 1 S1 dsRNA segment did not. It was particularly striking that a recombinant clone containing all types of 1 dsRNA segments, except for the S1 dsRNA segment derived from type 3, inhibited DNA synthesis by 75 to 85%. Because inactivated and live reoviruses suppress DNA synthesis, the S1 product and not the dsRNA appears to be the viral component responsible for this inhibition.[134] One interesting possibility from these data is that the inhibition of host DNA synthesis could be mediated through an interaction at the cell surface by the infecting virus particle, since the S1 gene codes for the viral hemagglutinin.[128] However, the possibility that the S1 product may exert an intracellular effect responsible for the inhibition of host DNA synthesis cannot be discounted.

2. *Negative-Strand RNA Virus*

a. Changes in the Structure or Function of Host Chromatin

Kranz and Reichmann[20] investigated inhibition of SV40 DNA synthesis by VSV in double-infected monkey kidney cells. Synchronized Vero cells were infected with SV40 and, at different times thereafter, the cells were coinfected with VSV and subsequently labeled with (^{3}H)-thymidine. Both SV40 and cellular DNA synthesis were inhibited at all times during the S phase. While gel electrophoretic profiles of SV40 (^{32}P)-DNA showed an accumulation of the monomeric forms at all stages of DNA supercoiling, the rate of DNA chain elongation did not vary under VSV superinfection. Under similar experimental conditions, the VSV-induced effect mimicks the effect of cycloheximide on the generation of SV40 DNA intermediates. An interpretation of these results is that inhibition of SV40 DNA synthesis in VSV-coinfected cells is an indirect result of the already reported VSV shut-off of host cell protein synthesis[136] in the sense that host translation shuf-off causes an incompletely supercoiled SV40 chromatin that converts the SV40 into a deficient template.

b. Suppression of Host Transcriptional Activity

Weck and Wagner[23] infected mouse myeloma cells (MPC-11) and found a marked reduction in host cellular RNA synthesis by measuring both the (^{3}H)uridine incorporation into total RNA (Figure 4A) and the (^{3}H)GTP incorporated to nuclear RNA in vitro. The authors reported no modifications on membrane permeability and no degradation of previously synthesized RNA. In fact, infected cells accumulated polyadenylated mRNA at the same rate as did uninfected cells (see Figure 4B). By the approach of differential α-amanitine sensitivities, it was found that only RNA polymerase II was affected progressively throughout infection and to a much greater extent than polymerase I and II.

More directly, Nishioka et al.[137] examined the effect on the accumulation of HSV thymidine kinase (tk)-specific cytoplasmic mRNA and the HSV-mediated degradation of host cell globin mRNA in VSV-coinfected FL cells. First, the HSV-induced host globin mRNA degradation was avoided in FL-coinfected cells. Second, by Northern blot hybridization of RNA samples extracted at various times after coinfection, less than 10% of those HSV-tk-specific mRNA levels found in cells infected with HSV alone were detected. This finding allowed the authors to postulate that the primary suppressor activity by VSV was at the level of transcription, rather than by a direct translation shut-off of preexisting HSV-1 mRNA.

Weck and Wagner,[138] studying VSV-induced host transcription suppression, reported that certain viral products must be made before drastic and prolonged suppression of cellular RNA metabolism takes place. UV-inactivation studies,[139] the use of temperature-sensitive (ts) mutants,[138] and defective interfering (DI) particles[140] have shown that primary transcription of VSV, but not amplification, is required for this inhibition.

Several mechanisms for the VSV suppressor effect of host cellular functions have been proposed, although a specific cellular target has yet to be identified. It has been reported that only transcription of the 3′ terminus of the virion RNA into a 47- to 48-nucleotide, the so-called leader RNA (1-RNA), was required for inhibition of cellular transcription.[21] This is coincident with reports by Kurilla et al.,[141] that demonstrated the presence of the 1-RNA in nucleus of VSV infected cells at the time when inhibition of cellular transcription was observed. Although there are differences between serotypes (New Jersey and Indiana), only the ability of both strains to synthesize this small RNA, in vivo and in vitro, correlated with the decreased rate of DNA-directed transcription. Following a different approach, McGowan et al.[21] have shown that in vitro transcription of SV40 and adenovirus DNA in whole HeLa cell extracts was inhibited by the addition of purified VSV 1-RNA. Further studies indicate that the inhibitory activity of the 1-RNA is dependent on its sequence-determinated secondary structure.[142]

Two kinds of evidence suggest that interaction of VSV 1-RNA with host proteins is probably involved in suppression: the primary structure of the RNA is reminiscent of polymerase III transcripts that are known to exist intracellulary as nucleoprotein complexes,[142,143] and the 1-RNA of VSV is associated with host protein *La*,[144] which is normally found associated with RNA polymerase III precursors.[143] Besides this requirement for the association of host cellular proteins with 1-RNA, it appears that viral protein synthesis is needed for the suppressor activity to take place. This is based on the evidence reported by several groups that demonstrated that the ability of VSV to suppress host cell RNA synthesis was markedly reduced in the presence of inhibitors of protein synthesis.[127,145] The data of Wu and Lucas-Lenard[146] suggest that suppression depends on the production of viral N protein in addition to the 1-RNA, and Blumberg et al.[147] found the 1-RNA complexed with VSV N protein. Finally, studies in rabies virus-infected cells support the hypothesis that the capacity of the 1-RNA to bind *La* protein may not be sufficient to suppress host RNA synthesis.[148,149] Rabies virus is a negative-strand rhabdovirus analogous to VSV in the sense of bullet-shaped morphology, genomic organization, and pattern of viral RNA and protein synthesis.[148] Although a 3′ terminal homology between VSV and rabies 1-RNAs may reflect the reported ability of both transcripts to bind the *La* protein,[148] a striking difference is that rabies has little or no apparent effect on host macromolecular synthesis.[149]

c. HnRNA and mRNA Stability

The effect of another negative-strand virus, the segmented human influenza virus, on host gene expression has been studied. A unique and interesting property of this virus is that viral mRNA synthesis occurs in the nucleus, utilizing capped RNA primers 10 to 13 nucleotides long. Since these primers are cleaved from newly synthesized polymerase II viral transcripts,[150,151] the replication of influenza virus is a RNA polymerase II dependent event.

Recently, Katze and Krug[33] examined the effects of influenza virus infection on the metabolism and expression of β-actin and ALV, in CEF and HeLa cells, respectively, (see also Section II). They found that within 1 to 2 hr, the rate of transcription decreased 40 to 60% (in a runoff assay with isolated nuclei) (see Figure 6), and that the remaining host transcripts that continued to be synthesized are degraded in the nucleus. This degradation is apparently performed by the influenza cap-dependent endonuclease. In contrast to the in vitro nuclear transcripts, the cytoplasmic β-actin, ALV, and three other HeLa cell mRNAs synthesized before infection remain stable. Degradation of HnRNA could be initiated by the cleavage of the 5′ end of polymerase II cellular transcripts by the influenza viral cap-dependent endonuclease. The resulting decapped host RNA would be likely to be more susceptible to degradation by cellular nucleases. This could be explained, as has been described in Section III, by the fact that the 5′ cap structure stabilizes HnRNA against nucleolytic degradation both in vivo and in vitro.[83,84]

3. Picornaviruses

As has been demonstrated for other RNA viruses, the inhibition of cellular transcription by picornaviruses [poliovirus, mengovirus, encephalomiocarditis (EMC) virus, Foot-and-Mouth disease virus (FMDV)] does not appear to be mediated through a decreased rate of RNA elongation or an increased rate of RNA degradation.[152-154] Instead, host RNA synthesis is probably inhibited due to diminished enzymatic activities of the three RNA polymerase, particularly polymerase II and, to a lesser extent, polymerases I and III.[2,155-158] Inhibition was first characterized in mengovirus-infected L cells[153,159] and subsequently reported in polio-infected HeLa cells,[156,160] in EMC-infected plasmocytoma cells,[161] and in FMDV-infected DHK cells.[162]

a. Changes in the Structure and Function of Cellular Chromatin

Infection of BHK cells with FMDV causes the disappearance of the histone H3 and the appearance of a new polypeptide (denominated Pi) that is detected at 2 hr p.i. coincidentally with host RNA synthesis inhibition. Polypeptide Pi shares common major peptides with histone H3 after controlled proteolysis with V8 protease or trypsin[163] (Figure 11). At present, it is not known if the transition of H3 to Pi is due to an enzymatic cleavage of histone H3 or a consequence of posttranslational modifications such as change in the degree of acetylation, phosphorylation, and/or methylation known to take place in histone.[164] Since Bohm et al.[165] have described that histone H3 is the first component of the nucleosomal core to be cleaved when chromatin preparations are subjected to exogenous proteolysis, Grigera and Tiaminetzky[163] speculate that the reported virus-codified,[166] or an endogenous induced proteolitic activity, modifies histone H3. Unfortunately these authors failed to obtain an in vitro processing of histone H3 with nuclear or cytoplasmic fractions of infected cells.

b. Suppression of Host Transcriptional Activity

Infection of mammalian cells with picornaviruses results in the rapid inhibition of normal host cell RNA synthesis (see Figure 1B). This effect has been characterized in whole cells and isolated nuclei.[2,158,161]

Apriletti and Penhoet,[158,167] studying the activity of the three major types of RNA polymerases solubilized from their RNA template, have shown that all the enzyme activities are recovered at control levels, and give DEAE-Sephadex® elution patterns indistinguishable from those enzymes isolated from mock-infected cells. Extending these studies, Schwartz

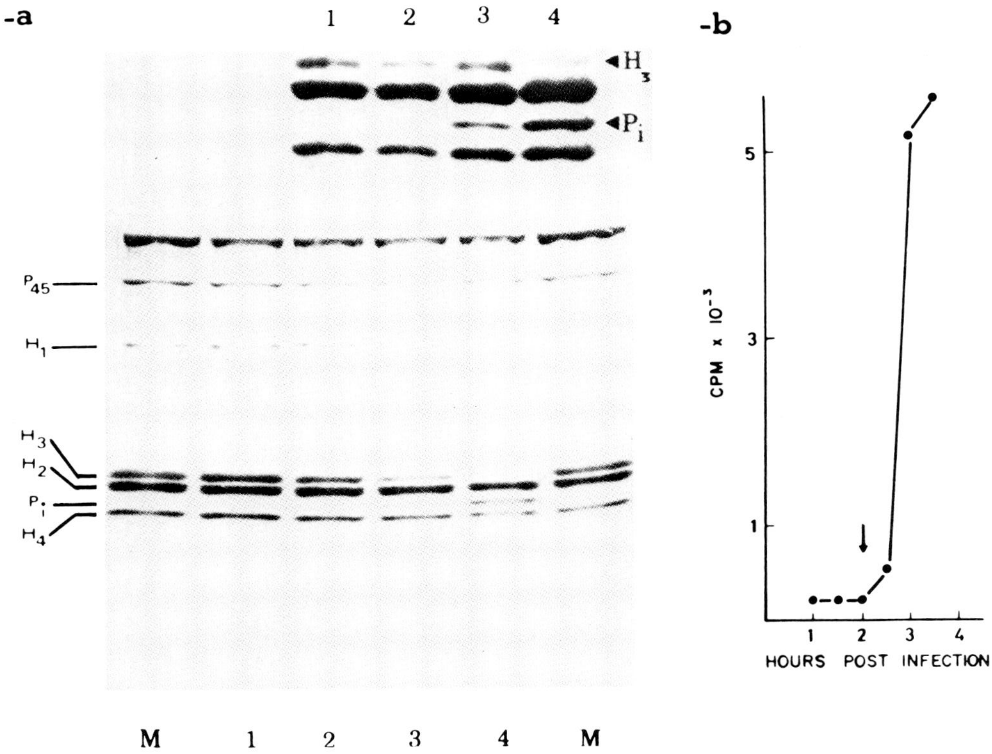

FIGURE 11. FMDV-induced modification of nuclear histone H3.[163] (a) Total nuclear proteins from infected and mock-infected cells were analyzed on PAGE-SDS gradient slab gels. Tracks 1 to 4 correspond to samples from infected cells at 1.5, 2, 2.5, and 3 hr p.i. (b) Rate of synthesis of FMDV-RNA analyzed in virus-infected BHK-21 cells in the presence of actinomycin D.

et al.[161] reported that endogenous RNA polymerase II activity was inhibited early p.i. and prior to RNA polymerases I and III, although the solubilized enzymes were fully active with exogenous DNA as template. Besides this evidence, it is important to keep in mind that the recent better-characterization on the polymerase II enzyme into three subspecies named IIO, IIA, and IIB (see Section III), and their putative role in transcription, leaves the nonmodification of RNA polymerase II during piconavirus infection open to question.

Several lines of evidence support the idea that suppression of transcription is at the level of initiation of RNA synthesis. Flores-Otero et al.[2] have shown that in nuclei isolated from polio-infected HeLa cells, RNA polymerase II was inhibited by 55% at 2 hr p.i. Following coupled strategies employing the anionic detergent Sarkosyl and ^{32}P-labeled cordycepin, it was reported that the number of polymerase II molecules engaged in chain elongation decreased as infection proceeded (see Figure 1). In agreement with this report, but using a different strategy, Crawford et al.[168] showed that at least one factor required for specific transcription by RNA polymerase II is deficient in extracts obtained from polio-infected HeLa cells. These authors used cell-free transcription systems with several eukaryotic promoters (major late adenovirus 2, SV40, early adeno and human β-globin gene) supplemented with whole cell extracts from polio and mock-infected cells. As can be seen in Figure 12, a rapid loss of specific polymerase II transcription (directed by the adeno-2 late promoter) was evident, reaching 75% inhibition at 2 hr p.i. By 3 hr p.i., polymerase II diminished by only 25% of the activity found in extracts from mock-infected cells. Further analysis of the transcription system demonstrated that polymerase II was not limiting in the poliovirus-

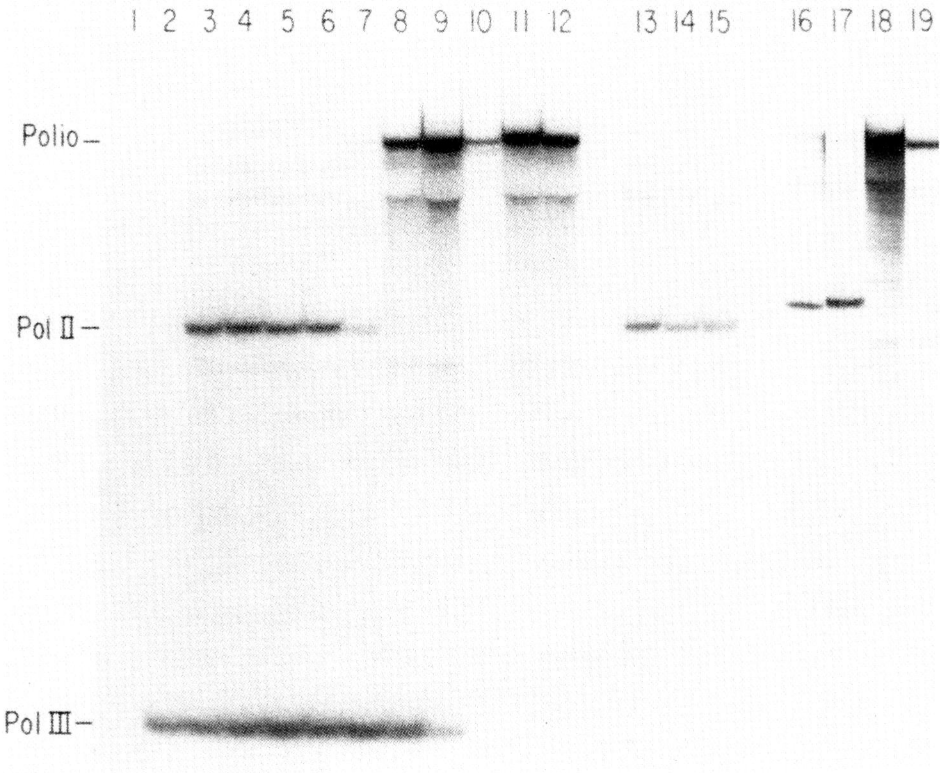

FIGURE 12. Inhibition of RNA polymerase II transcription in extracts from poliovirus-infected cells.[168] Whole HeLa cells extract prepared from untreated, mock-infected, and poliovirus-infected cells were assayed for polymerase II activity by the addition of DNA containing the major late promoter of adeno-2. Under these conditions, the polymerase II-initiated transcription produces a 2125 nucleotides run-off.[71] Lanes 1 to 3: uninfected HeLa cell extract with 5 μg/mℓ actinomicyn D, 0.5 μg/mℓ α-amanitin or no addition, respectively. Lane 4: mock-infected cell extract. Lanes 5 to 9: poliovirus-infected HeLa cell extracts made, respectively, 0, 1, 2, 3, and 4 p.i. Lane 10: (^{32}P)labeled poliovirion RNA. Lanes 11 and 12: polio-infected cell extracts with no DNA or 5 μg/mℓ actinomycin D. Lanes 13 to 15: mock-infected cell extracts incubated for 1 hr (lane 13) or incubated in the presence of 0.5 m*M* UTP and 0.5 μg/mℓ α-amanitin with mock-infected (lane 14) and polio-infected (lane 15) cell extracts. Lanes 16 to 18: assaying extracts prepared from mock-infected cells (lane 16), from mock-infected cells treated with 2 m*M* guanidine (lane 17), from polio-infected cells with (lane 19) or without (lane 18) guanidine.

infected cells exteact and that another component, necessary for transcription and present in a S100 extract,[169] was either missing or inactivated.

Although some workers[160,161,168,170] have failed to demonstrate any inhibition of RNA synthesis in vitro in extracts of uninfected cells when they are mixed with extracts from picornavirus-infected cells, transcription of isolated nuclei was suppressed when the transcription reaction was supplemented with cytoplasmic extracts from polio-infected cells.[171,172] This is coincident with the requirement of a functional viral genome for host transcription suppression,[159,173] although a productive infection need not occur.[170,174] Indirect evidence for the requirements of viral products has been published. Ho and Washington[175] have characterized a putative inhibitor as a lipoprotein since it was destroyed by phospholipase and heating and extracted into an organic solvent. On the other hand, Bienz et al.,[176] Bossart et al.,[172] and Fernandez-Tomas[177] reported the accumulation of virus-induced proteins in nuclei of polio-infected Hep-2 and HeLa cells. This accumulation coincides with the onset of host RNA synthesis inhibition.

C. Virus-Induced Cellular Structural Modifications

It is important to call attention to the fact that other cell alterations mediated by infection with animal viruses could affect host cellular mRNA production. This is the case of the well-documented effect of virus infection on the host cytoskeleton, since the major components of the cellular matrix have been implicated in many host functions.[178] Recently differences have been described between active and inactive genes in their association with the nuclear matrix. Also, it has been reported that supercoiled loops of DNA appear to be anchored to a salt-insoluble nuclear substructure to which a variety of transcribing genes is associated.[178]

Evidence in the literature suggests that viruses interact in a functionally significant manner with the cytoskeleton of host cells, and that some of them modify the genetic expression of some of the nuclear matrix structural polypeptides.[4,29,32-34] Although the possible relationship between cytoskeleton and host cellular genetic expression remains unclear, the effect of viral infection on cytoskeleton organization deserves a brief mention.

Upon infection, most viruses induce extensive morphological and functional alterations of the cytoskeleton.[179] Lenk and Penmann[180] have shown a gross morphological and biochemical alteration of the cytoskeleton in polio-infected HeLa cells, and suggest that changes are the result of an active viral modification of the normal cellular machinery. Significant changes were also observed in cells infected with VSV and NDV,[181] and in HSV-infected BHK-21 cells,[182] although their relationship with viral functions is not understood. Murti and Goorha[183] have shown progressive cytoskeletal alterations in all three filament types in frog virus-infected BHK cells; the inhibitory effect of this virus on host cell nucleic acid synthesis has been published also.[184] Recently[185] it has been reported that actin and VSV mRNAs, whose translation is inhibited by poliovirus, were released from the cytoskeletal framework when cells were coinfected with this picornavirus. Modifications in cytoskeletal organization have also been reported in adeno[186] and canine distemper virus[187] infected cells.

V. CONCLUDING REMARKS

Although knowledge of the viral suppressor effect on host gene expression is not all complete, at the present time it appears that the several viruses described here share some common features. It is clear that either DNA (adeno, HSV, SV40, Py) or RNA (polio, NDV) viruses have the capacity to induce a stimulation of host mRNA production, preceding viral genome replication. Also, it has been well-documented that with the exception of papova viruses, in all cases host cell mRNA declines late in infection in a way that viral mRNA becomes the dominant species during the postreplicative period. A third similarity is that the suppressor effect takes place, at least initially, at the transcriptional level, maybe in the initiation process of host mRNA synthesis.

Why do viral genes activate specific cellular transcription? Several possibilities are attractive. One is that certain genes require an activity that controls transcription in a manner similar to that found in the expression of DNA viruses. This could be localized at the level of promoter conformation, i.e., a host cell promoter that resembles a viral DNA promoter, and/or at the level of factor requirements which is shared under specific circumstances by viral cellular genes, as has been suggested for adeno Ela and HS.[103] Both situations could affect specific host genes. This type of regulation has been documented in β-tubulin[4] and HS proteins[9,14] during adenovirus infection with no modifications of other cellular genes. The fact that some kind of control by virus genome must be involved in such specific modifications has also been demonstrated by Babich et al.[13] in adenovirus cells, where only a small fraction of activated cellular genes contributes to the transcriptional pool of the cell.

Another possibility is that mimicking a cellular mechanism could have some benefits for the virus. For example, viral genes might make maximal use of a nonmodified, preexisting cellular machinery, as is the case for SV40, HSV, Py, etc. A third possibility is that some cellular functions could be of direct benefit to the virus, as in the case of some induced "host stress stimuli", which can favor the induction of early viral gene expression.[118] Some interesting observations have been reported. Zerbini et al.[188] showed that a HS of infected lymphoblastoid cells induced the appearance of Epstein-Barr virus and cytomegalovirus early antigen, and increased the percentage of cell expression viral capsid antigens. Similar results, although not very detailed, were reported by Coolins and Hightower[189] in NDV-infected CE cells, where accumulation of HS proteins and their functional mRNAs was detected. On the other hand, Imperiale et al.[103] reported that cells expressing a high level of HS mRNA also could partially provide Ela function for early adenovirus transcription.

Little is known about the mechanism of viral activation of cellular genes, except for the studies with the products of adenovirus Ela. Kao and Nevins,[9] Kao et al.,[14] and Velcich and Ziff[99] showed that the relative kinetics of Ela and HSP70 expression were consistent with Ela control. How could this coordinate induction of viral and host cell genes be explained? While some authors suggest that the activation by Ela appears to be indirect[9,91] in the sense that interaction with some host factor is involved, some evidence supports the hypothesis that products of Ela themselves are responsible for this phenomenon.[99] What is apparent is that the activation mechanism is at the level of primary transcription, although an increase in the half-life of host mRNA has also been reported in adenovirus-infected HeLa cells.[35]

While the mechanism of gene activation during viral infection has received attention recently, the suppressor effect has been more intensively studied. Infection of various cell types with RNA or DNA viruses, or possible defective interfering (DI) particles, frequently leads to a rapid and extensive inhibition of host macromolecular synthesis and, eventually, to cell death.

As has been described, the disruption of normal cell function during virus infection could affect synthesis of RNA and/or DNA and proteins. Although it is well-documented that viral infection suppresses overall cellular RNA production, it remains to be determined which of the multiple steps involved in mRNA production is altered during infection. Although the majority of viruses (adeno, HSV, VSV, influenza, picorna) exerts their inhibitory effect at the level of initiation of transcription,[2,9,113,114,124,126,137,138,168] it is obvious that suppression at this level causes concomitant and similar degrees of inhibition of the events that occur after transcription. One example is the altered steady-state level of nuclear mRNA described in adenovirus,[15,18,35] HSV,[25,29,32,114-116] and influenza[33] infected cells.

In conclusion, the growth of knowledge of the mechanisms that are involved in virus-infected suppression of host mRNA production has been exponential during the last 5 years. However, several important questions remain open, for instance: what is the role, if any, of virus penetration? Is the initial activation of the cell mRNA production a host stress response? How do RNA viruses that replicate in the cytoplasm suppress host mRNA transcription? These and other important considerations will be susceptible to study in the near future.

ACKNOWLEDGMENTS

I thank Dr. Grigera at the Centro de Virologia Animal, for the original Figure 11 and Dr. Khandjian at the Départment de Biologié Moleculaire, Université de Geneva, Switzerland, for the manuscript concerning the mouse p105 HS protein. I am grateful to Patricia Cortes for her help in the manuscript preparation.

REFERENCES

1. **Gariglio, P., Bellard, M., and Chambon, P.,** Clustering of RNA polymerase molecules in the 5′ moiety of the adult β-globin gene of hen erythrocytes, *Nucl. Acid Res.*, 9, 2589, 1981.
2. **Flores-Otero, G., Fernández-Tomás, C., and Gariglio-Vidal, P.,** DNA-bound RNA polymerase during poliovirus infection: reduction in the number of form II enzyme molecules, *Virology*, 116, 619, 1982.
3. **Gwo-Shu Lee, M., Lewis, S. A., Wilde, C. Deborah, and Cowan, N. J.,** Evolutionary history of a multigene family: an expressed human β-tubulin gene and three processed pseudogenes, *Cell*, 33, 477, 1983.
4. **Stein, R. and Ziff, E. B.,** HeLa cell β-tubulin gene transcription is stimulated by adenovirus 5 in parallel with viral early gene by an E1a-dependent mechanism, *Mol. Cell Biol.*, 4, 2792, 1984.
5. **Ashburner, M. and Bonner, J. J.,** The induction of gene activity in Drosophila by heat shock, *Cell*, 17, 241, 1979.
6. **Berger, E. M.,** The regulation and function of small heat-shock protein synthesis, *Dev. Genet.*, 4, 255, 1984.
7. **Pelham, H. R. B.,** A regulatory upstream promoter element in the Drosophilia Hsp 70 heat-shock gene, *Cell*, 30, 517, 1982.
8. **Ayme, A., Southgate, R., and Tissiers, A.,** Nucleotide sequences responsible for the thermal inducibility of the Drosophila small heat-shock protein genes in monkey Cos cells, *J. Mol. Biol.*, 182, 469, 1985.
9. **Kao, H. T. and Nevins, J. R.,** Transcriptional activation and subsequent control of the human heat shock gene during adenovirus infection, *Mol. Cell Biol.*, 3, 2058, 1983.
10. **Natarianni, E. L. and Preston, C. M.,** Activation of cellular stress proteins by herpes simplex virus temperature-sensitive mutants which overproduce immediate early polypeptides, *Virology*, 123, 113, 1982.
11. **La Thange, N. B., Shriver, K., Dawson, C., and Chan, W. L.,** Herpes simplex virus infection causes the accumulation of a heat-shock protein, *EMBO J.*, 3, 267, 1984.
12. **Khandjian, E. W. and Turler, H.,** Simian virus 40 and polyoma virus induces synthesis of heat shock proteins in permissive cells, *Mol. Cell Biol.*, 1, 807, 1983.
13. **Babich, A., Feldman, L., Nevins, J., Darnell, J., and Weinberg, C.,** Effects of adenovirus on metabolism of specific host mRNAs: transport control and specific translation discrimination, *Mol. Cell Biol.*, 3, 1212, 1983.
14. **Kao, H. T., Capasso, O., Heintz, N., and Nevins, J. R.,** Cell cycle control of the human HSP70 gene: implications for the role of a cellular E1A-like function, *Mol. Cell Biol.*, 5, 628, 1985.
15. **Yoder, S. S., Robberson, BL. L., Leys, E. J., Hook, A. G., Al-Ubaidi, M., Yeung, C., Kellems, R. E., and Berget, S. M.,** Control of cellular gene expression during adenovirus infection induction and shut-off of dihydrofolate reductase gene expression by adenovirus type 2, *Mol. Cell Biol.*, 3, 819, 1983.
16. **Masters, J. N., Yang, J. K., Cellini, A., and Attardi, G.,** A human dihydrofolate reductase pseudogene and its relationship to the multiple forms of specific messenger RNA, *J. Mol. Biol.*, 167, 23, 1983.
17. **Masters, J. N. and Attardi, G.,** Discrete human dihydrofolate reductase gene transcripts present in polysomal RNA map with their 5′ ends several hundred nucleotides upstream of the main mRNA start site, *Mol. Cell Biol.*, 5, 493, 1985.
18. **Yoder, S. S. and Bergett, S. M.,** Post-transcriptional control of DHFR gene expression during Adenovirus-2 infection, *J. Virol.*, 54, 72, 1985.
19. **Griffin, R. E.,** Structure and genomic organization of SV40 and polyoma virus, in *Molecular Biology of the Tumor Viruses* (Part 2), Tooze, J., Ed., Cold Spring Harbor Laboratory, Cold Spring Harbor, New York, 1980, 61.
20. **Kranz, D. and Reichmann, M. E.,** Inhibition of SV40 DNA synthesis by Vesicular Stomatitis virus in doubly infected monkey kidney cells, *Virology*, 128, 418, 1983.
21. **McGowan, J. J., Emerson, S. V., and Wagner, R. R.,** The strand leader RNA of VSV inhibits DNA-dependent transcription of adenovirus and SV40 genes in a soluble whole cell extract, *Cell*, 28, 325, 1982.
22. **Flint, S. J., Bettz, G. A., and Linzer, D. I. H.,** Synthesis and processing of simian virus 40-specific mRNA in adenovirus infected simian virus 40 transformed human cells, *J. Mol. Biol.*, 167, 335, 1983.
23. **Weck, P. K. and Wagner, R. R.,** Inhibition of RNA synthesis in mouse myeloma cells infected with vesicular stomatitis virus, *J. Virol.*, 25, 770, 1978.
24. **Pedley, S. and Cooper, R. J.,** The inhibition of HeLa cell RNA synthesis following infection with vaccinia virus, *J. Gen. Virol.*, 65, 1687, 1984.
25. **Nishioka, Y. and Silverstein, S.,** Degradation of cellular mRNA during infection by herpes simplex virus, *Proc. Natl. Acad. Sci. U.S.A.*, 74, 2370, 1977.
26. **Breathnach, R. and Chambon, P.,** Organization and expression of eukaryotic split genes coding for proteins, *Annu. Rev. Biochem.*, 50, 349, 1981.
27. **Carlson, D. P. and Ross, J.,** Humana β-globin promoter and coding sequences transcribed by RNA polymerase III, *Cell*, 34, 857, 1983.

28. **Nishioka, Y. and Silverstein, S.,** Requirement of protein synthesis for the degradation of host mRNA in Friend erythroleukemia cells infected with herpes simplex virus type 1, *J. Virol.*, 27, 619, 1978.
29. **Mayman, B. A. and Nishioka, Y.,** Differential stability of host mRNA in Friend erythroleukemia cells infected with herpes virus type 1, *J. Virol.*, 53, 1, 1985.
30. **Marty, Ph. F. L., Prechaczyk, M., Sabrouty, S. E., Dani, Ch., Jeantur, Ph., and Blanchard, J. M.,** Various rat adult tissues express only one major mRNA species from the glyceraldehyde-3-phosphate dehydrogenase multigenic family, *Nucl. Acid Res.*, 13, 1431, 1985.
31. **Hanauer, A. and Mandel, L.,** The GAPD gene family: structure of a human cDNA of an X chromosome linked pseudogene; amazing complexity of the gene family in mouse, *Cell*, 3, 2627, 1984.
32. **Inglis, S. C.,** Inhibition of host protein synthesis and degradation of cellular mRNA during infection by influenza and herpes simplex virus, *Mol. Cell Biol.*, 2, 1644, 1982.
33. **Katze, M. G. and Krug, R. M.,** Metabolism and expression of RNA polymerase II transcripts in influenza virus-infected cells, *Mol. Cell Biol.*, 4, 2198, 1984.
34. **Rice, A. P. and Roberts, B. E.,** Vaccinia virus induces cellular mRNA degradation, *J. Virol.*, 47, 529, 1983.
35. **Khalili, K. and Weinmann, R.,** Actin mRNA in HeLa cells. Stabilization after adenovirus infection, *J. Mol. Biol.*, 180, 1007, 1984.
36. **Shenk, T.,** Transcriptional control regions: nucleotide sequences requirement for initiation by RNA polymerase II and III, *Curr. Top. Microbiol. Immunol.*, 93, 25, 1981.
37. **Jongstra, J., Rendelhuber, T. L., Ondet, P., Benoist, C., Chae, C. B., Jeltch, J. M., Mathis, D. M., and Chambon, P.,** Induction of altered chromatin structures by simian virus 40 enhancer and promoter elements, *Nature (London)*, 307, 708, 1984.
38. **Larson, A. and Weintraub, H.,** An altered DNA conformation detected by S1 nuclease occurs at specific regions in active chicken globin chromatin, *Cell*, 29, 609, 1982.
39. **Mcknight, S. L., Kingsbury, R. C., Spence, A., and Smith, M.,** The distal transcription signals of the herpes virus tk gene share a common hexanucleotide control sequence, *Cell*, 37, 253, 1984.
40. **Rosenthal, N., Kress, M., Gruss, P., and Khony, G.,** BK viral enhancer element and human cellular homolog, *Science*, 222, 749, 1983.
41. **Mellon, P., Parker, V., Gluzman, Y., and Maniatis, T.,** Identification of DNA sequences required for transcription of the human α1-globin gene in a new SV40 host-vector system, *Cell*, 27, 279, 1981.
42. **Ghuzman, Y. and Shenk, T.,** Enhancers and eukaryotic gene expression, in *Current Communications in Molecular Biology*, Cold Spring Harbor Laboratory, Cold Spring Harbor, New York, 1983.
43. **Bannerji, J., Olson, L., and Schaffner, W.,** A lymphocyte specific cellular enhancer is located downstream of the joining region in immunoglobulin heavy chain genes, *Cell*, 33, 729, 1983.
44. **Walker, M. D., Edlund, T., Boulet, A. M., and Rutter, W. J.,** Cell-specific expression controlled by the 5′-flanking region of insulin and chromotrypsin genes, *Nature (London)*, 306, 557, 1983.
45. **Pelham, H. R. B.,** A regulatory upstream promoter element in the *Drosophila* Hsp heat-shock gene, *Cell*, 30, 517, 1982.
46. **Benoist, C. and Chambon, P.,** In vivo sequence requirements for the SV40 early promoter region, *Nature (London)*, 290, 304, 1981.
47. **Roberto, C., Will, H., Hernandez, N., and Schaller, H.,** Signals regulating hepatitis B surface antigen transcription, *Nature (London)*, 305, 336, 1983.
48. **Brady, J. N., Radonovich, M., Bodkin, M., Natarajan, V., Thoren, M., Das, G., Janik, J., and Salzman, N. P.,** Site-base subsitution and deletion mutations that enhance or suppress transcription of the SV40 major late RNA, *Cell*, 31, 624, 1982.
49. **Vigneron, M., Barrera-Saldaña, H. A., Baty, D., Everett, R. E., and Chambon, P.,** Effect of the 21bp repeat upstream element on the "in vitro" transcription from the early and late SV40 promoters, *EMBO J.*, 3, 2373, 1984.
50. **Dynan, W. S. and Tjian, R.,** The promoter specific transcription factor SP1 binds to upstream sequences in the SV40 early promoter, *Cell*, 35, 79, 1983.
51. **Mishoe, H., Brady, J. N., Radonovich, M., and Salzman, N. P.,** Simian virus 40 guanine-cytosine-rich sequences function as independent transcriptional control elements "in vitro", *Mol. Cell. Biol.*, 4, 2911, 1984.
52. **Hansen, V. and Sharp, P. A.,** Sequences controlling "in vitro" transcription of SV40 promoters, *EMBO J.*, 2, 2293, 1983.
53. **Hartzell, S. W., Byrne, B. J., and Subramanian, K. N.,** Mapping of the later promoter of simian virus 40, *Proc. Natl. Acad. Sci. U.S.A.*, 81, 23, 1984.
54. **Hamer, D. H. and Khoury, G.,** Enhancers and eukaryotic gene expression, in *Current Communications in Molecular Biology*, Gluzman, Y. and Shenk, T., Eds., Cold Spring Harbor Laboratory, Cold Spring Harbor, New York, 1983.
55. **Wasylyk, B., Wasylyk, C., Augerean, P., and Chambon, P.,** The SV40 72bp repeat preferentially potentiates transcription starting from proximal natural or substitute promoter elements, *Cell*, 32, 503, 1983.

56. **Gillies, S. D., Folsom, V., and Tonegawa, S.,** Cell type-specific enhancer element associated with a mouse MHC gene, Eβ, *Nature (London),* 310, 594, 1984.
57. **Moreau, P., Hena, R., Wasylyk, B., Everett, R., Gaub, M. P., and Chambon, P.,** The SV40 72 base pairs repeat has a striking effect on gene expression both in SV40 and other chimeric recombinants, *Nucl. Acid Res.,* 9, 6047, 1981.
58. **Treisman, R. and Maniatis, T.,** Simian virus 40 enhancer increases number of RNA polymerase II molecules on linked DNA, *Nature (London),* 315, 72, 1985.
59. **Weber, F. and Schaffner, W.,** Simian virus 40 enhancer increases RNA polymerase density within the linked gene, *Nature (London),* 315, 75, 1985.
60. **Hayward, W. S. and Neel, N. G.,** Retroviral gene expression, *Curr. Top. Microbiol. Immunol.,* 91, 217, 1981.
61. **Dahmus, M. E.,** Phosphorylation of eukaryotic DNA-dependent RNA polymerases, *J. Biol. Chem.,* 256, 3332, 1981.
62. **Christmann, J. L. and Dahmus, M. E.,** Monoclonal antibody specific for calf thymus RNA polymerase IIo and IIa, *J. Biol. Chem.,* 256, 11798, 1981.
63. **Dahmus, M. E. and Kedinger, C.,** Transcription of adenovirus-2 major late promoter inhibited by monoclonal antibody directed against RNA polymerase IIo and IIa, *J. Biol. Chem.,* 258, 2303, 1983.
64. **Garcia-Carranca, A., Gariglio, P., and Dahmus, M.,** Structure of RNA polymerase associated to SV40 transcription complex, *Arch. Biochem. Biophys.,* in press.
65. **Rangel, L. M., Fernández-Tomás, C., Dahmus, M., and Gariglio, P.,** Modification of RNA polymerase IIO subspecie after poliovirus infection, *J. Virol.,* in press.
66. **Matsui, T., Segall, J., Weil, P. A., and Roeder, R. G.,** Multiple factors required for accurate initiation of transcription by RNA polymerase II, *J. Biol. Chem.,* 255, 11992, 1981.
67. **Samuels, M., Fire, A., and Sharp, P. A.,** Separation and characterization of factors mediating accurate transcription by RNA polymerase II, *J. Biol. Chem.,* 257, 14419, 1982.
68. **Egly, J. M., Miyamoto, N. G., Moncollin, V., and Chambon, P.,** Is actin a transcription initiation factor for RNA polymerase B?, *EMBO J.,* 3, 2363, 1984.
69. **Parker, C. S. and Topol, J.,** A Drosophila RNA polymerase II transcription factor binds to regulatory site of an Hsp 70 gene, *Cell,* 37, 273, 1984.
70. **Tolunay, H. E., Yang, L., Anderson, W. F., and Safer, B.,** Isolation of an active transcription initiation complex from HeLa cell-free extract, *Proc. Natl. Acad. Sci. U.S.A.,* 81, 5916, 1984.
71. **Manley, J. L., Fire, A., Cano, A., Sharp, P. A., and Gefter, M. L.,** DNA-dependent transcription of adenovirus genes in a soluble whole cell extract, *Proc. Natl. Acad. Sci. U.S.A.,* 77, 3855, 1981.
72. **Dignam, J. D., Leibowitz, R. M., and Roeder, R. G.,** Accurate transcription initiation by RNA polymerase II in a soluble extract from isolated mammalian nuclei, *Nucl. Acid Res.,* 11, 1475, 1983.
73. **Sassone-Corsi, P., Wildeman, A., and Chambon, P.,** A transacting factor is responsible for the simian virus 40 enhancer activity "in vitro", *Nature (London),* 313, 458, 1985.
74. **Bogenhagen, D. F. and Brown, D. D.,** Nucleotide sequences in *Xenopus* 5S DNA required for transcription termination, *Cell,* 24, 261, 1981.
75. **Bakken, A., Morgan, G., Sollner-Webb, B., Roan, J., Busby, S., and Reeder, R. H.,** Mapping of transcription initiation and termination signals on *Xenopus laevis* ribosomal DNA, *Proc. Natl. Acad. Sci. U.S.A.,* 79, 56, 1982.
76. **Darnell, J. E., Jr.,** Variety in the level of gene control in eukaryotic cells, *Nature (London),* 297, 365, 1982.
77. **Price, D. H. and Parker, C. S.,** The 3′ end of Drosophila histone H3 mRNA is produced by a processing activity "in vitro", *Cell,* 38, 423, 1984.
78. **Hofer, E., Hofer, R. H. W., and Darnell, J. E., Jr.,** Globin RNA transcription: a possible termination site and demonstration of transcriptional control correlated with altered chromatin structure, *Cell,* 29, 887, 1982.
79. **Citron, B., Falck-Pedersen, E., Salditt-Georgieff, M., and Darnell, J. E., Jr.,** Transcription termination occurs within a 1000 base pair of the mouse β-globin (major) gene, *Nucl. Acid Res.,* 12, 8723, 1984.
80. **Falck-Pedersen, E., Logan, J., Shenk, T., and Darnell, J. E., Jr.,** Transcription termination within the E1A gene of adenovirus induced by insertion of the mouse β-major globin terminator element, *Cell,* 40, 897, 1985.
81. **Weinmann, R., Acherman, S., Bunick, D., Concino, M., and Zandomeni, R.,** *In vitro* transcription of adenovirus genes, *Curr. Top. Microbiol. Immunol.,* 109, 125, 1983.
82. **Zeeni, M., Nevins, J. R., and Darnell, J. E., Jr.,** Nuclear RNA is spliced in the absence of poly(A) addition, *Cell,* 26, 39, 1981.
83. **Furiuchi, Y., LaFiandra, A., and Shatkin, A. J.,** 5′ terminal structure and mRNA stability, *Nature (London),* 266, 235, 1977.
84. **Lockard, R. E. and Lane, C.,** Requirement for 7-methyl guanosine in translation of globin mRNA "in vivo", *Nucl. Acid Res.,* 5, 3237, 1978.

85. **Harpold, M. M., Wilson, M. C., and Darnell, J. E., Jr.,** Chinese hamster polyadenilated messenger ribonucleic acid: relationship to non-polyadenilated sequences and relative conservation during messenger ribonucleic acid processing, *Mol. Cell. Biol.*, 1, 188, 1981.
86. **Perez-Bercoff, R.,** But is the 5′ end of messenger RNA always involved in initiation?, in *Protein Biosynthesis in Eukaryotes*, (NATO Advanced Series), Perez-Bercoff, R., Ed., Plenum Press, New York, 1982, 41.
87. **Shatkin, A. J.,** mRNA cap binding proteins: essential factors for initiation translation, *Cell*, 40, 223, 1985.
88. **Adesnik, M. and Darnell, J. E., Jr.,** Biogenesis and characterization of histone mRNA in HeLa cells, *J. Mol. Biol.*, 67, 397, 1972.
89. **Gueyette, W. A., Matusik, R. J., and Rosen, J. M.,** Prolactin-mediated transcriptional and post-transcriptional control of casein gene expression, *Cell*, 17, 1013, 1979.
90. **Wiscocil, R., Bensky, P., Dower, W., Goldberger, R. F., Gordon, J. I., and Deeley, R. G.,** Coordinate regulation of two estrogen-dependent genes in avian liver, *Proc. Natl. Acad. Sci. U.S.A.*, 77, 4474, 1980.
91. **Nevins, J. R.,** Mechanism of activation of early viral transcription by the adenovirus E1a gene product, *Cell*, 26, 213, 1981.
92. **Feldman, L. T., Imperiale, M. J., and Nevins, J. R.,** Activation of early adenovirus transcription by the herpes virus immediate early gene: evidence for a common cellular control factor, *Proc. Natl. Acad. Sci. U.S.A.*, 79, 4952, 1982.
93. **Ricciardi, R., Jones, R., Cepko, C., Sharp, P., and Roberts, B.,** Expression of early adenovirus gene requires a viral encoded acidic polypeptide, *Proc. Natl. Acad. Sci. U.S.A.*, 76, 6121, 1981.
94. **Gaynor, R. B., Tsukamoto, A., Montell, C., and Berk, A. J.,** Enhanced expression of adenovirus transforming proteins, *J. Virol.*, 44, 276, 1982
95. **Leff, T., Elkaim, R., Godin, C. R., Jalinot, P., Sassone-Corsi, P., Perricaudet, M., Kedinger, C., and Chambon, P.,** Individual products of the adenovirus 12*S* and 13*S* mRNAs E1a stimulate viral EII and EIII expression at the transcriptional level, *Proc. Natl. Acad. Sci. U.S.A.*, 81, 4381, 1984.
96. **Montell, C., Courtois, G., Eng, C., and Berk, A.,** Complete transformation by Adenovirus 2 requires both E1a proteins, *Cell*, 36, 951, 1984.
97. **Svenson, C. and Akusjarvi, G.,** Adenovirus 2 early region 1A stimulates expression of both viral and cellular genes, *EMBO J.*, 3, 789, 1984.
98. **Flint, S. J., Plumb, M. A., Yang, U. C., Stein, G. S., and Stein, J. L.,** Effect of adenovirus infection on expression of human histone genes, *Mol. Cell. Biol.*, 4, 1363, 1984.
99. **Velcich, A. and Ziff, E.,** Adenovirus E1a proteins repress transcription from the SV40 early promoter, *Cell*, 40, 705, 1985.
100. **Nevins, J. R.,** Induction of the synthesis of a 70,000 dalton mammalian heat shock protein by the adenovirus E1A gene product, *Cell*, 29, 913, 1982.
101. **Green, M. M., Teisman, R., and Maniatis, T.,** Transcriptional activation of cloned human β-globin genes by viral immediate early gene products, *Cell*, 35, 137, 1983.
102. **Castiglia, C. L. and Flint, S. J.,** Effects of adenovirus infection on rRNA synthesis and maturation in HeLa cells, *Mol. Cell Biol.*, 3, 662, 1983.
103. **Imperiale, M. J., Kao, H. T., Feldman, L. T., Nevins, J. R., and Strickland, S.,** Common control of the heat shock gene and early adenovirus genes: evidence for a cellular E1A-like activity, *Mol. Cell. Biol.*, 4, 867, 1984.
104. **Braithwaite, A. W., Cheetman, B. F., Li, P., Parish, C. R., Waldron-Stevens, L. K., and Bellet, A. J. D.,** Adenovirus-induced alteration of the cell growth cycle: a requirement for expression of E1A but not E1B, *J. Virol.*, 45, 192, 1983.
105. **Borelli, E., Hen, R., and Chambon, P.,** Adenovirus-2 E1a product repress enhancer-induced stimulation of transcription, *Nature (London)*, 312, 608, 1984.
106. **Sarnow, R., Hearing, P., Anderson, C. W., Halberto, D. N., Shenk, T., and Levine, A. J.,** Adenovirus early region 1B 58,000-dalton tumor antigen is physically associated with an early region 4 25,000-dalton protein in productively infected cells, *J. Virol.*, 49, 692, 1984.
107. **Chow, K. C. and Pearson, G. D.,** Adenovirus infection elevates levels of topoisomerase I, *Proc. Natl. Acad. Sci. U.S.A.*, 82, 2247, 1985.
108. **Elkareh, A., Murphy, A. J. M., Fitcher, T., Efstrafiadis, A., and Silverstein, S.,** "Trans-activation" control signals in the promoter of the herpesvirus thymidine kinase gene, *Proc. Natl. Acad. Sci. U.S.A.*, 82, 1002, 1985.
109. **Flanagan, J. F.,** Virus-specific ribonucleic acic synthesis in KB cells infected with herpes simplex virus, *J. Virol.*, 1, 583, 1967.
110. **Fenwick, M. L. and Walker, M. J.,** Suppression of the synthesis of cellular macromolecules by herpes simplex virus, *J. Gen. Virol.*, 41, 37, 1978.
111. **Read, G. S. and Frenkel, N.,** Herpes simplex virus mutants defective in the virion-associated shut-off of host polypeptide synthesis and exhibiting abnormal synthesis of (immediate early) viral polypeptide, *J. Virol.*, 46, 498, 1983.

112. **Kaplan, A. S. and Ben-Porat, T.,** The pattern of viral and cellular DNA synthesis in pseudo-rabies virus-infected cells in logarithmic phase growth, *Virology,* 19, 205, 1963.
113. **Stenberg, R. M. and Pizer, L. I.,** Herpes simplex virus-induced changes in cellular and adenovirus RNA metabolism in an adenovirus type-5 transformed human cell line, *J. Virol.,* 42, 474, 1982.
114. **Pizer, L. I. and Beard, P.,** The effect of herpes simplex virus infection on mRNA polyoma virus-transformed cells, *Virology,* 75, 477, 1976.
115. **Nakai, H., Maxwell, I. H., and Pizer, L. I.,** Herpesvirus infection alters the steady-state levels of cellular polyadenilated RNA in polyoma virus-transformed cells, *J. Virol.,* 42, 1131, 1982.
116. **Spector, D. and Pizer, L. I.,** Herpes virus infection modifies adenovirus RNA metabolism in adenovirus transformed cells, *J. Virol.,* 27, 1, 1978.
117. **Reddy, V. B., Thimmappaya, B., Dhar, R., Subramanian, K. N., Zain, B. S., Pau, J., Ghosh, P. K., Celma, M. L., and Weissman, S. M.,** The genome of simian virus 40, *Science,* 200, 494, 1978.
118. **Li, G. C., Schrieve, D. C., and Zena, W.,** Correlation between synthesis of heat-shock proteins and development of tolerance to heat and to adriamycin in Chinese hamster fibroblast: heat shock and other inducers, in *Heat Shock: From Bacteria to Man,* Schlessinger, M. J., Ashburner, M., and Tissiers, A., Eds., Cold Spring Harbor Laboratory, Cold Spring Harbor, New York, 1982, 395.
119. **Keller, J. M. and Alwine, J. C.,** Activation of the SV40 late promoter: direct effects of T antigen in the absence of viral DNA replication, *Cell,* 36, 381, 1984.
120. **Brady, J., Bolen, J. B., Radonovich, M., Saltzman, N., and Khoury, G.,** Stimulation of simian virus 40 late gene expression by simian virus 40 tumor antigen, *Proc. Natl. Acad. Sci. U.S.A.,* 81, 2040, 1984.
121. **Tack, L. and Beard, P.,** Both trans-acting factors and chromatin structure are involved in the regulation of transcription from the early and late promoters in Simian virus 40 chromosomes, *J. Virol.,* 54, 207, 1985.
122. **Holowczak, J. A.,** Poxvirus DNA, *Curr. Top. Microbiol. Immunol.,* 97, 28, 1982.
123. **Bertholet, C., Drillien, R., and Wittek, R.,** One hundred base pairs of 5′ flanking sequence of a vaccinia virus late gene are sufficient to temporally regulate late transcription, *Proc. Natl. Acad. Sci. U.S.A.,* 82, 2096, 1985.
124. **Becker, Y. and Joklik, W. K.,** mRNA in cells infected with vaccinia virus, *Proc. Natl. Acad. Sci. U.S.A.,* 51, 577, 1964.
125. **Salzman, N. P., Shatkin, A. J., and Sebring, E. D.,** The synthesis of DNA-like RNA in the cytoplasm of HeLa cells infected with vaccinia virus, *J. Mol. Biol.,* 8, 405, 1964.
126. **Kit, S. and Dubbs, D. R.,** Biochemistry of vaccinia-infected mouse fibroblasts (strain L-M). Effects on nucleic acid and protein synthesis, *Virology,* 18, 274, 1962.
127. **Kaariainen, L. and Ranki, M.,** Inhibition of cell functions by RNA-virus infection, *Annu. Rev. Microbiol.,* 38, 91, 1984.
128. **Joklik, W. K.,** Structure and function of reovirus genome, *Microbiol. Abstr.,* 45, 483, 1981.
129. **Duncan, M. R., Stanish, S. M., and Cox, D. C.,** Differential sensitivity of normal and transformed human cells to reovirus infection, *J. Virol.,* 28, 444, 1978.
130. **Hand, R. and Kasupski, G. J.,** DNA and histone synthesis in reovirus-infected cells, *J. Gen. Virol.,* 39, 437, 1978.
131. **Hand, R. and Tamm, I.,** Initiation of DNA replication in mammalian cells and its inhibition by reovirus infection, *J. Mol. Biol.,* 82, 175, 1974.
132. **Chaly, N., Johnstone, M., and Hand, R.,** Alterations in the nuclear structure and function in reovirus-infected cells, *Clin. Invest. Med.,* 4, 141, 1979.
133. **Roner, M. R. and Cox, D. C.,** Cellular integrity is required for inhibition of initiation of cellular DNA synthesis by reovirus type 3, *J. Virol.,* 53, 350, 1985.
134. **Sharpe, A. H. and Fields, B. N.,** Reovirus inhibition of cellular DNA synthesis: role of S1 gene, *J. Virol.,* 38, 389, 1981.
135. **Ensminger, W. D. and Tamm, I.,** Inhibition of synchronized cellular deoxyribonucleic acid synthesis during Newcastle Disease Virus, Mengovirus or Reovirus infection, *J. Virol.,* 5, 672, 1970.
136. **Lodish, H. F. and Porter, M.,** Translational control of protein synthesis after infection by vesicular stomatitis virus, *J. Virol.* 36, 719, 1980.
137. **Nishioka, Y., Jones, G., and Silverstein, S.,** Inhibition of vesicular stomatitis virus of herpes simplex virus-directed protein synthesis, *Virology,* 124, 238, 1983.
138. **Weck P. K. and Wagner, R. R.,** Transcription of vesicular stomatitis virus is required to shut-off cellular RNA synthesis, *J. Virol.,* 30, 410, 1979.
139. **Yaoi, Y., Matsui, H., and Amano, M.,** Effect of U.V.-irradiated vesicular stomatitis virus on nucleic acid synthesis in chicken embryo cells, *J. Gen. Virol.,* 8, 165, 1970.
140. **Baxt, B. and Bablanian, R.,** Mechanism of vesicular stomatitis virus-induced cytophatic effects. II. Inhibition of macromolecular synthesis induced by infection and defective interfering particles, *Virology,* 72, 383, 1976.

141. **Kurilla, M. G., Piwnica-Worms, H., and Keene, J. D.,** Rapid and transient localization of the leader RNA and VSV in the nuclei of infected cells, *Proc. Natl. Acad. Sci. U.S.A.*, 79, 5240, 1982.
142. **Reddy, R., Henning, D., Tan, E., and Busch, H.,** Identification of a La protein binding site in a RNA polymerase III transcript (4.51 RNA), *J. Biol. Chem.*, 258, 8352, 1983.
143. **Rinke, J. and Steitz, J. A.,** Precursor molecules of both 5*S* ribosomal RNA and tRNAs are bound by a cellular protein reactive with anti-La lupus antibodies, *Cell*, 29, 149, 1982.
144. **Kurilla, M. G. and Keene, J. D.,** The leader RNA of vesicular stomatitis virus is bound by a cellular protein reactive with anti-La lupus antibodies, *Cell*, 34, 837, 1983.
145. **McGowan, J. J. and Wagner, R. R.,** Inhibition o cellular DNA synthesis by vesicular stomatitis virus, *J. Virol.*, 38, 356, 1981.
146. **Wu, F. S. and Lucas-Lenard, J.,** Inhibition of RNA accumulation in mouse L cells infected with VSV requires viral RNA transcription, *Biochemistry*, 19, 804, 1980.
147. **Blumberg, B. M., Giorgi, C., and Kolakofsky, D.,** N protein of vesicular stomatitis virus selectively encapsidates leader RNA "in vitro", *Cell*, 32, 837, 1983.
148. **Kurilla, M. G., Cabradilla, C. D., Holloway, B. P., and Keene, J. D.,** Nucleotide sequence and host La protein interactions of rabies virus leader RNA, *J. Virol.*, 50, 773, 1984.
149. **Matsumoto, S.,** Morphology of rabies virion and cytopathology of virus infected cells, *Symp. Ser. Immunobiol. Handb.*, 21, 25, 1974.
150. **Herz, C., Stavnezzer, E., Krug, R. M., and Gurney, T.,** Influenza virus, an RNA virus, synthesizes its messenger RNA in the nucleus of infected cells, *Cell*, 26, 391, 1981.
151. **Plotch, S. J., Bouloy, M., Ulmanen, I., and Krug, R. M.,** A unique cap (m^7GpppXm)-dependent influenza virion endonuclease cleaves capped RNAs to generate the primers that initiate viral RNA transcription, *Cell*, 23, 847, 1981.
152. **Ehrenfeld, E. and Lund, H.,** Unstralated vesicular stomatitis virus after poliovirus infection, *Virology*, 80, 297, 1977.
153. **Franklin, R. M. and Baltimore, D.,** Patterns of macromolecular synthesis in normal and virus-infected cells, *Cold Spring Harbor Symp. Quant. Biol.* 27, 175, 1962.
154. **Colby, D. S., Finnerty, V., and Lucas-Lenard, J.,** Fate of mRNA of L-cells infected with mengovirus, *J. Virol.*, 13, 858, 1974.
155. **Zimmerman, E. F., Hecter, M., and Darnell, J. E., Jr.,** RNA synthesis in poliovirus-infected cells, *Virology*, 19, 400, 1963.
156. **Contreras, G., Summers, D. F., Maizel, J. V., and Ehrenfeld, E.,** HeLa cell nucleolar RNA synthesis after poliovirus infection, *Virology*, 53, 120, 1973.
157. **Darnell, J. E., Girard, M., Baltimore, D., Summers, D. F., and Maizel, J. V.,** The synthesis and translation of poliovirus RNA, in *The Molecular Biology of Viruses*, Colter, J. and Paranchych, W., Eds., Academic Press, New York, 1967, 375.
158. **Apriletti, J. W. and Penhoet, E. E.,** Recovery of DNA-dependent RNA polymerase activities from L cells after Mengovirus infection, *Virology*, 61, 597, 1974.
159. **Baltimore, D. and Franklin, R. M.,** The effect of mengovirus infection on the activity of the DNA-dependent RNA polymerase of L-cells, *Proc. Natl. Acad. Sci. U.S.A.*, 48, 1383, 1962.
160. **Holland, J. J.,** Inhibition of DNA-primed RNA synthesis during poliovirus infection of human cells, *Biochem. Biophys. Res. Commun.*, 9, 556, 1962.
161. **Schwartz, L. B., Lawrence, C., Thach, R. E., and Roeder, R. G.,** Encephalomyocarditis virus infection of mouse plasmocytoma cells. II. Effect on host RNA synthesis and RNA polymerases, *J. Virol.*, 14, 611, 1974.
162. **Polatnick, J., Vande-Wonde, G. F., and Arlinghaus, R. B.,** Changes in protein and nucleic acid metabolism in BHK cells infected with FMDV, *Arch. Virusforsch.*, 23, 218, 1968.
163. **Grigera, P. R. and Tisminetzky, S. G.,** Histone H3 in BHK cells infected with Foot-and-Mouth disease virus, *Virology*, 136, 10, 1984.
164. **Weisrod, S.,** Active chromatin, *Nature (London)*, 297, 289, 1982.
165. **Bohm, L., Briand, G., Sautier, P., and Crane-Robinson, C.,** Proteolytic digestion studies of chromatin core-histones structure, *Eur. J. Biochem.*, 119, 67, 1981.
166. **Khorant, B. D., Langner, J., and Powers, J. C.,** Protein synthesis and cleavage in picornavirus-infected cells, in *Biosynthesis, Modification and Processing of Cellular and Viral Proteins*, Koch, G. and Riechter, D., Eds., Academic Press, New York, 1980, 277.
167. **Apriletti, J. W. and Penhoet, E. E.,** Cellular RNA synthesis in normal and mengovirus-infected L-929 cells, *J. Biol. Chem.*, 253, 603, 1978.
168. **Crawford, N., Fire, A., Samuels, M., Sharp, P. A., and Baltimore, D.,** Inhibition of transcription factor activity by poliovirus, *Cell*, 27, 555, 1981.
169. **Weil, P. A., Luse, D. S., Segall, J., and Roeder, R. G.,** Selective and accurate initiation of transcription at the Ade2 major late promoter in a soluble system dependent on purified RNA polymerase II and DNA, *Cell*, 18, 469, 1979.

170. **Martin, E. M. and Kerr, J. M.,** Virus-induced changes in host cell macromolecular synthesis, in *The Molecular Biology of Viruses*, Crawford, L. V. and Stokes, M. G. P., Eds., Cambridge University Press, New York, 1968, 15.
171. **Balandin, I. G. and Franklin, R. M.,** The effect of Mengovirus infection on the activity of the DNA-dependent RNA polymerase of L-cells. II. Preliminary data on the inhibitory factor, *Biochem. Biophys. Res. Commun.*, 15, 27, 1964.
172. **Bossart, W., Egger, D., Rasser, Y., and Bienz, K.,** Accumulation of poliovirus proteins in uninfected isolated HEp-2 cell nuclei "in vitro", *Intervirology*, 21, 150, 1984.
173. **Penmann, S. and Summers, D.,** Effects on host cell metabolism following synchronous infection with poliovirus, *Virology*, 27, 614, 1965.
174. **Collins, F. D. and Roberts, W. K.,** Mechanism of mengovirus-induced cell injury in L cells: use of inhibitors of protein synthesis to dissociate virus specific events, *J. Virol.*, 10, 969, 1972.
175. **Ho, P. P. K. and Washington, A. L.,** Evidence for a cellular ribonucleic acid synthesis inhibitor from poliovirus-infected HeLa cells, *Biochemistry*, 10, 3646, 1971.
176. **Bienz, K., Egger, D., Rasser, Y., and Bossart, W.,** Accumulation of poliovirus proteins in the host cell nucleus, *Intervirology*, 18, 189, 1982.
177. **Fernandez-Tomas, C.,** The presence of viral-induced proteins in nuclei from poliovirus-infected HeLa cells, *Virology*, 116, 629, 1982.
178. **Lazarides, E.,** Intermediate filaments as mechanical integrators of cellular space, *Nature (London)*, 283, 249, 1980.
179. **Luftig, R. B.,** Does the cytoskeleton play a significant role in animal virus replication?, *J. Theor. Biol.*, 99, 173, 1982.
180. **Lenk, R. and Penmann, S.,** The cytoskeletal framework and poliovirus metabolism, *Cell*, 16, 289, 1979.
181. **Rutter, G. and Mannweiler, K.,** Alterations of actin-containing structures in BHK 21 cells infected with Newcastle and vesicular stomatitis virus, *J. Gen. Virol.*, 37, 233, 1977.
182. **Bibor-Hardy, V., Bernard, M., and Simard, R.,** Nuclear matrix modifications at different stages of infection by herpes simplex virus type 1, *J. Gen. Virol.*, 66, 1095, 1985.
183. **Murti, K. G. and Goorha, R.,** Interaction of Frog virus-3 with the cytoskeleton. I. Altered organization of microtubules, intermediate filaments and microfilaments, *J. Cell Biol.*, 96, 1248, 1983.
184. **Maes, R. and Granoff, A.,** Viruses and renal carcinoma of *Rana pipiens*. IV. Nucleic acid synthesis in frog virus-3 infected BHK 21/13 cells, *Virology*, 33, 491, 1967.
185. **Bonneau, A. M., Darveu, A., and Sonenberg, N.,** Effect of viral infection on host protein synthesis and mRNA association with the cytoplasmic cytoskeletal structure, *J. Cell Biol.*, 100, 1209, 1985.
186. **van Venrooij, W. J., Sillekens, P. T. G., van Eekelen, C. B. G., and Reinders, R. J.,** On the association of mRNA with the cytoskeleton in uninfected and adenovirus-infected human KB cells, *Exp. Cell Res.*, 135, 79, 1981.
187. **Howard, J. M., Eckert, B. I., and Bourguignon, L. Y. W.,** Comparison of cytoskeletal organization in canine distemper virus-infected cells, *J. Gen. Virol.*, 64, 2379, 1983.
188. **Zerbini, M. L., Musiani, M., and La Placa, M.,** Effect of heat shock on Epstein-Barr virus and cytomegalovirus expression, *J. Gen. Virol.*, 66, 633, 1985.
189. **Collins, P. L. and Hightower, L. E.,** Newcastle disease virus stimulates the cellular accumulation of stress (heat shock) mRNAs and proteins, *J. Virol.*, 44, 703, 1982.
190. **Schrier, P. I., Bernards, R., Vaessen, R. T. M. J., Hooweling, A., and van der Erb, A. J.,** Expression of class I major histocompatibility antigens switched off by highly oncogenic adenovirus 12 in transformed cells, *Nature (London)*, 305, 771, 1983.
191. **Diaz-Moreno, A., Tapia-Ramirez, J., Gariglio, P., and Fernández-Tomás, C.,** IV Meeting of the European Group of Molecular Biology of Picornavirus, Seillac, France, September 1985.

Chapter 4

INHIBITION OF HOST PROTEIN SYNTHESIS BY DNA VIRUSES (ADENOVIRUS, HERPES VIRUS, AND POXVIRUS): MECHANISM OF ACTION

E. Paez and M. Esteban

TABLE OF CONTENTS

I. INTRODUCTION

Cytopathogenic viruses induce a decline in the rate of host macromolecular synthesis during infection of tissue culture cells. Cellular DNA, RNA, and protein synthesis become selectively inhibited. The virus-induced inhibition of host translation (shut-off) has been studied intensively and appears to be a complex phenomenon, which includes modifications of the host cell translational apparatus, competition between mRNAs, preferential degradation of cellular mRNAs, and inhibition of transcription, processing, and/or transport of cellular mRNAs.

We would like to provide a brief summary of some of the best characterized virus-cell systems to give an idea of the complexity and diversity of the mechanisms used by RNA and DNA viruses to inhibit cellular protein synthesis. Table 1 summarizes the early studies on the shut-off mechanism by some of the most representative groups of viruses. The shut-off by characteristic RNA viruses has been studied in great detail and will be described in another chapter.

One of the most extensively studied virus-host cell interaction is the control of translation in poliovirus infected cells.[14] In this system, the shut-off occurs at the initiation step of translation.[15] Cellular mRNA is functional, since it can be translated in cell-free systems.[16] These authors demonstrated that the failure of infected cells to translate cellular mRNA was in the crude initiation factor preparation. Addition of purified initiation factors showed that eIF-4B[17] or eIF-3[18] was inactivated in poliovirus infected cells. Sonenberg et al.[19] discovered a 24 kdalton protein associated with both initiation factors that binds to the 5′-cap structure of mRNAs. Poliovirus-induced block is reversed by a complex of five polypeptides that includes the 24 kdalton protein.[20,21] Poliovirus inactivates the cap binding proteins, allowing selective translation of the uncapped viral mRNAs.[22]

Competition between mRNAs has been proposed as the mechanism of shut-off vesicular stomatitis virus (VSV).[23] However, other studies indicate that VSV can induce shut-off when coinfected with defective-interfering particles.[24] Under this condition, a marked suppression of wild-type VSV mRNA synthesis occurs, so competition between mRNAs cannot explain this effect, and a mechanism of inhibition by viral transcripts has been proposed.[24,25] Thomas and Wagner[26] have shown an alteration in the fraction containing initiation factors eIF-3 and eIF-4B. This effect appears to be similar to that observed with poliovirus, although VSV mRNAs have a 5′-cap structure. On the other hand, a correlation between synthesis of leader mRNA and inhibition of cellular RNA synthesis has been shown.[27]

In reovirus infected cells, it was proposed that there be a transition of the translational machinery from cap-dependent to cap-independent.[28] The bulk of viral mRNAs isolated from infected cells at late times postinfection (p.i.) is uncapped, but they cannot be translated in lysates from uninfected cells. Therefore, it was proposed that a viral factor is necessary for translation of late uncapped mRNAs. Zarbl and Millward[29] have demonstrated that monoclonal antibodies to $\sigma 3$, one of the outer capsid polypeptides, inhibit the translation of uncapped reovirus mRNAs in lysates from infected cells. Contradictory results reported that late mRNAs are sensitive to m^7 GTP inhibition.[30] They proposed a mechanism of shut-off by mRNA competition for a discriminatory factor. Further work is needed to clarify the mechanism of shut-off induced by reovirus.

In this review, the shut-off of host protein synthesis caused by several cytopathogenic DNA viruses will be discussed more extensively. Briefly, adenovirus infection appears to alter the translational machinery by an unidentified protein[31,32] or by the small VAI-RNAs.[33] Degradation of cellular mRNA was proposed as the mechanism of herpes virus-induced shut-off.[34] However, recent reports support an alteration of the translational machinery dependent on viral gene expression.[35] In vaccinia virus infected cells, a discriminatory mechanism by small viral transcripts has been proposed.[36,37]

Table 1
SHUT-OFF OF HOST PROTEIN SYNTHESIS BY CYTOPATHIC VIRUSES

Virus	Cell type	50% Inhibition (hr p.i.)	Ref.
Picornavirus			
Polio 1	HeLa	1.5	1
Polio 2	HeLa	<1.5	2, 3
EMC	Krebs-II	4.5	4
Mengo	L	0.5	5
Togavirus			
Sindbis	Chick	2	6
Rhabdovirus			
VSV	L	2.5	7
Adenovirus			
Adeno 5	KB	9—12	8
Adeno 2	MK	>20	9
Herpes virus			
HSV-1	Hep-2	2—4	10
CDV			
FV-3	FHM	1—3	11
Poxvirus			
Vaccinia	HeLa	0.5—2	12
	L	<0.5	13

From the mechanisms proposed, it can be concluded that each group of animal viruses has developed a particular strategy to inhibit host protein synthesis. However, what these viruses have in common is that initiation of protein synthesis is the primary target for virus-induced shut-off. The critical step for the control of translation is likely to be at the level of formation of the initiation complex between the mRNA, initiation factors, and the 40S ribosomal subunit. This is most probably because initiation of translation is the rate-limiting step in protein synthesis.

II. ADENOVIRUS

A. General Properties

Adenoviruses were first isolated from human adenoids.[38] Several members have been identified in vertebrates. In humans, 41 different serotypes have been found, and human adenoviruses types 2 and 5 have been extensively used to understand the virus-cell interactions in this group of viruses.[39-44] An important characteristic of this virus-cell system is the oncogenic potential based on studies with rodent cells. This fact, together with the dependence on the transcriptional and translational machinery of the host cell, has stimulated the investigation in this field, making important contributions, as the discovery of mRNA splicing,[45-47] and more recently the contribution of the virus-associated (VA) RNAs first described by Reich et al.,[48] in what appears to be a new function for small RNAs, as positive effectors in translational control.[33,49-51]

The structure of adenoviruses is a nonenveloped icosahedron of 65 to 80 nm in diameter, containing 252 capsomeres arranged as hexons and pentons with fibers that project from the 12 vertices.[52] The virus contains a double-stranded DNA (dsDNA) of 20 to 25 megadaltons (Mdaltons) with a virus-coded 55 kdalton terminal polypeptide covalently linked to each 5′ end of the linear genome, which acts as a primer for initiation of DNA synthesis. DNA replication occurs exclusively in the nucleus of the host cell. Some types of adenovirus cause cytopathic effect, others persist for long periods, and certain types can induce malignant transformation.

In this review, we will focus on the productive cycle of adenovirus types 2 and 5, where an efficient shut-off of cellular protein synthesis occurs at late times after virus infection.[8] For convenience, the productive cycle is divided into an early phase of infection, followed by DNA replication that begins about 8 hr after infection, and a late phase starting with the onset of DNA replication, in spite of the fact that there are genes expressed at early and late times p.i.

Six transcriptional units (E1A, E1B, E2, E3, E4, and L1) are transcribed at early times. In fact, these transcripts are now classified as immediate-early (L1), pre-early (E1A), delayed-early (E1B, E2A, E2B, E3, and E4), and intermediate (IVa2 and IX). Each unit encodes for different polypeptides as a result of differential splicing of the mRNAs. These polypeptides are involved in different processes such as DNA replication, cellular transformation, and probably shut-off of host protein synthesis; products of the region E2 function in DNA replication. A 72 kdalton DNA binding protein is encoded by region E2A,[53] and region E2B encodes the 87 kdalton precursor of the 55 kdalton protein covalently attached to the termini of viral DNA[54] and the 140 kdalton DNA polymerase.[55,56] Products of both the E1A and E1B regions function in transformation of rat cells in culture,[57,58] but a product of E1A also facilitates transcription of all the other early genes,[59-61] and this product is structurally related to the protein products of the myc and myb oncogenes.[62] Recently, it has been reported that E1A gene products also enhance transcription by RNA polymerase III[63] and elevate the levels of cellular topoisomerase I.[64] A 55 kdalton protein of region E1B by making a complex with a 25 kdalton protein of region E4 is probably responsible for mediating shut-off of host protein synthesis. Regions E3 and L1 encode several early polypeptides of unknown function. Those encoded by E3 have no effect on virus growth in cell culture, and L1 is a late region that is also expressed at early times after infection.[65,66]

At about 8 hr p.i., viral DNA replication begins in the cell nucleus, while host DNA synthesis is reduced by more than 90%.[67] At this time, late mRNAs are transcribed and most of them are produced from the major late transcriptional unit with the promoter localized at 16.45 map units.[68,69] These transcripts can be polyadenylated at five different sites that define the five groups (L1 to L5). Differential splicing of these transcripts generates several mRNAs with a common tripartite leader sequence. These mRNAs encode structural polypeptides that form the viral capsid and core and several nonstructural polypeptides. Two small RNA species called virus-associated (VA) I and II RNAs are transcribed by the cellular RNA polymerase III, in contrast with most of the early and late genes that are transcribed by the cellular RNA polymerase II. They are encoded by two different genes that are located at about 30 map units.[70,71] VAI-RNA is preferentially transcribed at late times and is essential for efficient translation of late genes.[33]

The length of the infectious cycle is influenced by both the multiplicity of infection and the growth state of the cell, but it is usually about 36 hr. The final step is the assembly of the virion in the nucleus, which produces about 10,000 viral particles per infected cell, with an increase in the total DNA and protein content of the cell of about twofold.

B. Inhibition of Cellular DNA Synthesis

In productively-infected cells, cellular DNA synthesis is strongly inhibited after adenovirus infection.[72] This inhibition is not well understood; however, it might be a consequence of the shut-off of host protein synthesis.

Hodge and Scharff[73] showed that when viral DNA replication occurred during the G phase of the cell cycle, inhibition of the initiation of host DNA synthesis was observed. However, if viral DNA replication was in the S phase, no initiation of cellular DNA synthesis occurred, but those molecules that had already initiated were completed. These experiments suggested a block at the level of initiation of host DNA synthesis during viral DNA replication. Because adenovirus DNA synthesis is initiated by the 87 kdalton precursor to the terminal protein, initiation of viral DNA replication could be independent of cellular factors involved in this process. The terminal protein covalently binds to dCMP,[74] and this complex functions as a primer in viral DNA replication.

Most of the work on the mechanism of inhibition of host DNA synthesis is based on the generation of adenovirus mutants. Using temperature-sensitive (ts) mutants, it has been established that viral DNA replication is not necessary to the inhibition of cellular DNA synthesis.[75,76] Virus mutants in the DNA polymerase or DNA-binding protein do not synthesize viral DNA, but induce inhibition of cellular DNA synthesis under nonpermissive conditions. Two early regions of the adenovirus genome (E1A and E1B) appear to be involved in the inhibition of cellular DNA synthesis. A host range mutant (H5hr1) with a mutation that only affects the 13S mRNA from region E1A[77,78] fails to inhibit cellular DNA synthesis. However, the E1A region is important for the expression of other early genes, and a more extensive study of all the early genes has to be done before this effect can be attributed to the gene product(s) encoded by the largest mRNA of region E1A. On the other hand, degradation of DNA might also limit the extent of cellular DNA synthesis and can contribute to the cytopathogenicity of the virus as has been observed in cells infected with *cyt* mutants of adenovirus 12.[79] These mutants produce large plaques and are less oncogenic. Viral and cellular DNA were degraded in cells infected by the mutant H12cyt70, but not by wild-type virus.[80] This observation implies that a viral gene product prevents the degradation of DNA. A deletion mutant (H5dl331)[81] also induces degradation of DNA.[82] This defect has been mapped using deletion mutants and is related to the loss of E1B-21 kdalton polypeptide function.[83,84]

C. Inhibition of Cellular RNA Synthesis

Several mechanisms have been proposed to explain the inhibition of cellular RNA synthesis by adenovirus. Some of them are based in a differential effect on viral and cellular RNA metabolism, i.e, inhibition of processing and/or transport of the host mRNAs. However, none of them is sufficient to explain the shut-off of cellular protein synthesis (see Section II.D).

Early experiments indicated that cellular transcription continues during adenovirus infection.[85] More recent experiments have confirmed these results. Using cDNA clones from different cellular mRNAs, it seems that adenovirus infection does not inhibit transcription of cellular mRNAs.[31,86-88] An effect on processing of cellular mRNA is more difficult to establish. Nuclear poly(A)-containing RNA isolated after adenovirus infection contains species with the same length as the corresponding cytoplasmic poly(A)-containing RNA from uninfected cells.[31,87] However, these experiments do not measure specific precursors, and it is not possible to determine whether nuclear RNA processing continued.

Inhibition of the appearance in the cytoplasm of cellular RNA at late times of adenovirus infection has been reported by several groups,[31,86,89,90] and this was proposed as the cause for the inhibition of host protein synthesis.[86] At the beginning of the late phase during adenovirus infection (12 hr p.i.), few cellular mRNAs and rRNAs enter the cytoplasm, and a 96% inhibition occurs by 18 hr p.i.[89] The failure of cellular RNA to appear in the cytoplasm could be due to failure of nuclear processing or transport of cellular RNAs, but this question is unanswered at present. Yoder and Berget[88] have shown that nuclear steady-state levels of dihydrofolate reductase (DHFR) sequences changed in correspondence to the changing rate of appearance of DHFR mRNA in the cytoplasm. Steady-state DHFR-specific nuclear RNA levels initially increased and subsequently decreased during the course of infection, indicating a differential stabilization of DHFR nuclear sequences. With rRNA, it appears to occur as a differential effect on the 28S and 18S. While the effects of adenovirus infection on 28S rRNA metabolism resemble those described above, the inhibition of appearance in the cytoplasm of 18S rRNA takes place at a later time than that of 28S rRNA.[89] In run-off transcription experiments, using cloned human rDNA containing the 5′ end of the 45S precursor of 18S rRNA, it has been shown that extracts from adenovirus-infected cells are much less efficient in transcribing this rDNA than extracts from uninfected cells and that the effect was specific of RNA polymerase I.[91] It seems that the inhibition of transcription is due to deficiency of some cellular factors important to initiation, and this effect could be secondary to the inhibition of host protein synthesis.

D. Inhibition of Cellular Protein Synthesis

Adenovirus produces a dramatic shut-off of cellular protein synthesis at late times after infection.[8] This inhibition is most likely due to a discriminatory effect on translation of viral and cellular mRNAs.[31,32,92] Using cDNA probes against several HeLa abundant mRNAs, including the actin mRNA, it has been shown that the steady-state levels of functional host mRNAs in the cytoplasm of infected cells are indistinguishable from those of uninfected cells at the time shut-off is complete.[31,92] Thus, the reduced levels of transport[86] do not account for the shut-off of host protein synthesis. However, a short half-life of some cellular mRNAs can reduce the steady-state levels in the cytoplasm, as has been reported for the SV40 early mRNA produced in SV80 cells[87] or histone mRNA.[93] At the time when the synthesis of cellular proteins is reduced to more than 90% compared to uninfected cells, there are still normal amounts of several cellular mRNAs present in the cytoplasm; furthermore, these cellular mRNAs can be translated in vitro.[31,92,94] The lack of activity is not due to degradation since the cellular mRNA remains with the same size and approximately equal abundance.[31,92] More than 90% inhibition of translation of actin mRNA is obtained when mRNAs from uninfected cells are translated in a message-dependent homologous HeLa cell extract from adenovirus infected cells compared to uninfected cells.[92] This inhibition of translation of cellular mRNAs in infected cell extracts suggests the presence of a discriminatory factor that is either virus-induced or virus-coded.

Another possibility of explaining the shut-off of cellular protein synthesis is competition between viral and cellular mRNAs.[95] In fact, a preferential translation of viral mRNAs in vitro is consistent with this hypothesis,[92] and might be due to differences in the 5′ leader sequences of viral and cellular mRNAs, or might be a consequence of their relative abundance. Most of the late viral mRNAs share the same leader sequence,[45-47] which might confer translational advantages, but there are three different transcriptional units (1A, 1B, and protein IX) that are translated at late times and do not contain these leader sequences. If the efficiency of translation of these mRNAs was slower than those that contain the tripartite leader sequence, a competition mechanism would explain the shut-off phenomenon. Indeed, an intact tripartite leader sequence has been shown to play an essential role in the translation of SV40 T antigen in adenovirus/SV40 recombinant viruses.[96] It has been postulated that inhibition of translation of cellular mRNA is not due to competition, based on the observation that in H5dl331 infected cells, neither viral nor cellular proteins is synthesized. This deletion mutant fails to produce VAI-RNA. Cells infected with this mutant produce late viral mRNA, but are not translated. In the absence of viral protein synthesis, one could expect synthesis of cellular proteins if a competition mechanism is responsible for shut-off. However, it has been recently demonstrated that this complete block of viral and cellular protein synthesis is due to the phosphorylation of the α subunit of initiation factor eIF-2.[49-51] VAI-RNA plays a role in suppressing eIF-2α kinase activity during adenovirus infection of HeLa cells (see Section V.F). Thus, a mechanism of shut-off by competition of viral and cellular mRNAs cannot be excluded at the moment.

The cellular mRNA is associated with small polyribosomes in infected cells, indicating that inhibition is at the level of initiation of translation.[31,92] Since shut-off is prevented in the presence of inhibitors of DNA synthesis, the inhibitor is a late function.[31]

Babiss and Ginsberg[32] have recently reported a number of mutants in the E1B region of adenovirus 5 that fail to shut-off cellular protein synthesis. These mutants are defective for the production of the 55 kdalton polypeptide. H5dl118 also synthesized a truncated form of the 21 kdalton polypeptide, but all of them are defective of efficiently shutting off cellular protein synthesis, suggesting that the 55 kdalton protein is involved in this process. The 55 kdalton protein is produced at early times, and as we described before, shut-off of cellular protein synthesis is prevented in the presence of inhibitors of DNA synthesis, implicating late viral functions. Furthermore, 293 and KB-18 cells produce the 55 kdalton protein, but are viable, and cellular proteins are synthesized normally. In this regard, it is interesting that monoclonal antibodies directed against the 55 kdalton protein immunoprecipitated, in addition to this protein, a polypeptide of 25 kdalton that is encoded by E4 region of adenovirus genome.[97] Because of mutants that have been engineered into the E4 region resembling the E1B mutants,[98] it is tempting to speculate that a 55 to 25 kdalton protein complex may indirectly induce shut-off by altering the expression of some late viral mRNA or polypeptide. In fact, there is a threefold reduction in the cytoplasmic steady-state levels of viral mRNA, indicating that 55 kdalton protein affects maturation, stability, or transport of viral mRNA.[32] It has been shown that the 55 kdalton protein can also form a complex with the p53 cellular protein in adenovirus transformed cells.[99] A comparative study about the function of the 55 kdalton in these protein complexes may give some clues on how the mechanism of shut-off is activated in adenovirus-infected cells.

III. HERPES VIRUS

A. General Properties

Herpes viruses are similar to adenovirus in that they also have a dsDNA as genome and replicate in the nucleus. Among the several members isolated from vertebrates, there are five human herpes virus distributed in three subfamilies: Alphaherpesvirinae [herpes simplex virus type 1 (HSV-1), herpes simplex virus type 2 (HSV-2), and varicella-zoster virus]; Betaherpesvirinae [cytomegalovirus]; and Gammaherpesvirinae [Epstein-Barr virus].[100,101] Most of the work has been concentrated with members of the subfamily Alphaherpesvirinae that also include pseudorabies virus from pigs and equine herpes virus 1. This is probably due to the wide host range and the short infectious cycle.[102-104] A general characteristic of herpes virus is the latency, responsible for the phenomenon of recurrent herpes infections.[105]

The structure of herpes virus is an enveloped icosahedron composed of 162 capsomeres. The virion consists of four structural components: core, capsid, tegument, and envelope.[106] The size varies from 120 to 300 nm depending on the thickness of the tegument and the state of the envelope. Herpes virus contains a linear dsDNA of 80 to 150 Mdalton, with minor variations in a specific virus.[104] The 100 Mdalton HSV genome is arranged in two regions L and S (L represents about 80% of the viral genome), and both regions are flanked by inverted repetitions.[107,108] The L and S regions invert relative to each other, giving rise to four isomers.[109] These inversions appear to be mediated by recombination in the inverted repeats region.[110,111] A deletion mutant that has lost the internal inverted repeats can grow normally but does not invert.[112]

The host RNA polymerase II is responsible for transcription of viral genes,[113] but this process is sequentially regulated by viral products in a cascade fashion.[106,114] At least five groups of proteins have been defined (alpha, $beta_1$, $beta_2$, $gamma_1$, and $gamma_2$).[115,116] The alpha or immediate-early genes codifies for five polypeptides named 0, 4, 22, 27, and 47. Their synthesis starts soon after infection and reaches a peak between 2 to 4 hr. Functional alpha polypeptides are required for the synthesis of beta- or early polypeptides. ts mutants accumulate in the alpha 4 gene; a role of this gene in the regulation of beta genes has been proposed. The alpha genes map at or near the inverted repetitions and only two of them (alpha 22 and alpha 47) produce spliced mRNAs.[117] The beta groups are synthesized thereafter and before the onset of DNA synthesis. Beta proteins are involved in DNA replication (i.e., DNA-binding protein, thymidine kinase, DNA polymerase) in the shut-off of synthesis of alpha and host proteins, and in the induction of gamma genes.[115,116] The gamma or late polypeptides are mostly made after DNA replication; however, $gamma_1$ genes can be expressed before DNA replication, but at reduced amounts. Most of the gamma proteins are structural components of the virion. Some of them appear to function early after infection in shut-off [35,118,119] and induction of alpha genes.[120] Beta and gamma genes map in both L and S regions, but in comparison with adenovirus, gene overlapping and splicing do not occur very frequently.[121] Furthermore, genes with multiple initiation sites for transcription have been shown.

Herpes virus encodes about 100 proteins, and some proteins are processed after their synthesis. This includes cleavage, phosphorylation, glycosilation, and others.[122,123] These viral proteins can be easily detected because of the shut-off of host protein synthesis by 2 to 4 hr after infection.[106] The infectious cycle of HSV is about 12 to 18 hr.

B. Inhibition of Cellular DNA Synthesis

Early after infection, herpes virus blocks the synthesis of cellular DNA.[124-126] Herpes virus DNA can be separated from cellular DNA by CsCl gradient centrifugation because of the different buoyant density due to high G-C content.[125,127] The viral product responsible for the shut-off of cellular DNA is most probably an early viral protein, because the block of host DNA synthesis is observed in the absence of viral DNA synthesis,[128] but not in the presence of cycloheximide.[129] DNA synthesis was inhibited soon after removal of the drug.

ts mutants of HSV-1 and HSV-2 defective in shut-off of host DNA synthesis have been isolated.[130,131] Two different genes related in their function appear to be involved; one is the viral DNA polymerase. It was suggested that an interaction between these viral products is required for viral DNA replication and may lead to a block of cellular DNA synthesis. Fenwick and Walker[132] have presented evidence that cellular DNA synthesis can be inhibited with UV-inactivated virus, but this is probably an indirect effect related to the early shut-off of protein synthesis produced by a component of the virion.

C. Inhibition of Cellular RNA Synthesis

Transcription of host cell RNA is inhibited after HSV infection, but to a lesser extent than that observed with DNA and protein synthesis.[126,133] Pulse-chase experiments in the presence of actinomycin D show an inhibition of synthesis of cellular RNA, greater on smaller RNAs. Ribosomal RNA synthesis is also inhibited, most probably by an effect on processing of precursors to produce rRNA.[134] Using BHK cells transformed with polyoma virus and measuring the rate of polyoma mRNA synthesis after HSV-1 infection showed that both the synthesis and the cytoplasmic steady-state levels were strongly inhibited.[135] An increase in nonadenylated RNA was also observed, suggesting an effect on processing or deadenylation of cellular mRNA.[136] Similar experiments with cells transformed with adenovirus confirmed that both synthesis and transport to the cytoplasm were inhibited after infection with HSV-1.[137,138] Another two steps that can decrease the cytoplasmic pool of active mRNA are the stability and functional activity to direct protein synthesis. Nishioka and Silverstein[34] have shown that globin mRNA sequences were degraded early after infection. This has been confirmed by Inglis[139] for the actin and glyceraldehyde-3-phosphate dehydrogenase mRNAs. Recently, this result has been extended,[35] although another report presented evidences for a differential stability of cellular mRNAs.[140] They have found that unlike globin and actin mRNAs, the steady-state level of histone mRNA remains unchanged, although the transcription of the histone H_3 gene was also inhibited. It was proposed that this apparent stabilization could be due to the lack of poly(A) at the 3′ end of histone mRNAs or inhibition of a host protein responsible for activating a cellular RNase specific for these mRNAs. Schek and Bachenheimer,[35] however, have also observed degradation of histone mRNAs. On the other hand, cellular mRNA activity was lost after infection with HSV, as determined by translation in vitro.[141] Fenwick and McMenamin[142] have shown that this effect occurs in the presence of actinomycin or cycloheximide or with UV-inactivated virus, suggesting that a virion-associated component causes a structural change in the host mRNA to the extent that is is no longer functional in a translation system in vitro.

D. Inhibition of Cellular Protein Synthesis

Different strains of herpes virus vary in the extent of inhibition of host protein synthesis, but usually translation declines soon after infection. Roizman et al.[10] showed an inhibition of total protein synthesis of about 70% at 3 hr p.i. with HSV-1. The shut-off of cellular protein synthesis appears to be a multistep process.[34,119,132] At least there are two different stages, an early or virion-associated shut-off that is due to a virion component, and a delayed or expression-dependent shut-off that requires the synthesis of viral proteins.

Fenwick and Walker[132] and Nishioka and Silverstein[143] have shown that the early shut-off of cellular protein synthesis is mediated by a virion component. Infection with UV-inactivated virus or in the presence of actinomycin does not prevent the inhibition of cellular protein synthesis. Furthermore, HSV shuts off host protein synthesis in enucleated cells,[132] and this effect was prevented by infecting with virus inactivated by heating[143] or neutralized with HSV-specific antiserum.[132] Earlier investigations demonstrated a rapid disaggregation of HSV infected cell polyribosomes.[144,145] More recently it has been shown that a virion component is responsible for this effect and correlated with the early shut-off.[132,143] This phenomenon is further supported by the results obtained during mixed infections with HSV and VSV.[146] They found that VSV prevented the production of HSV proteins, probably by inhibiting transcription, but disaggregation of polyribosomes occurs, an effect specific for HSV because it is known that VSV does not dissociate polyribosomes during infection. ts mutants have been isolated that fail to cause early shut-off at the nonpermissive temperature.[119,147] These mutants also fail to suppress the synthesis of host polypeptides during infections in the presence of actinomycin, confirming the presence of an abnormal virion component. Because HSV-2 shuts off cellular protein synthesis faster than HSV-1, recombinants of these viruses have permitted mapping of this function between 0.52 and 0.59 map units, but the gene products are unknown.[118,148] Ribosomes are also altered during the process of HSV infection. After the initial disaggregation, a substantial fraction of the polyribosomes formed late in infection was inactive in protein synthesis, indicating an alteration of the protein synthetic machinery.[149] Other observations have shown phosphorylation of the S6 protein of the small ribosomal subunit soon after infection.[150,151]

On the other hand, Nishioka and Silverstein[34,143] have shown that after the initial disaggregation of polyribosomes, the cellular mRNA was degraded by a process that required viral protein synthesis. This secondary or delayed shut-off appears to be specific, or at least there is a differential rate of degradation.[35,140] In this regard, Isom et al.[152] reported that in rat hepatoma cells, HSV-2 infection inhibits the synthesis and secretion of specific plasma proteins, a process that correlates with a decrease in the amount of RNA and requires viral protein synthesis. However, other plasma proteins remain unchanged. Delayed shut-off of cellular protein synthesis is probably mediated by viral beta proteins, but it does not exclude gamma proteins possibly being involved. In HSV-infected cells in the presence of cycloheximide, cellular protein synthesis is inhibited soon after removal of the drug, a process that was prevented in part if actinomycin was added at the time of cycloheximide removal.[115] This experiment may suggest that some alpha proteins could be involved in the shut-off phenomenon, because there was no complete prevention in the presence of actinomycin. However, Read and Frenkel,[119] using a mutant (vhs 1 from virus-associated host shut-off) deficient in early shut-off, have shown a complete prevention of shut-off by actinomycin, excluding a role of alpha proteins in delayed shut-off. A recent report has shown that cellular mRNA degradation is induced by a factor associated with the infecting HSV-1 and HSV-2 virions.[35] Because ts mutants induce cellular mRNA degradation, but have little effect on cellular protein synthesis at the nonpermissive temperature and because wild-type virus induce shut-off before an extensive degradation of cellular mRNA occurs, an additional viral function requiring viral expression is needed to fully inhibit host cell protein synthesis.

The inhibition of alpha proteins at the time that beta proteins are synthesized is another area of interest.[115] As we described before, in HSV infected cells in the presence of cycloheximide, there was a block in host cell protein synthesis after removal of the drug, but this block also affected alpha proteins. Beta or gamma proteins were suggested to induce this shut-off of both host and viral alpha proteins, because actinomycin partially blocks this inhibition. In this regard, ts mutants that affect the beta 8 protein (a DNA-binding protein) accumulate more alpha mRNA, suggesting a role of this protein in the decline of alpha proteins, probably by inhibiting transcription.[153] Other reports have proposed a requirement for alpha 4 protein in the regulation of HSV-1 gene expression, using alpha 4 ts-mutants.[154,155] These studies demonstrated that alpha 4 protein is involved in the negative regulation of alpha genes, as well as for expression of early and late genes. A recent report approached the effect of isolated alpha 4 alleles on the expression of isolated HSV-1 alpha, beta, and gamma regulatory elements, using constructs of different HSV-1 promoters fused to the gene for chloramphenicol acetyltransferase gene.[156] They concluded that low levels of alpha 4 protein have a stimulatory effect on alpha promoters, and that higher concentrations have an inhibitory effect on these promoters. However, a number of other HSV-1 specific polypeptides can affect the level of synthesis of viral and cellular polypeptides in infected cells. Thus, virion components have been implicated in expression of both cellular and viral alpha genes,[114,119,157] and the immediate-early polypeptides alpha 0 and alpha 27 may also be involved in modulating the expression of HSV-1 genes.[158-160]

IV. POXVIRUS

A. General Properties

Poxviruses are the largest animal viruses that have been isolated. They have a complex structure and contain a linear dsDNA as genome, but unlike adenoviruses and herpes viruses, replication occurs in the cytoplasm of infected cells. This characteristic is made possible because all of the enzymes required for viral mRNA synthesis are present within the virion, and the DNA also encodes for their own enzymes required for DNA replication.[161,162]

Vaccinia virus, a member of the *Orthopoxvirus* genus, is the prototype of this group of viruses and is a highly cytocidal virus that shuts off host cell protein synthesis very soon after infection.[163,164] The structure is a brick-shaped ellipsoid and measures approximately 200 × 300 nm. It contains a core where the genome is located, two lateral bodies of unknown function at both sides of the core, and an envelope. Extracellular virus contains an additional envelope obtained from the cellular membrane modified by the inclusion of viral proteins, although most of the virus remains cell-associated.[165]

The genome is a large 122 Mdalton linear dsDNA, cross-linked at both ends.[166-168] While the central region is well conserved among isolates or during serial cell passage, genetic changes such as deletions or rearrangements occur spontaneously at the ends.[169-173] Vaccinia genome also contains an inverted terminal repetition of 10 kilobases (kb),[174,175] with sets of tandem repeats located at both ends.[176-178] A role of the inverted terminal repetition in DNA replication has been proposed. However, it has been observed that more than half of the repetition can be deleted from the left end of vaccinia genome without altering viral infectivity in cell culture.[173,179] Upon penetration into cells, viral cores are released in the cytoplasm, and there follows a rapid synthesis of viral mRNAs by the enzymes present within the core.

These include DNA-dependent RNA polymerase, poly(A) polymerase, an enzyme complex for capping (RNA triphosphatase, guanylyl, and methyl transferases), an endoribonuclease (probably involved in RNA processing), nucleoside triphosphatases, and a DNA topoisomerase with a suggested role in transcription.

Viral mRNAs can be classified into three groups: immediate-early, delayed-early, and late.[13,180] DNA-RNA hybridization studies indicate that early mRNA sequences compete with half of the late mRNAs.[181,182] This is now explained because late transcripts are heterogeneous in size as a consequence of undefined 3′ ends,[183] and they run through the early genes. When these genes are transcribed in opposite directions they produce dsRNA that accumulates at late times of infection.[184,185] Several early mRNAs have now been analyzed, and lack of splicing and processing at the 5′ end has been observed.[186-189] These viral mRNAs also revealed a lack of consensus of regulatory sequences with the initiation and termination sites of prokaryotic and eukaryotic mRNAs. Recently, a functional analysis of a promoter region has been reported.[190] This promoter has both early and late regulatory signals within 31 nucleotides upstream of the early and late RNA start sites, respectively. Among the early proteins there are several enzymes involved in viral DNA synthesis. Replication of the viral DNA occurs in the cytoplasm and begins at the ends of the genome.[191,192] After viral DNA replication, a shift in gene expression takes place. Late viral mRNAs are transcribed and hybridized with almost the whole genome.[193] Most of these mRNAs encode for virion structural proteins. At this time, the synthesis of early proteins is inhibited,[181,194,195] although early mRNAs remain functional in the cytoplasm of infected cells.[196,197] Several late genes have been precisely mapped.[183,198-204] The characterization of early and late genes will permit an understanding of the unique signal sequences evolved by poxviruses.

The cytopathic effects caused by these viruses are frequently very rapid and extensive. An early cell rounding is observed between 1.5 to 2 hr p.i., and a later cytopathic effect occurs after the onset of DNA synthesis that is characterized by cell fusion and other cellular changes producing cellular death.[163,164] Poxviruses also induce a strong inhibition of cellular metabolism. The effect on host macromolecular synthesis has been investigated more extensively, and several hypotheses have been proposed.[164]

B. Inhibition of Cellular DNA Synthesis

Early reports show an inhibition of cellular DNA synthesis soon after infection with vaccinia virus.[205-208] A virion component was suggested as responsible for this inhibition. Cellular DNA is degraded in infected cells, a process probably mediated by the virion-associated DNases.[209] In cells infected with UV-inactivated virus, the rapidly labeled 4S DNA is not incorporated into higher molecular weight DNA, because the nascent single stranded fragments become degraded.[210]

C. Inhibition of Cellular RNA Synthesis

Vaccinia virus inhibits the synthesis of cellular mRNA after infection.[212] If this effect is due to an inhibition of transcription or processing, it remains to be elucidated. Rice and Roberts[211] have examined the cytoplasmic steady-state levels of cellular mRNAs by Northern blot analysis using cDNAs of beta-actin and alpha-tubulin as probes. They have shown that these mRNAs are degraded during the course of infection. However, no attempts have been made to discover if this effect is caused by a virion-associated component or a viral product synthesized after infection. Another possibility is the activation of a latent cellular RNase. The implication of this effect on the shut-off of cellular protein synthesis will be discussed in Section IV.D.

On the other hand, inhibition of processing and transport to the cytoplasm of ribosomal RNA have been reported.[212,213] This inhibition occurs by 2 to 3 hr p.i., infection, but since the synthesis of ribosomal proteins is inhibited before this time, it is unclear if the effects observed on rRNA metabolism are secondary to the shut-off of cellular protein synthesis. Similarly to adenovirus and herpes virus, generations of ts mutants of vaccinia virus defective in the inhibition of cellular macromolecular synthesis may provide useful information about the inhibitory mechanism, but no such mutants have yet been reported.

D. Inhibition of Cellular Protein Synthesis

Vaccinia virus induces a rapid and drastic shut-off of cellular protein synthesis.[12,13,206,214,215] Greater than 50% reduction in protein synthesis can be observed by 0.5 to 1 hr p.i. This very early shut-off occurs concomitantly with degradation of polyribosomes.[216]

There are different hypotheses about the viral product(s) responsible for this shut-off. One postulates a virion component;[12,214,217] the second hypothesis proposes small viral transcripts synthesized early after infection.[163,218] This work has been recently reviewed extensively by Bablanian,[164] so we will only describe the most significant data about the viral products involved in shut-off and how this shut-off is produced.

Early reports using actinomycin D suggested that shut-off occurs in the presence of this drug, and for that reason a virion component would be involved in shut-off.[12,214] However, it has since been shown that the concentration of this drug is critical, and 20 to 30 μg/mℓ of actinomycin D are needed to completely prevent viral gene expression.[36,219] A similar effect was observed using UV-irradiated virus.[220] Virus irradiated with low doses of UV induces shut-off in the absence of viral protein synthesis, but RNA is synthesized under these conditions. Higher doses of UV are needed to prevent shut-off. Results from in vitro experiments with cell-free systems have shown an inhibition of protein synthesis by viral cores.[217] This inhibition appears to be at the level of initiation of protein synthesis.[221] An 11 kdalton phosphoprotein has been isolated from viral cores that inhibit protein synthesis without discrimination.[222] However, other reports have shown that this inhibition only occurs when high concentrations of viral cores are added to the cell-free system,[223] and that boiled cores can also inhibit protein synthesis, excluding a viral protein.[224] The surface tubules have also been implicated in shut-off.[225] However, the significance of these results in the mechanism of cellular shut-off is unclear because it has been demonstrated that high concentrations of actinomycin D do not affect virus penetration and uncoating, yet inhibit viral RNA synthesis and prevent shut-off from occurring.[36,219] Furthermore, Raczynski and Condit[226] have shown an inhibition of vaccinia virus growth by the drug 2′-*O*-methyladenosine. This is a ribonucleoside analogue that inhibits early viral protein synthesis and prevents cellular shut-off. The effect was virus specific because a drug-resistant mutant was isolated. From the chemical nature of the drug, the most obvious explanation for its effect is that it inhibits production of functional viral mRNA. Thus, these results, together with those obtained with actinomycin D, indicate that viral gene expression is needed to induce shut-off.

An interesting phenomenon that has remained unexplained for a long time is that observed by the use of cycloheximide. Cellular protein synthesis is severely inhibited when cycloheximide is removed a few hours p.i. in spite of the fact that this drug is reversible in uninfected cells.[219,227] A correlation between this shut-off and the extent of viral RNA synthesis has been reported.[228] Our recent results have shown that viral and ribosomal RNAs are degraded in cycloheximide-treated infected cells, most likely by activation of the 2-5A synthetase/RNase L system, since a characteristic cleavage pattern of ribosomal RNA can be observed.[229,230] An indiscriminate degradation of RNA may explain why protein synthesis continues inhibited after cycloheximide removal. A protein kinase that phosphorylates the alpha subunit of the initiation factor eIF-2 and inhibits protein synthesis can also be involved in this effect, as we have observed activation of this sytem in cycloheximide-treated cells both in vitro and in vivo.[231,232] The activator of the 2-5A synthetase/RNase L and protein kinase systems is dsRNA. It is known that dsRNA is produced by vaccinia virus as a result of symmetrical transcription.[184,185] However, dsRNA probably does not play a role in shut-off as it inhibits viral and cellular protein synthesis with no selectivity.[233]

The second hypothesis has proposed vaccinia virus-induced RNA as the viral product responsible for shut-off.[163,218] In infected cells in the presence of 5 μg/mℓ of actinomycin D, small RNAs were synthesized, host polyribosomes were dissociated, and host shut-off occurred.[218] It was proposed that the small RNAs may compete with cellular RNAs and induce shut-off. As mentioned before, in cycloheximide-treated infected cells, a correlation was established between viral RNA synthesized and the extent of shut-off.[163,219,228]

Bablanian et al.[36] have shown that RNAs 50 to 100 nucleotides long were associated with shut-off. Small RNAs were synthesized in cells infected with vaccinia virus inactivated with low doses of UV, and these RNAs hybridized to vaccinia DNA. Shut-off occurred under these conditions, but it was prevented if 20 to 30 μg/mℓ of actinomycin D was added at the time of infection. Other studies in vitro have supported this hypothesis.[234,235]

Coppola and Bablanian[37] have extended these results using in vitro transcripts made by viral cores. These transcripts can selectively inhibit translation of cellular mRNAs in a cell-free system. They also inhibit translation of uncapped mRNAs, but have no effect on translation of cytoplasmic RNA obtained from vaccinia virus infected cells. Cytoplasmic RNA obtained from cycloheximide-treated vaccinia virus-infected cells also inhibits cellular mRNA translation in a cell-free system. It appears that this effect is not due to competition because there was a total inhibition of cellular protein synthesis without an effect on viral protein synthesis. Several controls were made to rule out the possibility of a polypeptide as responsible for the inhibition. Furthermore, RNA isolated from viral cores that have been preincubated for 30 min can also inhibit cellular protein synthesis in a cell-free system. However, the authors suggested that this virion RNA is not sufficient to block protein synthesis because, as mentioned before, actinomycin D (30 μg/mℓ) prevents the shut-off, and transcription is necessary to accumulate the viral RNA inhibitor. This result may explain why high concentrations of viral cores inhibit protein synthesis in cell-free systems.[217,223]

How does vaccinia virus inhibit protein synthesis? Few reports have been published on the mechanism of action of vaccinia virus-induced shut-off. In vitro experiments with cell-free systems showed an inhibition at the level of the 40S-methionyl-tRNA complex formation.[221] Disaggregation of host polyribosomes is consistent with this effect. Two different viral products can be involved: the 11 kdalton phosphoprotein isolated from viral cores[222] or early small viral RNAs.[36,37]

A mechanism proposed initially by Carrasco[236] suggested that alteration of the cellular membrane after infection induces an imbalance of sodium ions; this, in turn, might discriminate between cellular and vira mRNAs. Vaccinia virus infection leads to a modification in the permeability of the cell membrane.[237] This modification is observed very early during infection, and a gradual resealing follows. Although this mechanism could be important in the shut-off by other viruses,[238] the contribution to the shut-off induced by vaccinia virus is unclear. Changes of the intracellular ion concentrations have been observed only late in infection.[239]

Another mechanism proposed phosphorylation of ribosomal proteins in vaccinia virus infected cells.[240] They found an enhanced phosphorylation of the S6 protein of the small ribosomal subunit after infection. Although this mechanism might explain an alteration of the ribosomes, consistent with the disaggregation of host polyribosomes, the correlation with host shut-off is still unknown.

A recent report has demonstrated that vaccinia viruses induce degradation of beta-actin and alpha-tubulin mRNAs.[211] Other reports have shown a relatively low level of host mRNAs, several hours after infection.[195,241] In this regard, a similarity with the shut-off induced by herpes virus can be established. Herpes virus also induces degradation of cellular mRNAs, and this effect has been implicated in host shut-off.[34] It appears that this degradation is induced by a virion-associated factor, but an additional viral function requiring viral expression is needed to completely shut-off host protein synthesis.[35] Vaccinia virus induced shut-

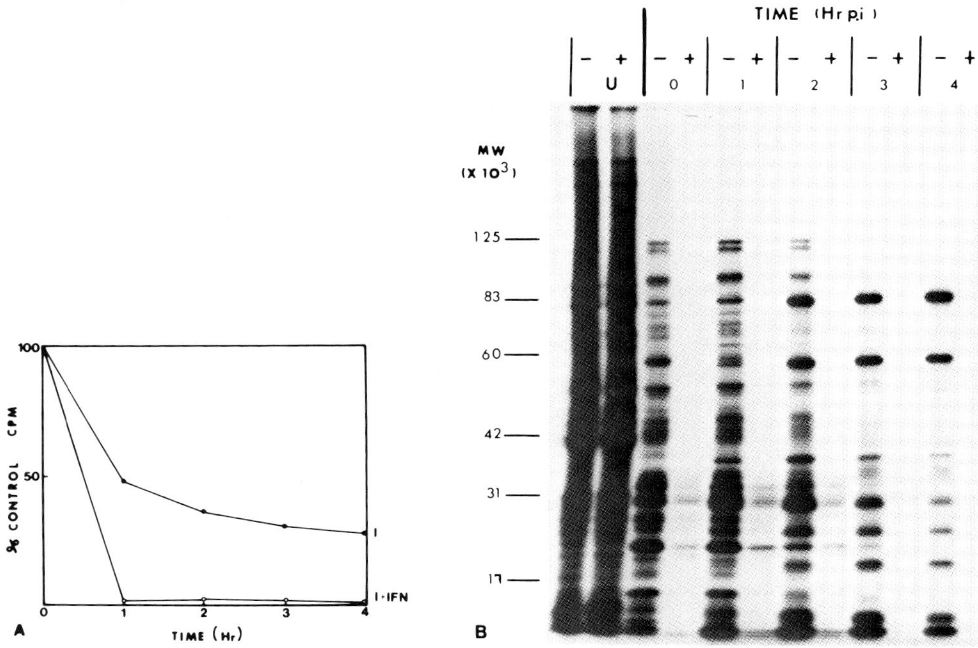

FIGURE 1. Effect of interferon on protein synthesis and extent of shut-off in vaccinia infected cells. Cells were pulse labeled with ^{35}S-methionine for 20 min after the 15 min period for virus adsorption (zero time) and at 1, 2, 3, and 4 hr p.i. To the left, kinetics of protein synthesis. To the right, SDS-PAGE analysis and autoradiography. (−), untreated cells; (+), interferon-treated cells.

off occurs after reversal of cycloheximide and is prevented in the presence of high concentrations of actinomycin D. Thus, if degradation of cellular mRNA and shut-off are related processes, the most likely mechanism to explain this effect is the activation of a cellular nuclease by early viral mRNAs. In interferon-treated vaccinia virus infected cells, a latent cellular RNase L is activated by a series of 2′ to 5′ linked oligonucleotides.[229] However, this RNase L degrades both viral and cellular mRNAs with no selectivity.[242] Furthermore, vaccinia virus induced shut-off occurs very early during infection (Figure 1).[13,229] At this time it is unlikely that an extensive degradation of cellular mRNA, a process that occurs gradually after infection, may fully explain the drastic shut-off of host protein synthesis.

V. MECHANISMS OF TRANSLATIONAL CONTROL

Gene expression is controlled to a large extent at the transcriptional level, but there is increasing evidence that translational control mechanisms also regulate the level of expression of specific genes. These mechanisms are important during normal cell growth and differentiation, and after virus infection.[243,244] A better understanding of the different mechanisms of translational control may help toward deciphering the molecular basis of the virus-induced shut-off of host protein synthesis.

A. Degradation of RNA

Mechanisms that affect the steady-state levels of cytoplasmic RNA should have an important effect on translation. As mentioned in Sections III and IV, herpes virus and vaccinia virus induce a specific degradation of cellular mRNAs after infection.[34,211] Although shut-off occurs very early during infection (and is most likely produced by other mechanisms) degradation of cellular mRNAs might contribute to the inhibition of host protein synthesis at a later time p.i. Different mechanisms have been proposed: the activation of a latent cellular nuclease, the synthesis of a viral or cellular nuclease, the release of a nuclease present in the virion, and recompartmentalization of mRNA, making it more susceptible to enhanced turnover or alteration of cellular mRNA-protein complexes, rendering the mRNA accessible to a cellular nuclease. However, the mechanism by which this degradation occurs is still unknown. Inglis[139] has also shown that influenza virus induces degradation of cellular mRNA, and suggests that a common cellular nuclease might be involved as a response to some types of viral infection because very different viruses can induce mRNA degradation.

RNA degradation has also been proposed as a mechanism of inhibition of viral replication by interferon (IFN).[233,245] The interaction of IFN to the cell receptors induces by unknown mechanism the synthesis of three cellular proteins: dsRNA-dependent 2-5A synthetase, endoribonuclease (RNase L), and 2-5A phosphodiesterase (Figure 2). The activator of this mechanism is most likely dsRNA that is produced after virus infection as a replicative intermediate or by symmetrical transcription. Upon infection, the 2-5A synthetase is activated by dsRNA and using ATP as substrate catalyzes the synthesis of a series of oligonucleotides ppp(A2′p)nA collectively referred to as 2-5A,[246] which in turn activates a latent RNase L that degrades mRNA as well as rRNA.[247-249] A correlation between activation of 2-5A synthesis with degradation of viral RNAs and inhibition of protein synthesis has been observed in IFN-treated, encephalomyocarditis (EMC), reovirus, and vaccinia virus infected cells.[229,250,251] It appears that the degradation is not selective for viral mRNAs; however, it is possible that an activation of this system at the site of viral replication complexes renders viral mRNA more susceptible to degradation.[252] Indeed, in extracts of IFN treated cells, VSV or EMC mRNAs by generating dsRNA at their poly(A) tail or poly(C) tract, respectively, are more rapidly degraded than the corresponding controls. Moreover, the RNase L is activated near the viral replication site because the dsRNA synthesized induces the synthesis of 2-5A in its vicinity, and viral mRNA may be more accessible to degradation.

On the other hand, we have observed that the 2-5A synthetase/RNase L system is blocked in IFN-treated vaccinia virus infected cells.[231] In several cell lines of different origins, vaccinia virus is relatively insensitive to IFN. This resistance is related to an interference phenomenon between the virus and the IFN-mediated enzyme activities (Figure 2). Two vaccinia virus induced enzymes, ATPase causing degradation of ATP and phosphatase which dephosphorylates 2-5A, have been implicated in this phenomenon.[253] However, at late times after infection, 2-5A is produced in high levels in intact cells.[230,254] This is most likely because of the accumulation of dsRNA and the inhibition of the 2-5A phosphodiesterase that occur at late times after infection. In spite of the presence of 2-5A, cleavage of rRNA does not occur or is delayed in IFN-treated infected cells. We have shown that the failure of high levels of 2-5A to activate the 2-5A dependent RNase L during vaccinia virus infection of tissue culture cells appears to be related to an interference phenomenon between viral products and the 2-5A system.[230] Several lines of evidence indicate that vaccinia virus proteins modulate the 2-5A system. rRNA cleavage occurred early in infection when translation was inhibited with cycloheximide, but not when transcription was blocked with actinomycin. However, when viral proteins are synthesized, it took longer for rRNA cleavage to develop. This was also documented with a ts22 mutant of vaccinia virus.[255] This mutant has an abortive phenotype late in infection at the nonpermissive temperature. Cleavage of rRNA was only observed in the abortive state, but if viral proteins were synthesized, cleavage was prevented. It was proposed that vaccinia proteins act as biological modifiers of the 2-5A system, and that by limiting the extent of rRNA cleavage, they might contribute to the resistance of vaccinia virus replication to IFN. The availability of vaccinia virus mutants might provide the means to identify viral functions involved in the interference between this cytoplasmic DNA virus and the 2-5A system. In this regard, ts and deletion mutants of

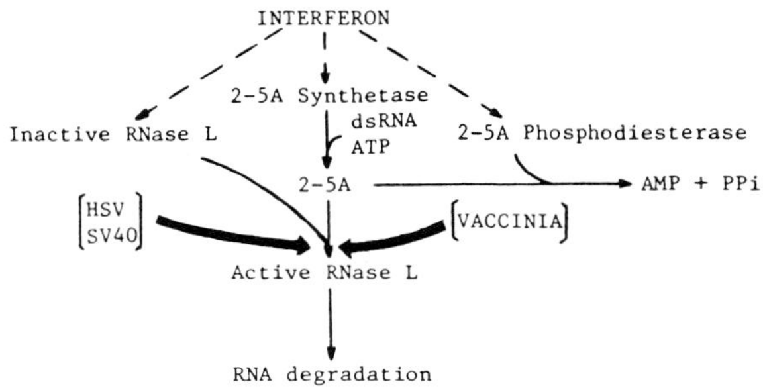

FIGURE 2. Mechanism of activation of the dsRNA dependent 2-5A synthetase/RNase L system and interference by DNA viruses.

vaccinia virus have been isolated,[173,255] and some of these mutants have different degrees of sensitivity to IFN.[179]

This phenomenon is not exclusive of vaccinia virus infection (Figure 2), as 2-5A products that do not activate the RNase L have been found in IFN-treated SV40 infected cells.[256] Furthermore, in IFN-treated cells infected with HSV, 2-5A products that inhibit the activation of the RNase L have been isolated.[257] These viruses might have evolved different strategies to overcome the antiviral action of IFN to prevent the activation of the RNase L.

B. Phosphorylation of the Alpha Subunit of eIF-2

Several conditions that affect the activity of eIF-2 and block protein synthesis have been reported. Studies conducted with reticulocyte lysates have shown that heme deficiency, dsRNA, and high temperature (heat shock) induce phosphorylation of the small alpha subunit of the initiation factor eIF-2, inhibiting protein synthesis. This phosphorylation has also been observed in IFN-treated virus-infected cells and in extracts from cells infected with an adenovirus mutant that does not synthesize small VAI-RNAs. A role for these small RNAs in the shut-off has been proposed in this virus-cell system. There is increasing evidence that this process may be under more complex control mechanisms.

Protein synthesis in reticulocyte cell-free systems declines within a few minutes unless hemin is added.[258] Heme deficiency activates an inhibitor of initiation of protein synthesis. Low concentrations of dsRNA activate a similar inhibitor.[259] In both cases, cyclic AMP-independent protein kinases are activated by phosphorylation, which in turn, phosphorylate the small alpha subunit of the initiation factor eIF-2, interfering with its function.[259] Both kinases phosphorylate the eIF-2 at the same sites in serine and threonine residues,[260] however, they are different proteins with molecular weights of 90 and 68 kdaltons for the heme deficiency and for the dsRNA-dependent kinases, respectively. A 68 kdalton protein kinase is induced in IFN-treated cells, and is also activated by dsRNA in cell extracts[261,262] and intact cells.[263] This activated protein kinase is capable of phosphorylating eIF-2 and histones.[262,264] Galabru and Hovanessian[265] have presented new evidence that this protein kinase exists as a complex with at least two subunits of 68 (p68) and 48 kdaltons (p48). This complex has been isolated by an immunoaffinity column prepared with monoclonal antibodies specific for the p68 protein kinase. They have shown that the p48 is responsible for the phosphorylate exogenous eIF-2 and histones (Figure 3). A phosphatase that removes the phosphate moiety has been described.[266]

The process of initiation of protein synthesis and the mechanism of action of the protein kinase are described in Figure 4. The initiation factor eIF-2 is necessary for the ternary complex formation with met-tRNA$_f$ and GTP. During the initiation process, GTP is hydrolized and eIF-2.GDP must be reconverted to eIF-2.GTP. A guanosine nucleotide exchange factor (GEF or eIF-2B)[267] catalyzes GTP/GDP exchange. It was demonstrated that phos-

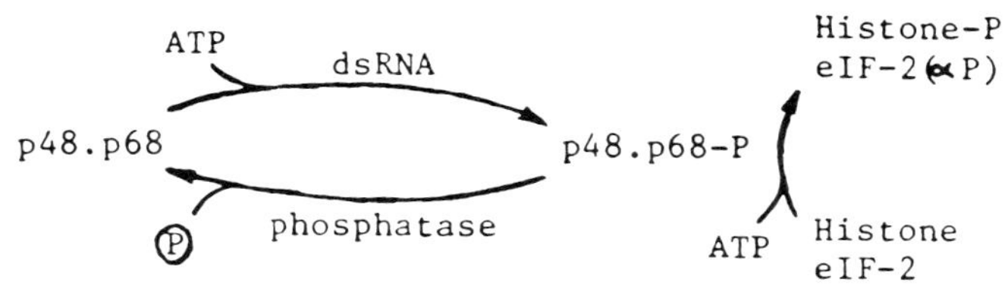

FIGURE 3. Mechanism of activation of the dsRNA dependent protein kinase.

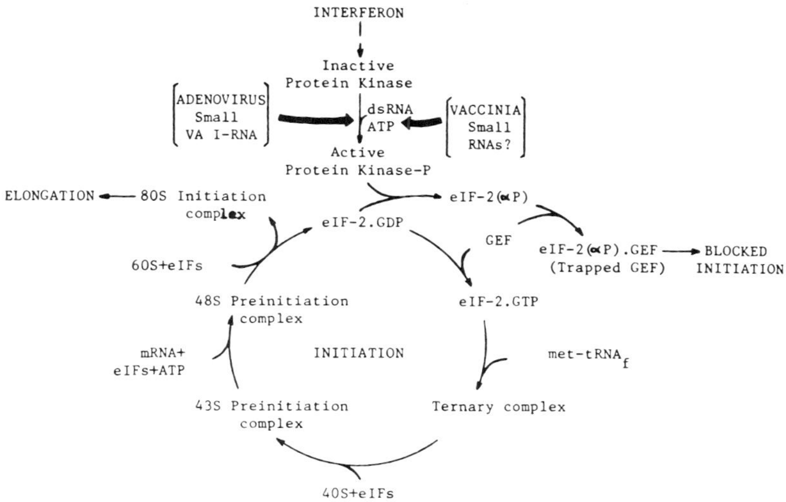

FIGURE 4. Role of the dsRNA dependent protein kinase in the inhibition of protein synthesis and blockage of interferon action by DNA viruses.

phorylation of the alpha subunit of eIF-2, but not of the beta subunit, specifically blocks the ability of GEF to exchange GDP for GTP in eIF-2.[268] One hypothesis is that GEF is present in limiting amounts, and it is trapped by phosphorylated eIF-2, blocking initiation of protein synthesis.[269,270] This model might explain why protein synthesis is inhibited when a 20 to 30% of eIF-2 is only phosphorylated.[271] The dsRNA dependent protein kinase system appears not to be selective; however, mengovirus and coxsackie virus mRNAs are resistant to the dsRNA-mediated inhibition in a reticulocyte lysate under conditions that inhibit cellular mRNA translation.[272] These authors proposed a model based on competition between dsRNA and mRNA for eIF-2. Viral mRNA with high affinity binds to eIF-2 and blocks the formation of a complex of active protein kinase, dsRNA, and eIF-2 that should lead to phosphorylation of eIF-2 and inhibition of protein synthesis.

An inactivation of the protein kinase system by vaccinia virus and adenovirus has been reported.[49-51,231,273,274] The vaccinia virus product(s) responsible for the inactivation of the protein kinase system appears to be different from the ones that inactivate the 2-5A synthetase/RNase L system.[231] Both the p68 protein kinase and the initiation factor eIF-2 are not phosphorylated (Figure 4), neither in IFN-treated vaccinia virus cell extracts[231,273,274] nor in intact cells.[232] This inhibition is specific because it does not affect to a similar extent the phosphorylation of any other cellular protein. Furthermore, the effect is reversed after binding the cell-extracts to poly(I):(C) agarose. This finding established that inactivation of the IFN-mediated protein kinase is not the result of enzyme degradation or competitive binding for dsRNA. The vaccinia virus products have not been characterized, but from the kinetics of synthesis it appears to be an early viral protein. However, the contribution of viral RNA cannot be ruled out, since there is no complete prevention by cycloheximide. In this regard, it is tempting to speculate a similar pathway to that observed with VAI-RNA mutant adenovirus-infected cells (Figure 4). As mentioned earlier, this mutant fails to synthesize the small VAI-RNAs and grows poorly.[33] VAI-RNA is required for efficient translation by preventing activation of the protein kinase system.[49-51] Extracts of mutant-infected cells showed enhanced phosphorylation of the p68 protein kinase and of the alpha subunit of exogenously added eIF-2. This effect does not occur in cells infected with the wild-type virus, indicating that VAI-RNA prevents inactivation of eIF-2 function. Viral products that block the activation of the protein kinase system could be useful for complex viruses like adenovirus and vaccinia virus that produce dsRNA as a result of symmetrical transcription and deal with IFN during their normal life cycle. In this regard, adenovirus and vaccinia virus are very resistant to the antiviral effects of IFN.[233,245] An inhibition of the antiviral host response might facilitate persistence of the virus for a long time after the primary infection.

C. Phosphorylation of Ribosomal Proteins

Different viruses induce disaggregation of polyribosomes early during infection. An alteration in the ribosomes might affect the affinity for cellular mRNA. Vaccinia virus and herpes virus enhanced phosphorylation of the S6 protein of the small ribosomal subunit.[150,151,240] An increase in phosphorylation of S6 protein has been observed in uninfected cells treated with cycloheximide.[275] In other cell systems, there is a correlation between the extent of phosphorylation of protein S6 and the rate of initiation of protein synthesis.[275,276] The enhanced phosphorylation appears to induce an apparent selective advantage to form polyribosomes. The significance of this phenomenon in shut-off, if any, has not yet been determined.

D. Competition Between Different mRNAs

A competition mechanism between viral and cellular mRNAs has been proposed to explain the shut-off of cellular protein synthesis induced by VSV.[23] At 4 hr p.i., there was a threefold increase in poly(A)-containing mRNA. Although VSV mRNA has the same efficiency to translate as cellular mRNAs, the production of higher amounts of viral mRNA would compete with diluted cellular mRNA to form initiation complexes. However, competition does not appear to cause VSV induced shut-off of host protein synthesis, because this shut-off also occurs under conditions that limit viral transcription.[24]

Viruses that produce mRNA with a high affinity for the initiation factor eIF-2 might induce shut-off by a competition mechanism. This phenomenon has been observed by competition experiments between mengovirus RNA and globin mRNA in a reticulocyte cell-free system.[277] These authors have shown that viral RNA has a 30-fold higher affinity for binding to eIF-2 than globin mRNA. However, the significance of this mechanism in virus induced shut-off is unclear, because a complete inhibition of cellular protein synthesis usually occurs after infection with different viruses. Furthermore, competition cannot explain the observation that cell-free systems from virus-infected cells rendered message-dependent by nuclease treatment are unable to translate cellular mRNAs. Competition among mRNAs is most likely involved in the differential expression of cellular mRNAs, as has been demonstrated for alpha and beta-globin mRNAs.[278] Beta-globin mRNA is more efficiently translated as a result of an increase in the extent of initiation.

As mentioned earlier, Carrasco[236] proposed that an alteration of the membrane permeability after viral infection induces an increase in the intracellular ionic concentration, and that viral mRNA has a higher affinity to initiate translation under these conditions. It has been demonstrated that several viruses induce membrane leakiness after infection, and that viral mRNA has a wider range of ionic requirements for translation. However, this mechanism does not appear to be responsible for the shut-off induced by DNA viruses, and may be restricted to some RNA virus[238] (see Chapter 6).

It is also interesting to discuss the results of mixed infections between different viruses. As noted previously, coinfection with herpes virus and VSV inhibits protein synthesis of the former virus, probably by an inhibition of transcription mediated by VSV, but the polyribosomes become disaggregate, an effect specific of herpes virus infection.[146] When VSV is coinfected with vaccinia virus, synthesis of proteins of both viruses occurs, although the extent of specific viral polypeptides is reduced for both viruses when compared to single infections.[231] That vaccinia virus proteins are synthesized can be concluded by the analysis of the pattern of polypeptides synthesized after infection with both viruses and because typical vaccinia virus-induced early cell rounding is observed, a phenomenon that needs previous synthesis of viral proteins to occur.[219] On the other hand, whereas VSV replication is inhibited by IFN, it has been shown that vaccinia virus can rescue VSV from the antiviral action of IFN during mixed infections in IFN-treated cells.[231,273] A similar mechanism of shut-off of host protein synthesis has been proposed for both VSV and vaccinia, based on small viral transcripts,[24,36] and it is tempting to suggest that competition between VSV and vaccinia mRNAs occurs during mixed infections.

E. Inactivation of Cap-Binding Proteins

This mechanism, first discovered during the study of the mechanism of inhibition of host protein synthesis by poliovirus,[19,22] has been proposed in other groups of virus. An inactivation of cap-binding proteins appears to occur at late times of reovirus infection when viral mRNA is uncapped.[28] Semliki Forest virus infection apparently induces a deficiency in cap-binding proteins at late times,[281] and an alteration of initiation factors has been observed in VSV infected cells.[26]

As mentioned earlier, eIF-3 and eIF-4B were shown to be defective in extracts of poliovirus infected cells.[17,18] During studies on the proteins that cross-link to the cap structure of mRNA, a 24 kdalton protein associated with both initiation factors was isolated,[19] and an inactivation of this cap-binding protein was proposed to explain the mechanism of shut-off induced by poliovirus.[22] However, it was observed that the poliovirus induced block is not reversed by this cap-binding protein, but by a complex of five polypeptides that includes the 24 kdalton protein.[20,21] The mechanism by which these cap-binding proteins regulate the initiation of translation appears to be more complex. In fact, a dependence of uncapped poliovirus mRNA with one of these polypeptides has been described.[282] The observation that binding to the cap requires ATP and magnesium and the correlation between ATP requirement and secondary structure of the mRNA suggested that the cap-binding proteins may function by unwinding the secondary structure of the 5′ end of mRNA,[283] but the binding of the mRNA to the translational machinery appears to be a function of the initiation factor eIF-2.[284]

This mechanism of shut-off may be restricted to only some RNA viruses. Here, it is also interesting to point out that during mixed infection between poliovirus and vaccinia virus, a block of vaccinia protein synthesis occurs.[279] In this case, vaccinia virus-induced early cell rounding is also inhibited. However, both viral mRNAs can be translated in a reticulocyte cell-free system.[280] This result can be explained because of the capacity of poliovirus to inactivate the cap-binding proteins and to inhibit translation of vaccinia virus mRNAs that contain a 5′ cap structure.

F. Regulation by Small Viral Transcripts

During the last few years, several reports on the regulation of protein synthesis after viral infection have raised the hypothesis that small viral transcripts have a role in the selective discrimination between viral and cellular mRNA translation. Although the strongest evidence for this mechanism is provided by a vaccinia virus-infected cell system,[36,37] several reports have been published recently for such different viruses as adenovirus,[33] Epstein-Barr virus,[285] VSV,[24,25] and influenza.[286]

Small virus-associated VA-RNAs are synthesized in large amounts during the late phase of adenovirus infection.[48] VAI-RNA is necessary for efficient translation of late viral mRNAs,[33] and phosphorylation of the alpha subunit of eIF-2 appears to be responsible for the inhibition of protein synthesis during infection with a deletion mutant that fails to synthesize this viral RNA.[49-51] The most likely explanation of this effect is that VAI-RNA prevents the phosphorylation of the initiation factor eIF-2 because it can form dsRNA in high amounts, and it is known that high concentrations of dsRNA block the activity of the protein kinase system.[259,287] This proposal would argue against a discriminatory role of the VA-RNAs between viral and cellular mRNAs. Further work will be necessary to clarify whether small VA-RNAs can differentially affect expression of viral and cellular functions during the normal process of infection. In this regard, a recent, and in some way surprising, report has shown that influenza viral mRNA can escape from both wild-type and VAI-RNA deletion mutant adenovirus-induced block of protein synthesis, suggesting that influenza virus has established a system that permits translation of its own mRNAs under conditions that block protein synthesis with or without selectivity.[286] A mechanism of selective binding of specific mRNAs might explain the translational control exerted by different viruses. In fact, Epstein-Barr virus encodes two small RNAs (EBER-1 and 2)[285] that can functionally substitute for VA-RNAs,[288] and a correlation has been established between synthesis of small viral transcripts and VSV-induced shut-off[24] where two VSV-encoded RNAs appeared to be involved;[25] however, the mechanism of this selectivity has not yet been clarified.

Vaccinia virus-induced shut-off of host protein synthesis seems to be mediated by viral transcripts.[36] Vaccinia virus transcripts synthesized by the virion-associated RNA polymerase can discriminate in a reticulocyte cell-free system between translation of viral and cellular mRNA.[37] This study has been extended to demonstrate that this in vitro discrimination resides exclusively in the small (100 to 300 nucleotides) nontranslating poly(A) containing viral RNA transcripts. Analysis of this inhibitory viral RNA has shown that the selective inhibition appears to be due to free poly(A) present in this fraction.[289] In fact, addition of poly(A) to the in vitro cell-free system inhibited selectively HeLa cell mRNA translation, with little effect on vaccinia virus mRNA translation. This effect was eliminated by the addition of poly(dT). Furthermore, poly(G), poly(C), and poly(dA) did not have any effect, only poly(U) having a moderate effect that is most probably related to the formation of dsRNA with poly(A) and the activation of the protein kinase system. Poly(A) also inhibits translation of L cell, EMC, and reovirus mRNAs; however, VSV mRNA was more resistant. It will be interesting if poly(A) is present in the VSV mRNAs that have been proposed as likely mediators of the VSV-induced shut-off.

G. Regulation by Antisense RNA

By probably producing RNA:RNA duplexes, nucleic acids complementary to a given mRNA can inhibit synthesis of the protein product. This phenomenon has been found to control translation in prokaryotes, where it was first described. For example, in *Escherichia coli,* expression of the IS 10 transposable gene[290] and the gene for the outer membrane protein (Omp F) is apparently regulated at the translational level by antisense RNAs.[291] Similar mechanisms have not yet been described in eukaryotes. However, plasmid DNA constructions that transcribed the noncoding DNA strand, synthetic antisense RNA, and synthetic oligonucleotides have been used to inhibit expression of several genes after introduction into animal cell or bacteria. An inhibition by antisense RNA has been shown for several genes transiently or constitutively expressed as herpes thymidine kinase (TK),[292,293] beta-galactosidase,[294] beta-globin,[295] chicken TK, chloramphenicol acetyl transferase (CAT) and actin.[296] Furthermore, it has been shown that antisense RNA directed against viral proteins inhibits viral replication.[297] The application of this kind of immune system to both animal and plant cells is very promising and permits establishing transgenic organisms carrying genes that encode for antisense RNA against different viruses. The inhibition can be observed with as little as 52 bp of homology to the 5′ leader sequence of a mRNA and does not have to include the initiator AUG. A 3′ antisense RNA seems to be less effective. In fact, naturally occurring antisense RNA in prokaryotes is complementary to the 5′ regions of the corresponding mRNA.[290,291] There are evidences that this mechanism can occur in the cytoplasm[295] or in the nucleus.[293] Very little is known about the mechanism of action and the parameters that affect this process, but it provides a suggestive mechanism to regulate protein synthesis and hence, gene expression. This system offers an important tool for studying biological functions of specific viral and cellular genes by inhibiting their expression in vivo.

VI. CONCLUDING REMARKS

The understanding of the mechanism of shut-off by different viruses has been accelerated in the last 10 years because of the important advances in molecular biology. Synthesis of viral and cellular DNA, RNA, and proteins can now be screened with specific probes that permit measuring a specific viral or cellular product during the process of viral infection.

Different mechanisms have been proposed and some of them, such as the regulation by small transcripts, phosphorylation of eIF-2, or the inactivation of the cap-binding proteins, are very suggestive; future work will provide important findings in this exciting area of research. Far less is known about the possible role of the antisense RNA in controlling cellular metabolism in eukaryotes, although this is already an alternative and more direct method for studying the biological function of specific genes. It has to be determined whether these RNAs function directly or by interacting with viral or cellular RNA or proteins.

ACKNOWLEDGMENTS

This investigation was supported by NIH grant AI 16780 to Mariano Esteban Eduardo Paez is a recipient of a fellowship from the Consejo Superior de Investigaciones Cientificas, Spain.

REFERENCES

1. **Zimmerman, E. F., Heeter, M., and Darnell, J. E.,** RNA synthesis in poliovirus-infected cells, *Virology*, 19, 400, 1963.
2. **Bablanian, R., Eggers, H. J., and Tamm, I.,** Studies on the mechanism of poliovirus-induced cell damage. I. The relation between poliovirus-induced metabolic and morphological alterations in cultured cells, *Virology*, 26, 100, 1965.
3. **Bablanian, R., Eggers, H. J., and Tamm, I.,** Studies on the mechanism of poliovirus-induced cell damage. II. The relationship between poliovirus growth and virus-induced morphological changes in cells, *Virology*, 26, 114, 1965.
4. **Martin, E. M., Malec, J., Sved, S., and Work, T. S.,** Studies on protein and nucleic acid metabolism in virus-infected mammalian cells. I. Encephalomyocarditis virus in Krebs-II mouse ascites tumor cells, *Biochem. J.*, 80, 585, 1961.
5. **Baltimore, D., Franklin, R. M., and Callender, J.,** Mengovirus induced inhibition of host ribonucleic acid and protein synthesis, *Biochem. Biophys. Acta*, 76, 425, 1963.

6. **Strauss, J. H., Burge, B. W., and Darnell, J. E.,** Effect of sindbis virus on host protein synthesis, *Virology,* 37, 367, 1969.
7. **McAllister, P. E. and Wagner, R. R.,** Differential inhibition of host protein synthesis in L cells infected with RNA^- temperature-sensitive mutants of vesicular stomatitis virus, *J. Virol.,* 18, 550, 1976.
8. **Bello, L. J. and Ginsberg, H. S.,** Inhibition of host protein synthesis in type 5 adenovirus-infected cells, *J. Virol.,* 1, 843, 1967.
9. **Ledinko, N.,** Changes in metabolic and enzymatic activities of monkey kidney cells after infection with adenovirus 2, *Virology,* 15, 173, 1966.
10. **Roizman, B., Borman, G.S., and Rousta, M. K.,** Macromolecular synthesis in cells infected with herpes simplex virus, *Nature (London),* 206, 1374, 1965.
11. **Goorha, R. and Granoff, A.,** Macromolecular synthesis in cells infected by frog virus 3. I. Virus-specific protein synthesis and its regulation, *Virology,* 60, 237, 1974.
12. **Moss, B.,** Inhibition of HeLa cell protein synthesis by the vaccinia virion, *J. Virol.,* 2, 1028, 1968.
13. **Esteban, M. and Metz, D. H.,** Early virus protein synthesis in vaccinia virus-infected cells, *J. Gen. Virol.,* 19, 201, 1973.
14. **Ehrenfeld, E.,** Picornavirus inhibition of host cell protein synthesis, in *Comprehensive Virology,* Vol. 19, Fraenkel-Conrat, H. and Wagner, R. R., Eds., Plenum Press, New York, 1984, 177.
15. **Willems, M. and Penman, S.,** The mechanism of host cell protein synthesis inhibition by poliovirus, *Virology,* 30, 355, 1966.
16. **Kaufmann, Y., Goldstein, E., and Penman, S.,** Poliovirus-induced inhibition of polypeptide initiation *in vitro* on native polyribosomes, *Proc. Natl. Acad. Sci. U.S.A.,* 73, 1834, 1976.
17. **Rose, J. K., Trachsel, M., Leong, K., and Baltimore, D.,** Inhibition of translation by poliovirus inactivation of a specific initiation factor, *Proc. Natl. Acad. Sci. U.S.A.,* 75, 2732, 1978.
18. **Helentjaris, T., Ehrenfeld, E., Brown-Luedi, M. L., and Hershey, J. W. B.,** Alternations in initiation factor activity from poliovirus-infected HeLa cells, *J. Biol. Chem.,* 254, 10973, 1979.
19. **Sonenberg, N., Morgan, M., Merrick, W., and Shatkin, A. J.,** A polypeptide in eukaryotic initiation factors that crosslinks specifically to the 5′-terminal cap in mRNA, *Proc. Natl. Acad. Sci. U.S.A.,* 75, 4843, 1978.
20. **Tahara, S. M., Morgan, M. A., and Shatkin, A. J.,** Two forms of purified m^7G-cap binding protein with different effects on capped mRNA translation in extracts of uninfected and poliovirus-infected HeLa cells, *J. Biol. Chem.,* 256, 7691, 1981.
21. **Grifo, J. A., Tahara, S. M., Morgan, M. A., Shatkin, A. J., and Merrick, W. C.,** New initiation factor activity required for globin mRNA translation, *J. Biol. Chem.,* 258, 5804, 1983.
22. **Trachsel, H., Sonenberg, N., Shatkin, A. J., Rose, J. K., Leong, K., Bergman, J. E., Gordon, J., and Baltimore, D.,** Purification of a factor that restores translation of VSV mRNA in extracts from poliovirus-infected HeLa cells, *Proc. Natl. Acad. Sci. U.S.A.,* 77, 770, 1980.
23. **Lodish, H. F. and Porter, M.,** Translational control of protein synthesis after infection by vesicular stomatitis virus, *J. Virol.,* 36, 719, 1980.
24. **Schnitzlein, W. M., O'Barrion, M. K., Poirot, M. K., and Reichman, M. B.,** Effect of intracellular vesicular stomatitis virus mRNA concentration on the inhibition of host cell protein synthesis, *J. Virol.,* 45, 206, 1983.
25. **Dunnigan, D. D. and Lucas-Lenard, J. M.,** Two transcription products of the vesicular stomatitis virus genome may control L-cell protein synthesis, *J. Virol.,* 45, 618, 1983.
26. **Thomas, J. R. and Wagner, R. R.,** Inhibition of translation in lysates of mouse L cells infected with vesicular stomatitis virus: presence of a defective ribosome-associated factor, *Biochemistry,* 22, 1540, 1983.
27. **Grinnel, B. W. and Wagner, R. R.,** Comparative inhibition of cellular transcription by vesicular stomatitis virus serotypes New Jersey and Indiana: role of each viral leader RNA, *J. Virol.,* 48, 88, 1983.
28. **Skup, D., Zarbl, H., and Millward, S.,** Regulation of translation in L-cells infected with reovirus, *J. Mol. Biol.,* 151, 35, 1981.
29. **Zarbl, H. and Millward, S.,** The reovirus multiplication cycle, in *The Viruses, The Reoviridae,* Fraenkel-Conrat, H. and Wagner, R. R., Eds., Plenum Press, New York, 1983, 107.
30. **Detjen, B. M., Walden, W. E., and Thach, R. E.,** Translational specificity in reovirus-infected mouse fibroblasts, *J. Biol. Chem.,* 257, 9855, 1982.
31. **Babich, A., Feldman, L. T., Nevins, J. R., Darnell, J. E., and Weinberger, C.,** Effects of adenovirus on metabolism of specific host mRNAs: transport control and specific translational discrimination, *Mol Cell. Biol.,* 3, 1212, 1983.
32. **Babiss, L. E. and Ginsberg, H. S.,** Adenovirus type 5 early region 1b gene product is required for efficient shut-off of host protein synthesis, *J. Virol.,* 50, 202, 1984.
33. **Thimmappaya, B., Weinberger, C., Schneider, R. J., and Shenk, T.,** Adenovirus VAI RNA is required for efficient translation of viral mRNAs at late times after infection, *Cell,* 31, 543, 1982.
34. **Nishioka, Y. and Silverstein, S.,** Degradation of cellular mRNA during infection by herpes simplex virus, *Proc. Natl. Acad. Sci. U.S.A.,* 74, 2370, 1977.

35. **Schek, N. and Bachenheimer, S. L.,** Degradation of cellular mRNAs induced by a virion-associated factor during herpes simplex virus infection of Vero cells, *J. Virol.*, 55, 601, 1985.
36. **Bablanian, R., Coppola, G., Scribani, S., and Esteban, M.,** Inhibition of protein synthesis by vaccinia virus. IV. The role of low molecular weight virus RNA in the inhibition of protein synthesis, *Virology*, 112, 13, 1981.
37. **Coppola, G. and Bablanian, R.,** Discriminatory inhibition of protein synthesis in cell-free systems by vaccinia virus transcripts, *Proc. Natl. Acad. Sci. U.S.A.*, 80, 75, 1983.
38. **Rowe, W. F., Huebner, R. J., Gilmore, L. K., Parrot, R. H., and Ward, T. G.,** Isolation of a cytogenic agent from human adenoids undergoing spontaneous degradation in tissue culture, *Proc. Soc. Exp. Biol. Med.*, 84, 570, 1953.
39. **Ziff, E.,** Transcription and RNA processing by the DNA tumor viruses, *Nature (London)*, 287, 491, 1980.
40. **Tooze, J., Ed.,** *The Molecular Biology of Tumor Viruses 2. DNA Tumor Viruses*, 2nd ed., Cold Spring Harbor Laboratory, Cold Spring Harbor, New York, 1980.
41. **Logan, J. S. and Shenk, T.,** Transcriptional and translational control of adenovirus gene expression, *Microbiol. Rev.*, 46, 377, 1982.
42. **Persson, H. and Philipson, L.,** Regulation of adenovirus, *Curr. Top. Microbiol. Immunol.*, 97, 157, 1982.
43. **Flint, S. J.,** Adenovirus cytopathology, in *Comprehensive Virology*, Vol. 19, Fraenkel-Conrat, H. and Wagner, R. R., Eds., Plenum Press, New York, 1984, 297.
44. **Horwitz, M. S.,** Adenoviruses and their replication, in *Virology*, Fields, B. N., Knipe, D. M., Chanock, R. M., Melnick, J., Roizman, B., and Shope, R., Eds., Raven Press, New York, 1985, 433.
45. **Berget, S. M., Moore, C., and Sharp, P. A.,** Spliced segments at the 5′-terminus of adenovirus 2 late mRNA, *Proc. Natl. Acad. Sci. U.S.A.*, 74, 3171, 1977.
46. **Chow, L. T., Gelines, R. E., Broker, T. R., and Roberts, R. J.,** An amazing sequence arrangement at the 5′ end of adenovirus 2 messenger RNA, *Cell*, 12, 1, 1977.
47. **Klessing, D. F.,** Two adenovirus mRNAs have a common 5′ terminal leader sequence encoded at least 10 kb upstream from their main coding regions, *Cell*, 12, 9, 1977.
48. **Reich, P. R., Forget, B., Rose, J., and Weissman, S. M.,** RNA of low molecular weight in KB cells infected with Ad 2, *J. Mol. Biol.*, 17, 428, 1966.
49. **Reichel, P. A., Merrick, W. C., Siekierka, J., and Mathews, M.,** Regulation of a protein synthesis initiation factor by adenovirus virus-associated RNA_I, *Nature (London)*, 313, 196, 1985.
50. **Siekierka, J., Mariano, T. M., Reichel, P. A., and Mathews, M. B.,** Translational control by adenovirus: lack of virus-associated RNA_I during adenovirus infection results in phosphorylation of initiation factor eIF-2 and inhibition of protein synthesis, *Proc. Natl. Acad. Sci. U.S.A.*, 82, p. 1959, 1985.
51. **Schneider, R. J., Safer, B., Munemitsu, S. M., Samuel, C. E., and Shenk, T.,** Adenovirus VAI RNA prevents phosphorylation of the eukaryotic initiation factor 2α subunit subsequent to infection, *Proc. Natl. Acad. Sci. U.S.A.*, 82, 4321, 1985.
52. **Horne, R. W., Bonner, S., Waterson, A. P., and Wildy, P.,** The icosahedral form of an adenovirus, *J. Mol. Biol.*, 1, 84, 1959.
53. **Van der Vliet, P. C. and Levine, A. J.,** DNA binding proteins specific for cells infected by adenovirus, *Nature (London), New Biol.*, 246, 170, 1973.
54. **Robinson, A. J., Younghusband, H. B., and Bellet, A. J. D.,** A circular DNA-protein complex from adenoviruses, *Virology*, 56, 54, 1973.
55. **Lichy, J. H., Field, J., Horwitz, M. S., and Hurwitz, J.,** Separation of the adenovirus terminal protein precursor from its associated DNA polymerase: role of both proteins in the initiation of adenovirus DNA replication, *Proc. Natl. Acad. Sci. U.S.A.*, 79, 5225, 1982.
56. **Stillman, B. W., Tamanoi, F., and Mathews, M. B.,** Purification of an adenovirus-encoded DNA polymerase that is required for initiation of DNA replication, *Cell*, 31, 613, 1982.
57. **Sambrook, J., Botchan, M., Gallimore, P., Ozanne, B., Petterson, U., Williams, J. F., and Sharp, P. A.,** Viral DNA sequences in cells transformed by simian virus 40, adenovirus type 2 and adenovirus type 5, *Cold Spring Harbor Symp. Quant. Biol.*, 39, 615, 1975.
58. **Graham, F. L., Abrahams, P. J., Mulder, C., Heijneker, H. L., Warnaar, S. O., DeVries, F. A. J., Fiers, W., and Van der Eb, A. J.,** Studies on *in vitro* transformation by DNA and DNA fragments of human adenovirus and simian virus 40, *Cold Spring Harbor Symp. Quant. Biol.*, 39, 637, 1975.
59. **Berk, A. J., Lee, F., Harrison, T., Williams, J., and Sharp, P. A.,** Pre-early Ad 5 gene product regulates synthesis of early viral mRNAs, *Cell*, 17, 935, 1979.
60. **Jones, N. and Shenk, T.,** An adenovirus type 5 early gene function regulates expression of other early viral genes, *Proc. Natl. Acad. Sci. U.S.A.*, 76, 3665, 1979.
61. **Nevins, J. R.,** Mechanism of activation of early viral transcription by adenovirus E1A gene product, *Cell*, 26, 213, 1981.
62. **Ralston, R. and Bishop, J. M.,** The protein products of the *myc* and *myb* oncogenes and adenovirus E1A are structurally related, *Nature (London)*, 306, 803, 1983.

63. **Hoeffler, W. K. and Roeder, R. G.,** Enhancement of RNA polymerase III transcription by the E1A gene product of adenovirus, *Cell,* 41, 955, 1985.
64. **Chow, K. C. and Pearson, G. D.,** Adenovirus infection elevates levels of cellular topoisomerase I, *Proc. Natl. Acad. Sci. U.S.A.,* 82, 2247, 1985.
65. **Chow, L. T., Lewis, J. B., and Broker, T. R.,** RNA transcription and splicing at early and intermediate times after adenovirus-2 infection, *Cold Spring Harbor Symp. Quant. Biol.,* 44, 401, 1980.
66. **Shaw, A. B. and Ziff, E. B.,** Transcripts from the adenovirus 2 major later promoter yield a single early family of 3′ coterminal mRNA and five late families, *Cell,* 22, 905, 1980.
67. **Horwitz, M. S.,** Intermediates in the replication of type 2 adenovirus DNA, *Virology,* 8, 675, 1971.
68. **Evans, R. M., Fraser, N., Ziff, E., Weber, J., Wilson, M., and Darnell, J. E.,** The initiation sites for RNA transcription in Ad 2 DNA, *Cell,* 12, 733, 1977.
69. **Nevins, J. R. and Darnell, J. E.,** Groups of adenovirus type 2 mRNAs derived from a large primary transcript: probable nuclear origin and possible common 3′ ends, *J. Virol.,* 25, 811, 1978.
70. **Petterson, U. and Philipson, L.,** Location of sequences on the adenovirus genome coding for the 5.5S RNA, *Cell,* 6, 1, 1975.
71. **Mathews, M.,** Genes for VA-RNA in Ad 2, *Cell,* 6, 223, 1975.
72. **Ginsberg, H. S., Bello, L. J., and Levine, A. J.,** Control of biosynthesis of host macromolecules in cells infected with adenovirus, in *The Molecular Biology of Viruses,* Colter, J. S. and Paranchych, W., Eds., Academic Press, New York, 1967, 547.
73. **Hodge, L. D. and Scharff, M. D.,** Effect of adenovirus on host cell DNA synthesis in synchronized cells, *Virology,* 37, 554, 1969.
74. **Lichy, J. H., Horwitz, M. S., and Hurwitz, J.,** Formation of a covalent complex between the 80,000 dalton adenovirus terminal protein and 5′ dCMP *in vitro, Proc. Natl. Acad. Sci. U.S.A.,* 78, 3678, 1981.
75. **Ensinger, M. and Ginsberg, H. S.,** Selection and preliminary characterization of temperature-sensitive mutants of type 5 adenovirus, *J. Virol.,* 10, 328, 1972.
76. **Wilkie, N. M., Ustacelabi, S., and Williams, J. F.,** Characterization of temperature-sensitive mutants of adenovirus type 5: nucleic acid synthesis, *Virology,* 51, 499, 1973.
77. **Harrison, T. J., Graham, F., and Williams, J. F.,** Host range mutants of adenovirus type 5 defective for growth on HeLa cells, *Virology,* 77, 319, 1977.
78. **Ricciardi, R. P., Jones, R. L., Cepko, C. T., Sharp, P. A., and Roberts, B. E.,** Expression of early adenovirus genes requires a viral encoded acidic polypeptide, *Proc. Natl. Acad. Sci. U.S.A.,* 78, 6121, 1981.
79. **Takemori, N., Riggs, J. L., and Aldrich, C.,** Genetic studies with tumorigenic adenoviruses. I. Isolation of cytocidal *(cyt)* mutants of adenovirus type 12, *Virology,* 36, 575, 1968.
80. **Ezoe, H., Lai Fatt, R. B., and Mak, S.,** Degradation of intracellular DNA in KB cells infected with *cyt* mutants of human adenovirus type 12, *J. Virol.,,* 40, 20, 1981.
81. **Jones, N. and Shenk, T.,** Isolation of adenovirus type 5 host-range deletion mutants defective for transformation of rat embryo cells, *Cell,* 17, 683, 1979.
82. **Lai Fatt, R. B. and Mak, S.,** Mapping of an adenovirus function involved in the inhibition of DNA degradation, *J. Virol.,* 42, 969, 1982.
83. **Subramanian, T., Kuppuswamy, M., Mak, S., and Chinnadurai, G.,** Adenovirus cyt^+ locus, which controls cell transformation and tumorigenicity, is an allele of lp^+ locus, which codes for a 19-kilodalton tumor antigen, *J. Virol.,* 52, 336, 1984.
84. **Pilder, S., Logan, J., and Shenk, T.,** Deletion of the gene encoding the adenovirus 5 early region 1B 21,000-molecular-weight polypeptide leads to degradation of viral and host cell DNA, *J. Virol.,* 52, 664, 1984.
85. **Price, R. and Penman, S.,** Transcription of the adenovirus genome by an α-amanitin-sensitive RNA polymerase in HeLa cells, *J. Virol.,* 9, 621, 1972.
86. **Beltz, G. and Flint, S. J.,** Inhibition of HeLa cell protein synthesis during adenovirus infection: restriction of cellular messenger RNA sequences to the nucleus, *J. Mol. Biol.,* 131, 353, 1979.
87. **Flint, S. J., Beltz, G. A., and Linzer, D. I. H.,** Synthesis and processing of simian virus 40-specific RNA in adenovirus-infected, simian virus 40-transformed human cells, *J. Mol. Biol.,* 167, 335, 1983.
88. **Yoder, S. S. and Berget, S. M.,** Posttranscriptional control of DHFR gene expression during adenovirus 2 infection, *J. Virol.,* 54, 72, 1985.
89. **Castiglia, C. L. and Flint, S. J.,** Effects of adenovirus infection on rRNA synthesis and maturation in HeLa cells, *Mol. Cell. Biol.,* 3, 662, 1983.
90. **Yoder, S. S., Robberson, B. L., Leys, E. J., Hook, A. G., Al-Ubaidi, M., Yeung, C.-Y., Kellems, R. E., and Berget, S. M.,** Control of cellular gene expression during adenovirus infection: induction and shut-off of dihydrofolate reductase gene expression by adenovirus type 2, *Mol. Cell. Biol.,* 3, 819, 1983.
91. **Learned, R. M., Smale, S. T., Haltiner, M. M., and Tjian, R.,** Regulation of human ribosomal RNA transcription, *Proc. Natl. Acad. Sci. U.S.A.,* 80, 3558, 1983.

92. **Khalili, K. and Weinmann, R.,** Shut-off of actin biosynthesis in adenovirus serotype-2-infected cells, *J. Mol. Biol.*, 175, 453, 1984.
93. **Flint. S. J., Plumb, M. A., Yang, U.-C., Stein, G. S., and Stein, J. L.,** Effect of adenovirus infection on expression of human histone genes, *Mol. Cell. Biol.*, 4, 1363, 1984.
94. **Anderson, C. W., Lewis, J. B., Atkins, J. F., and Gesteland, R. F.,** Cell-free synthesis of adenovirus 2 proteins programmed by fractionated mRNA: a comparison of polypeptide products and mRNA lengths, *Proc. Natl. Acad. Sci. U.S.A.*, 71, 2576, 1974.
95. **Ray, B. K., Brendler, T. G., Adya, S., Daniels-McQueen, S., Miller, J. K., Hershey, J. W. B., Grifo, J. A., Merrick, W. C., and Thach, R. E.,** Role of mRNA competition in regulation translation: further characterization of mRNA discriminatory factors, *Proc. Natl. Acad. Sci. U.S.A.*, 80, 663, 1983.
96. **Thummel, C., Tijian, R., Hu, S.-L., and Grodzicker, T.,** Translational control of SV40 T antigen expressed from the adenovirus late promoter, *Cell*, 33, 455, 1983.
97. **Sarnow, P., Hearing, P., Anderson, C. W., Halbert, D. N., Shenk, T., and Levine, A. J.,** Adenovirus early region 1B-58,000 dalton tumor antigen is physically associated with an early region 4 25,000-dalton protein in productively infected cells, *J. Virol.*, 49, 692, 1984.
98. **Challberg, S. S. and Ketner, G.,** Deletion mutants of adenovirus 2: isolation and initial characterization of virus carrying mutations near the right end of the viral genome, *Virology*, 114, 196, 1981.
99. **Sarnow, P., Sullivan, C. A., and Levine, A. T.,** A monoclonal antibody detecting the adenovirus type 5 E1b-58kd tumor antigen: characterization of the E1b-58kd tumor antigen in adenovirus-infected and transformed cells, *Virology*, 120, 510, 1982.
100. **Mathews, R. E. F.,** Classification and nomenclature of viruses. Fourth report of the International Committee on Taxonomy of Viruses, *Intervirology*, 17, 1, 1982
101. **Roizman, B.,** The family Herpesviridae. General description, taxonomy and classification, in *The Viruses: Herpesvirus*, Vol. 1, Roizman, B., Ed., Plenum Press, New York, 1982, 1.
102. **Spear, P. G. and Roizman, B.,** Herpes simplex viruses, in *DNA Tumor Viruses*, 2nd ed., Tooze, J., Ed., Cold Spring Harbor Laboratory, Cold Spring Harbor. New York, 1980, 615.
103. **Fenwick, M. L.,** The effects of herpesviruses on cellular macromolecular synthesis, in *Comprehensive Virology*, Vol. 19, Fraenkel-Conrat, H. and Wagner, R. R., Eds., Plenum Press, New York, 1984, 359.
104. **Roizman, B. and Batterson, W.,** Herpesviruses and their replication, in *Virology*, Fields, B. N., Knipe, D. M., Chanock, R. M., Melnick, J., Roizman, B., and Shope, R., Eds., Raven Press, New York, 1985, 497.
105. **Burnet, F. M. and Williams, S. W.,** Herpes simplex: a new point of view, *Med. J. Aust.*, 1, 637, 1939.
106. **Roizman, B. and Furlong, D.,** The replication of herpesviruses, in *Comprehensive Virology*, Vol. 3, Fraenkel-Conrat, H. and Wagner, R. R., Eds., Plenum Press, New York, 1974, 229.
107. **Sheldrick P. and Berthelot, N.,** Inverted repetitions in the chromosome of herpes simplex virus, *Cold Spring Harbor Symp. Quant. Biol.*, 39, 667, 1975.
108. **Wadsworth, S., Jacob, R. J., and Roizman, B.,** Anatomy of herpes simplex virus DNA. II. Size, composition, and arrangement of inverted terminal repetitions, *J. Virol.*, 15, 1487, 1975.
109. **Hayward, G. S., Jacob, R. J., Wadsworth, S. C., and Roizman, B.,** Anatomy of herpes simplex virus DNA: evidence for four populations of molecules that differ in the relative orientation of their long and short segments, *Proc. Natl. Acad. Sci. U.S.A.*, 74, 4243, 1975.
110. **Smiley, J., Fong, B., and Leung, W. C.,** Construction of a double jointed herpes simplex virus DNA molecule: inverted repeats are required for segment inversion and direct repeat promote deletions, *Virology*, 111, 345, 1981.
111. **Mocarski, E. S. and Roizman, B.,** The structure and role of the herpes simplex virus DNA termini in inversion, circularization and generation of virion DNA, *Cell*, 31, 89, 1982.
112. **Poffenberger, K. L., Tabares, E., and Roizman, B.,** Characterization of a viable, non-inverting herpes simplex virus 1 genome derived by insertion of sequences at the L-S component junction, *Proc. Natl. Acad. Sci. U.S.A.*, 80, 2690, 1983.
113. **Constanzo, F., Campadelly-Fiume, G., Foa-Tomas, L., and Cassai, E.,** Evidence that herpex simplex virus DNA is transcribed by cellular RNA polymerase II, *J. Virol.*, 21, 996, 1977.
114. **Post, L. E., Mackem, S., and Roizman, B.,** The regulation of α genes of herpes simplex virus: expression of chimeric genes produced by fusion of thymidine kinase with α gene promoters, *Cell*, 24, 555, 1981.
115. **Honess, R. W. and Roizman, B.,** Regulation of herpesvirus macromolecular synthesis. I. Cascade regulation on the synthesis of three groups of viral proteins, *J. Virol.*, 14, 8, 1974.
116. **Honess, R. W. and Roizman, B.,** Regulation of herpesvirus macromolecular synthesis: sequential transition of polypeptide synthesis requires functional viral polypeptides, *Proc. Natl. Acad. Sci. U.S.A.*, 72, 1276, 1975.
117. **Mackem, S. and Roizmann, B.,** Regulation of herpesvirus macromolecular synthesis: transcription-initiation sites and domains of α genes, *Proc. Natl. Acad. Sci. U.S.A.*, 77, 7122, 1980.

118. **Fenwick, M., Morse, L. S., and Roizman, B.,** Anatomy of herpes simplex virus DNA. XI. Apparent clustering of functions effecting rapid inhibition of host DNA and protein synthesis, *J. Virol.,* 29, 825, 1979.
119. **Read, G. S. and Frenkel, N.,** Herpes simplex virus mutants defective in the virion associated shut-off of host polypeptide synthesis and exhibiting abnormal synthesis of α (immediate early) viral polypeptides, *J. Virol.,* 46, 498, 1983.
120. **Batterson, W. and Roizman, B.,** Characterization of the herpes simplex virion-associated factor responsible for the induction of α genes, *J. Virol.,* 46, 371, 1983.
121. **Wagner, E. K.,** Individual HSV transcripts: characterization of specific genes, in *Herpesviruses,* Vol. 3, Roizman, B., Ed., Plenum Press, New York, 1984.
122. **Gibson, W. and Roizman, B.,** Proteins specified by herpes simplex virus. X. Staining and radiolabeling properties of B-capsid and virion proteins in polyacrylamide gels, *J. Virol.,* 13, 155, 1974.
123. **Honess, R. W. and Roizman, B.,** Proteins specified by herpes simplex virus. XIII. Glycosylation of viral polypeptides, *J. Virol.,* 16, 1308, 1975.
124. **Newton, A. A., Dendy, P. P., Smith, C. L., and Wildy, P.,** A pool size problem associated with the use of titriated thymidine, *Nature (London),* 194, 886, 1962.
125. **Roizman, B. and Roane, P. H. J.,** The multiplication of HSV. II. The relation between protein synthesis and the duplication of viral DNA in infected Hep-2 cells, *Virology,* 22, 262, 1964.
126. **Aurelian, L. and Roizman, B.,** Abortive infection of canine cells by HSV. II. The alternative suppression of synthesis of interferon and viral constituents, *J. Mol. Biol.,* 11, 539, 1965.
127. **Kaplan, A. S. and Ben-Porat, T.,** The pattern of viral and cellular DNA synthesis in pseudorabies virus-infected cells in the logarithmic phase of growth, *Virology,* 19, 205, 1963.
128. **Ben-Porat, T. and Kaplan, A. S.,** Mechanism of inhibition of cellular DNA synthesis by pseudorabies virus, *Virology,* 25, 22, 1965.
129. **Ben-Porat, T., Jean, J.-H., and Kaplan, A. S.,** Early functions of the genome of herpesvirus. IV. Fate and translation of immediate-early RNA, *Virology,* 59, 524, 1974.
130. **Halliburton, I. W. and Timbury, M. C.,** Temperature-sensitive mutants of herpes simplex virus type 2: description of three new complementation groups and studies on the inhibition of host cell DNA synthesis, *J. Gen. Virol.,* 30, 207, 1976.
131. **Daksis, J. I., Priemer, M. M., and Chan, V.-L.,** Isolation and preliminary characterization of a phosphonoacetic acid-resistant and temperature-sensitive mutant of herpes simplex virus type 1, *J. Virol.,* 42, 20, 1982.
132. **Fenwick, M. L. and Walker, M. J.,** Suppression of the synthesis of cellular macromolecules by HSV, *J. Gen. Virol.,* 41, 37, 1978.
133. **Hay, J., Koteles, G. J., Keir, H. M., and Subak-Sharpe, H.,** Herpesvirus-specified ribonucleic acids, *Nature (London),* 210, 387, 1966.
134. **Wagner, E. K. and Roizman, B.,** Ribonucleic acid synthesis in cells infected with HSV, *J. Virol.,* 4, 36, 1969.
135. **Pizer, L. I. and Beard, P.,** The effect of herpesvirus infection on mRNA in polyoma virus-transformed cells, *Virology,* 75, 477, 1976.
136. **Nakai, H., Maxwell, I. H., and Pizer, L. I.,** Herpesvirus infection alters the steady state levels of cellular polyadenylated RNA in polyoma virus-transformed BHK cells, *J. Virol.,* 42, 1131, 1982.
137. **Spector, D. and Pizer, L. I.,** Herpesvirus infection modified adenovirus RNA metabolism in adenovirus type 5-transformed cells, *J. Virol.,* 27, 1, 1978.
138. **Stenberg, R. M. and Pizer, L. I.,** HSV induced changes in cellular and adenovirus RNA metabolism in an adenovirus type 5-transformed human cell line, *J. Virol.,* 42, 474, 1982.
139. **Inglis, S. C.,** Inhibition of host protein synthesis and degradation of cellular mRNAs during infection by influenza and hepes simplex virus, *Mol. Cell. Biol.,* 2, 1644, 1982.
140. **Mayman, B. A. and Nishioka, Y.,** Differential stability of host mRNAs in Friend erythroleukemia cells infected with herpes simplex virus type 1, *J. Virol.,* 53, 1, 1985.
141. **Inglis, M. M. and Newton, A. A.,** Comparison of the activities of HSV-1 and cellular mRNAs as templates for *in vitro* translation, *Virology,* 110, 1, 1981.
142. **Fenwick, M. L. and McMenamin, M. M.,** Early virion-associated suppression of cellular protein synthesis by herpes simplex virus is accompanied by inactivation of mRNA, *J. Gen. Virol.,* 65, 1225, 1984.
143. **Nishioka, Y. and Silverstein, S.,** Requirement of protein synthesis for the degradation of host mRNA in Friend erythroleukemia cells infected with HSV-1, *J. Virol.,* 27, 619, 1978.
144. **Sydiskis, R. J. and Roizman, B.,** Polysomes and protein synthesis in cells infected with a DNA virus, *Science,* 153, 76, 1966.
145. **Sydiskis, R. J. and Roizman, B.,** The disaggregation of host polyribosomes in productive and abortive infection with HSV, *Virology,* 32, 678, 1967.
146. **Nishioka, Y., Jones, G., and Silverstein, S.,** Inhibition by vesicular stomatitis virus of HSV-directed protein synthesis, *Virology,* 124, 238, 1983.

147. **Knipe, D. M., Batterson, W., Nosal, C., Roizman, B., and Buchan, A.,** Molecular genetics of HSV. VI. Characterization of temperature-sensitive mutant defective in the expression of all early viral gene products, *J. Virol.*, 38, 539, 1981.
148. **Morse, L. S., Pereira, L., Roizman, B., and Schaffer, P. A.,** Anatomy of HSV DNA. X. Mapping of viral genes by analysis of polypeptides and functions specified by HSV-1 x HSV-2 recombinants, *J. Virol.*, 26, 389, 1978.
149. **Silverstein, S. and Engelhardt, D. L.,** Alternations in the protein synthetic apparatus of cells infected with HSV, *Virology*, 95, 334, 1979.
150. **Fenwick, M. L. and Walker, M. J.,** Phosphorylation of a ribosomal protein and of virus-specific proteins in cells infected with HSV, *J. Gen. Virol.*, 45, 397, 1979.
151. **Kennedy, I. M., Stevely, W. S., and Leader, D. P.,** Phosphorylation of ribosomal proteins in hamster fibroblasts infected with pseudorabies virus or herpes simplex virus, *J. Virol.*, 39, 359, 1981.
152. **Isom, H. C., Lico, W. S. L., Taylor, J. M., Wilworth, G. E., and Eadline, T. S.,** Rapid and selective shutoff of plasma protein production in HSV-2-infected hepatoma cells, *Virology*, 126, 548, 1983.
153. **Godowski, P. J. and Knipe, D. M.,** Mutations in the major DNA-binding protein gene of HSV-1 result in increased levels of viral gene expression, *J. Virol.*, 47, 478, 1983.
154. **Dixon, R. A. F. and Schaffer, P. A.,** Fine structure mapping and functional analysis of temperature-sensitive mutants in the gene encoding the HSV-1 immediate-early protein VP175, *J. Virol.*, 36, 189, 1980.
155. **Preston, C. M.,** Abnormal properties of an immediate early polypeptide in cells infected with the herpes simplex virus type 1 mutant *ts*K, *J. Virol.*, 32, 357, 1979.
156. **DeLuca, N. A. and Schaffer, P. A.,** Activation of immediate-early, early, and late promoters by temperature-sensitive and wild-type forms of herpes simplex virus type 1 protein ICP4, *Mol. Cell. Biol.*, 5, 1997, 1985.
157. **Campbell, M. E. M., Palfreyman, J. W., and Preston, C. M.,** Identification of herpes simplex virus DNA sequences which encode a trans-acting polypeptide responsible for stimulation of immediate early transcription, *J. Mol. Biol.*, 180, 1, 1984.
158. **Everett, R. D.,** Transactivation of transcription by herpes virus product: requirements for two HSV-1 immediate-early polypeptides for maximum activity, *EMBO J.*, 3, 3135, 1984.
159. **O'Hare, P. and Hayward, G. S.,** Evidence for a direct role for both the 175,000- and 110,000-molecular-weight immediate-early proteins of herpex simplex virus in the transactivation of delayed-early promoters, *J. Virol.*, 53, 751, 1985.
160. **Quinlan, M. P. and Knipe, D. M.,** Stimulation of expression of a herpes simplex virus DNA-binding protein by two viral functions, *Mol. Cell. Biol.*, 5, 957, 1985.
161. **Moss, B.,** Reproduction of poxviruses, in *Comprehensive Virology*, Vol. 3, Fraenkel-Conrat, H. and Wagner, R. R., Eds., Plenum Press, New York, 1974, 405.
162. **Moss, B.,** Replication of poxviruses, in *Virology*, Fields, B. N., Knipe, D. M., Chanock, R. M., Melnick, J., Roizman, B., and Shope, R., Eds., Raven Press, New York, 1985, 685.
163. **Bablanian, R.,** Structural and functional alterations in cultured cells infected with cytocidal viruses, *Prog. Med. Virol.*, 19, 40, 1975.
164. **Bablanian, R.,** Poxvirus cytopathogenicity: effects on cellular macromolecular synthesis, in *Comprehensive Virology*, Vol. 19, Fraenkel-Conrat, H. and Wagner, R. R., Eds., Plenum Press, New York, 1984, 391.
165. **Dales, S. and Pogo, B. G. T.,** Biology of poxviruses, in *Virology Monographs*, Vol. 18, Kingsbury, D. W. and Zur Hausen, H., Eds., Springer-Verlag, New York, 1981.
166. **Berns, K. I. and Silverman, C.,** Natural occurrence of cross-linked vaccinia virus deoxyribonucleic acid, *J. Virol.*, 45, 452, 1970.
167. **Geshelin, P. and Berns, K. I.,** Characterization and localization of the naturally occurring cross-links in vaccinia virus DNA, *J. Mol. Biol.*, 88, 785, 1974.
168. **McCarron, R. J., Cabrera, C. V., Esteban, M., McAllister, W. T., and Holowczak, J. A.,** Structure of vaccinia DNA: analysis of the viral genome by restriction endonucleases, *Virology*, 86, 88, 1978.
169. **Moyer, R. W. and Rothe, C. T.,** The white pock mutants of rabbit poxvirus. I. Spontaneous host range mutants contain deletions, *Virology*, 102, 119, 1980.
170. **Esposito, J. J., Cabradilla, C. D., Nakano, J. H., and Obijeski, J. F.,** Intracellular genomic sequence transposition in monkeypox virus, *Virology*, 109, 231, 1981.
171. **Panicali, D., Davis, S. W., Mercer, S. R., and Paoletti, E.,** Two major DNA variants present in serially propagated stocks of the WR strain of vaccinia virus, *J. Virol.*, 37, 1000, 1981.
172. **Moss, B., Winters, E., and Cooper, J. A.,** Deletion of a 9,000 base pair segment of the vaccinia virus genome that encodes non-essential polypeptides, *J. Virol.*, 40, 387, 1981.
173. **Paez, E., Dallo, S., and Esteban, M.,** Generation of a dominant 8-MDa deletion at the left terminus of vaccinia virus DNA, *Proc. Natl. Acad. Sci. U.S.A.*, 82, 3365, 1985.
174. **Garon, C. F., Barbosa, E., and Moss, B.,** Visualization of an inverted terminal repetition in vaccinia virus DNA, *Proc. Natl. Acad. Sci. U.S.A.*, 75, 4863, 1978.

175. **Wittek, R., Menna, A., Muller, K., Schumperli, D., Bosley, P. G., and Wyler, A.,** Inverted terminal repeats in rabbit poxvirus and vaccinia virus DNA, *J. Virol.*, 28, 171, 1978.
176. **Pedrali-Noy, G. and Weisbach, A.,** Evidence of a repetitive sequence in vaccinia virus DNA, *J. Virol.*, 24, 406, 1977.
177. **Grady, L. J. and Paoletti, E.,** Molecular complexity of vaccinia DNA and the presence of reiterated sequences in the genome, *Virology*, 79, 337, 1977.
178. **Wittek, R. and Moss, B.,** Tandem repeats within the inverted terminal repetition of vaccinia virus DNA, *Cell*, 21, 277, 1980,
179. **Paez, E. and Esteban, M.,** Interferon prevents the generation of spontaneous deletions at the left terminus of vaccinia virus DNA, *J. Virol.*, 56, 75, 1985.
180. **Moss, B. and Salzman, N. P.,** Sequential protein synthesis following vaccinia virus infection, *J. Virol.*, 2, 1016, 1968.
181. **Oda, K. and Joklik, W. K.,** Hybridization and sedimentation studies on "early" and "late" vaccinia messenger RNA, *J. Mol. Biol.*, 27, 395, 1967.
182. **Kaverin, N. V., Varich, N. L., Surgay, V. V., and Chernos, V. I.,** A quantitative estimation of poxvirus genome fraction transcribed as early and late mRNA, *Virology*, 65, 112, 1975.
183. **Cooper, J. A., Wittek, R., and Moss, B.,** Extension of the transcriptional and translational map of the left end of the vaccinia virus genome to 21 kilobase pairs, *J. Virol.*, 39, 733, 1981.
184. **Colby, C. and Duesberg, P. H.,** Double-stranded RNA in vaccinia virus infected cells, *Nature (London)*, 222, 940, 1969.
185. **Colby, C., Jurale, C., and Kates, J. I.,** Mechanism of synthesis of vaccinia virus double-stranded RNA *in vivo* and *in vitro*, *J. Virol.*, 7, 71, 1971.
186. **Venkatesan, S., Baroudy, B. M., and Moss, B.,** Distinctive nucleotide sequences adjacent to multiple initiation and termination sites of an early vaccinia virus gene, *Cell*, 125, 805, 1981.
187. **Venkatesan, S., Gershowitz, A., and Moss, B.,** Complete nucleotide sequences of two adjacent early vaccinia virus genes located within the inverted terminal repetition, *J. Virol.*, 44 637, 1982.
188. **Weir, J. P. and Moss, B.,** Nucleotide sequence of the vaccinia virus thymidine kinase gene and the nature of spontaneous frameshift mutations, *J. Virol.*, 46, 530, 1983.
189. **Hruby, D. E., Maki, R. A., Miller, D. B., and Ball, L. A.,** Fine structure analysis and nucleotide sequences of the vaccinia virus thymidine kinase gene, *Proc. Natl. Acad. Sci. U.S.A.*, 80, 3411, 1983.
190. **Cochran, M. A., Puckett, C., and Moss, B.,** *In vitro* mutagenesis of the promoter region for a vaccinia virus gene: evidence for tandem early and late regulatory signals, *J. Virol.*, 54, 30, 1985.
191. **Esteban, M., Flores, L., and Holowczak, J. A.,** Model for vaccinia virus DNA replication, *Virology*, 83, 467, 1978.
192. **Pogo, B. G. T. and O'Shea, M. T.,** The mode of replication of vaccinia virus DNA, *Virology*, 86, 1, 1978.
193. **Belle Isle, H., Venkatesan, S., and Moss, B.,** Cell-free translation of early and late mRNAs selected by hybridization to cloned DNA fragments derived from the left 14 million to 72 million daltons of the vaccinia virus genome, *Virology*, 112, 306, 1981.
194. **Paoletti, E. and Grady, L. J.,** Transcriptional complexity of vaccinia virus *in vivo* and *in vitro*, *J. Virol.*, 23, 608, 1977.
195. **Boone, R. F. and Moss, B.,** Sequence complexity and relative abundance of vaccinia virus mRNAs synthesized *in vivo* and *in vitro*, *J. Virol.*, 26, 554, 1978.
196. **Hruby, D. E. and Ball, L. A.,** Control of expression of the vaccinia virus thymidine kinase gene, *J. Virol.*, 40, 456, 1981.
197. **Vassef, A., Ben-Hamida, F., Dru, A., and Beaud, G.,** Translational control of early protein synthesis at the late stage of vaccinia virus infection, *Virology*, 118, 45, 1982.
198. **Wittek, R., Cooper, J. A., and Moss, B.,** Transcriptional and translational mapping of a 6.6-kilobase-pair DNA fragment containing the junction of the terminal repetition and unique sequence at the left end of the vaccinia virus genome, *J. Virol.*, 39, 722, 1981.
199. **Wittek, R., Hanggi, M., and Hiller, G.,** Mapping of a gene coding for a major late structural polypeptide on the vaccinia virus genome, *J. Virol.*, 49, 371, 1984.
200. **Wittek, R., Richner, B., and Hiller, G.,** Mapping of the gene coding for the two major vaccinia virus core polypeptides, *Nucl. Acids Res.*, 12, 4835, 1984.
201. **Mahr, A. and Roberts, B. E.,** Arrangement of late RNAs transcribed from a 7.1-kilobase *Eco RI* vaccinia virus DNA fragment, *J. Virol.*, 49, 510, 1984.
202. **Weir, J. P. and Moss, B.,** Regulation of expression and nucleotide sequence of a late vaccinia virus gene, *J. Virol.*, 51, 662, 1984.
203. **Bertholet, C., Drillien, R., and Wittek, R.,** One hundred base pairs of 5′ flanking sequence of a vaccinia virus late gene are sufficient to temporally regulate late transcription, *Proc. Natl. Acad. Sci. U.S.A.*, 82, 2096, 1985.

204. **Weinrich, S. L., Niles, E. G., and Hruby, D. E.,** Transcriptional and translational analysis of the vaccinia virus late geene L65, *J. Virol.*, 55, 450, 1985.
205. **Hanafusa, T.,** Alteration of nucleic acid metabolism by active and inactive forms of vaccinia virus, *Biken J.*, 3, 313, 1960.
206. **Kit, S. and Dubbs, D. R.,** Biochemistry of vaccinia-infected mouse fibroblasts (strain L-M). Effects on nucleic acid and protein synthesis, *Virology*, 18, 274, 1962.
207. **Joklik, W. K. and Becker, Y.,** The replication and uncoating of vaccinia DNA, *J. Mol. Biol.*, 10, 452, 1964.
208. **Jungwirth, C. and Launer, J.,** Effect of poxvirus infection on host cell DNA synthesis, *J. Virol.*, 2, 401, 1968.
209. **Pogo, G. B. T. and Dales, S.,** Biogenesis of poxvirus. Inactivation of host DNA polymerase by a component of the invading inoculum particle, *Proc. Natl. Acad. Sci. U.S.A.*, 70, 1726, 1973.
210. **Olgiati, D., Pogo, B. G. T., and Dales, S.,** Biogenesis of vaccinia: specific inhibition of rapidly labeled host DNA in vaccinia inoculated cells, *Virology*, 71, 325, 1976.
211. **Rice, A. P. and Roberts, B. E.,** Vaccinia virus induces cellular mRNA degradation, *J. Virol.*, 47, 529, 1983.
212. **Salzman, N. P., Shatkin, A. J., and Sebring, E. D.,** The synthesis of a DNA-like RNA in the cytoplasm of HeLa cells infected with vaccinia virus, *J. Mol. Biol.*, 8, 405, 1964.
213. **Jefferts, E. R. and Holowczak, J. A.,** RNA synthesis in vaccinia-infected L cells. Inhibition of ribosome formation and maturation, *Virology*, 46, 730, 1971.
214. **Shatkin, A. J.,** Actinomycin D and vaccinia virus infection of HeLa cells, *Nature (London)*, 199, 357, 1963.
215. **Salzman, N. P. and Sebring, E.,** Sequential formation of vaccinia virus proteins and viral deoxyribonucleic acid, *J. Virol.*, 1, 16, 1967.
216. **Joklik, W. K. and Merigan, T. C.,** Concerning the mechanism of action of interferon, *Proc. Natl. Acad. Sci. U.S.A.*, 56, 558, 1966.
217. **Ben-Hamida, F. and Beaud, G.,** *In vitro* inhibition of protein synthesis by purified cores from vaccinia virus, *Proc. Natl. Acad. Sci. U.S.A.*, 75, 175, 1978.
218. **Rosemond-Hornbeak, H. and Moss, B.,** Inhibition of host protein synthesis by vaccinia virus. Fate of cell mRNA and synthesis of small poly (A)-rich polyribonucleotides in the presence of actinomycin D, *J. Virol.*, 16, 34, 1975.
219. **Bablanian, R., Esteban, M., Baxt, B., and Sonnabend, J. A.,** Studies on the mechanisms of vaccinia virus cytopathic effects. I. Inhibition of protein synthesis in infected cells is associated with virus-induced RNA synthesis, *J. Gen. Virol.*, 39, 391, 1978.
220. **Bablanian, R., Coppola, G., Scribani, S., and Esteban, M.,** Inhibition of protein synthesis by vaccinia virus. III. The effect of UV-irradiated virus on inhibition of protein synthesis, *Virology*, 112, 1, 1981.
221. **Person, A., Ben-Hamida, F., and Beaud, G.,** Inhibition of 40S-Met-tRNAmet ribosomal initiation complex formation by vaccinia virus, *Nature (London)*, 287, 355, 1980.
222. **Ben-Hamida, F., Person, A., and Beaud, G.,** Solubilization of a protein synthesis inhibitor from vaccinia virus, *J. Virol.*, 45, 452, 1983.
223. **Cooper, J. A. and Moss, B.,** Transcription of vaccinia virus mRNA coupled to translation *in vitro*, *Virology*, 88, 149, 1978.
224. **Pelham, H. R. B., Sykes, J. M. M., and Hunt, T.,** Characteristics of a coupled cell-free transcription and translation system directed by vaccinia cores, *Eur. J. Biochem.*, 82, 199, 1978.
225. **Mbuy, G. N., Morris, R. E., and Bubel, H. C.,** Inhibition of cellular protein synthesis by vaccinia virus surface tubules, *Virology*, 116, 137, 1982.
226. **Raczynski, P. and Condit, R. C.,** Specific inhibition of vaccinia virus growth by 2′-O-methyladenosine. Isolation of a drug-resistant mutant, *Virology*, 128, 458, 1983.
227. **Moss, B. and Filler, R.,** Irreversible effects of cycloheximide during the early period of vaccinia virus replication, *J. Virol.*, 5, 99, 1970.
228. **Schrom, M. and Bablanian, R.,** Inhibition of protein synthesis by vaccinia virus. II. Studies on the role of virus-induced RNA synthesis, *J. Gen. Virol.*, 44, 625, 1979.
229. **Esteban, M., Benavente, J., and Paez, E.,** Effect of interferon on integrity of vaccinia virus and ribosomal RNA in infected cells, *Virology*, 134, 40, 1984.
230. **Paez, E. and Esteban, M.,** Resistance of vaccinia virus to interferons: modulation of the 2-5A system in interferon-treated, vaccinia virus infected cells, submitted, 1986.
231. **Paez, E. and Esteban, M.,** Resistance of vaccinia virus to interferon is related to an interference phenomenon between the virus and the interferon system, *Virology*, 134, 12, 1984.
232. **Esteban, M., Benavente, J., and Paez, E.,** Mode of sensitivity and resistance of vaccinia virus replication to interferon, *J. Gen. Virol.*, 67, 801, 1986.
233. **Lengyel, P.,** Biochemistry of interferons and their actions, *Annu. Rev. Biochem.*, 51, 251, 1982.

234. **Gershowitz, A. and Moss, B.,** Abortive transcription products of vaccinia virus are guanylylated, methylated, and polyadenylated, *J. Virol.*, 31, 849, 1979.
235. **Paoletti, E., Lipinskas, B. R., and Panicali, D.,** Capped and polyadenylated low-molecular weight RNA synthesized by vaccinia virus *in vitro*, *J. Virol.*, 33, 208, 1980.
236. **Carrasco, L.,** The inhibition of cell functions after viral infection, a proposed general mechanism, *FEBS Lett.*, 76, 11, 1977.
237. **Carrasco, L. and Esteban, M.,** Modification of membrane permeability in vaccinia-virus infected cells, *Virology*, 117, 62, 1982.
238. **Alonso, M. A. and Carrasco, L.,** Reversion by hypotonic medium of the shut-off of protein synthesis induced by encephalomyocarditis virus, *J. Virol.*, 37, 535, 1981.
239. **Norrie, D. H., Wolstenholme, J., Howcroft, H., and Stephen, J.,** Vaccinia virus-induced changes in (Na^+) and (K^+) in HeLa cells, *J. Gen. Virol.*, 62, 127, 1982.
240. **Kaerlein, M. and Horak, I.,** Phosphorylation of ribosomal proteins in HeLa cells infected with vaccinia virus, *Nature (London)*, 259, 250, 1976.
241. **Cooper, J. A. and Moss, B.,** *In vitro* translation of immediate early, early, and late classes of RNA from vaccinia virus-infected cells, *Virology*, 96, 368, 1979.
242. **Benavente, J., Paez, E., and Esteban, M.,** Indiscriminate degradation of RNAs in interferon-treated, vaccinia virus infected cells, *J. Virol.*, 51, 866, 1984.
243. **Ochoa, S. and de Haro, C.,** Regulation of protein synthesis in eukaryotes, *Annu. Rev. Biochem.*, 48, 549, 1979.
244. **Kaempfer, R.,** Regulation of eukaryotic translation, in *Comprehensive Virology*, Vol. 19, Fraenkel-Conrat, H. and Wagner, R. R., Eds., Plenum Press, New York, 1984, 99.
245. **Esteban, M. and Paez, E.,** Antiviral and antiproliferative properties of interferons: mechanism of action, *Prog. Med. Virol.*, 32, 159, 1985.
246. **Kerr, I. M. and Brown, R. W.,** pppA2′p5′A: an inhibitor of protein synthesis synthesized with an enzyme fraction from interferon-treated cells, *Proc. Natl. Acad. Sci. U.S.A.*, 75, 256, 1978.
247. **Ratner, L., Wiegand, R. C., Farrell, J., Sen, G. C., Cabrer, B., and Lengyel, P.,** Interferon, double stranded RNA and RNA degradation. Fractionation of the endonuclease INT system into two macromolecular components: role of a small molecule in nuclease activation, *Biochem. Biophys. Res. Commun.*, 81, 947, 1978.
248. **Baglioni, C., Minks, M. A., and Maroney, P. A.,** Interferon action may be mediated by activation of a nuclease by pppA2′p5′A2′p5′A, *Nature (London)*, 273, 684, 1978.
249. **Wreschner, D. H., James, T. C., Silverman, R. H., and Kerr, I. M.,** Ribosomal RNA cleavage, nuclease activation and 2-5A (ppp(A2′p)nA) in interferon-treated cells, *Nucl. Acids Res.*, 9, 1571, 1981.
250. **Williams, B. R. G., Golgher, R. R., Brown, R. W., and Kerr, I. M.,** Natural occurrence of 2-5A in interferon-treated, EMC virus-infected L cells, *Nature (London)*, 282, 582, 1979.
251. **Nilsen, T. W., Maroney, P. A., and Baglioni, C.,** Synthesis of 2-5A oligoadenylate and activation of an endoribonuclease in interferon-treated HeLa cells infected with reovirus, *J. Virol.*, 42, 1039, 1982.
252. **Nilsen, T. W. and Baglioni, C.,** Mechanism for discrimination between viral and host mRNA in interferon-treated cells, *Proc. Natl. Acad. Sci. U.S.A.*, 76, 2600, 1979.
253. **Paez, E. and Esteban, M.,** Nature and mode of action of vaccinia virus products that block activation of the interferon-mediated ppp(A2′p)nA-synthetase, *Virology*, 134, 29, 1984.
254. **Rice, A. P., Roberts, W. K., and Kerr, I. M.,** 2-5A accumulates to high levels in interferon-treated, vaccinia virus-infected cells in the absence of any inhibition of virus replication, *J. Virol.*, 50, 220, 1984.
255. **Condit, R. C., Motyczka, A., and Spizz, G.,** Isolation, characterization, and physical mapping of temperature-sensitive mutants of vaccinia virus, *Virology*, 128, 429, 1983.
256. **Hersh, C. L., Brown, R. E., Roberts, W. K., Swyryd, E. A., Kerr, I. M., and Stark, G. R.,** Simian virus 40-infected, interferon-treated cells contain 2′,5′-oligoadenylates which do not activate cleavage of RNA, *J. Biol. Chem.*, 259, 1731, 1984.
257. **Cayley, P. J., Davies, J., McGullagh, K. G., and Kerr, I. M.,** Activation of the ppp(A2′p)nA system in interferon-treated, herpes simplex virus-infected cells and evidence for novel inhibitors of the ppp(A2′p)nA-dependent RNase, *Eur. J. Biochem.*, 143, 165, 1984.
258. **Zucker, W. V. and Schulman, H. M.,** Stimulation of globin chain initiation by haemin in the reticulocyte cell free system, *Proc. Natl. Acad. Sci. U.S.A.*, 59, 582, 1968.
259. **Farrell, P. J., Balkow, K., Hunt, T., Jackson, R. J., and Trachsel, H.,** Phosphorylation of initiation factor eIF-2 and the control of reticulocyte protein synthesis, *Cell*, 11, 187, 1977.
260. **Samuel, C. E.,** Mechanism of interferon action: phosphorylation of protein synthesis initiation factor eIF-2 in interferon treated human cells by a ribosome associated kinase processing site specificity similar to hemin-regulated rabbit reticulocyte kinase, *Proc. Natl. Acad. Sci. U.S.A.*, 76, 600, 1979.
261. **Lebleu, B., Sen, G. C., Shaila, S., Cabrer, B., and Lengyel, P.,** Interferon, double-stranded RNA and protein phosphorylation, *Proc. Natl. Acad. Sci. U.S.A.*, 73, 3107, 1976.

262. **Roberts, W. K., Hovanessian, A., Brown, R. E., Clemens, M. J., and Kerr, I. M.,** Interferon mediated protein kinase and low-molecular weight inhibitor of protein synthesis, *Nature (London),* 264, 477, 1976.
263. **Gupta, S. L.,** Specific protein phosphorylation in interferon-treated uninfected and virus infected mouse L929 cells: enhancement by double-stranded RNA, *J. Virol.,* 29, 301, 1979.
264. **Sen, G. C., Taira, H., and Lengyel, P.,** Interferon, double-stranded RNA and protein phosphorylation. Characteristics of a double-stranded RNA-activated protein kinase system partially purified from interferon-treated Ehrlich ascites tumor cells, *J. Biol. Chem.,* 253, 5915, 1978.
265. **Galabru, J. and Hovanessian, A. G.,** Two interferon-induced proteins are involved in the protein kinase complex dependent on double-stranded RNA, *Cell,* 43, 685, 1985.
266. **Kimchi, A., Zilberstein, A., Schmidt, A., Shulman, L., and Revel, M.,** The interferon induced protein kinase PK-i from mouse cells, *J. Biol. Chem.,* 254, 9846, 1979.
267. **De Haro, C. and Ochoa, S.,** Mode of action of the hemin-controlled inhibitor of protein synthesis: studies with factors from rabbit reticulocytes, *Proc. Natl. Acad. Sci. U.S.A.,* 75, 2713, 1978.
268. **Siekierka, J., Mauser, L., and Ochoa, S.,** Mechanism of polypeptide chain intiation in eukarotes and its control by phosphorylation of the σ subunit of initiation factor 2, *Proc. Natl. Acad. Sci. U.S.A.,* 79, 2537, 1982.
269. **Safer, B.,** 2B or not 2B: regulation of the catalytic utilization of eIF-2, *Cell,* 33, 7, 1983.
270. **Ochoa, S.,** Regulation of protein synthesis initiation in eukaryotes, *Arch. Biochem. Biophys.,* 223, 325, 1983.
271. **Leroux, A. and London, I. M.,** Regulation of protein synthesis by phosphorylation of eukaryotic initiation factor 2α in intact reticulocytes and reticulocyte lysates, *Proc. Natl. Acad. Sci. U.S.A.,* 79, 2147, 1982.
272. **Rosen, H., Knoller, S., and Kaempfer, R.,** Messenger RNA specificity in the inhibition of eukaryotic translation by double-stranded RNA, *Biochemistry,* 20, 3011, 1981.
273. **Whitaker-Dowling, P. and Youngner, J.,** Vaccinia rescue of VSV from interferon-induced resistance: reversal of translation block and inhibition of protein kinase activity, *Virology,* 131, 128, 1983.
274. **Rice, A. P. and Kerr, I. M.,** Interferon-mediated, double-stranded RNA dependent protein kinase is inhibited in extracts from vaccinia virus-infected cells, *J. Virol.,* 50, 229, 1984.
275. **Thomas, G., Siegman, M., Kubler, A. M., Gordo, J., and Jimenez de Asua, L.,** Regulation of 40S ribosomal protein S6 phosphorylation in Swiss mouse 3T3 cells, *Cell,* 19, 1015, 1980.
276. **Gressner, A. M. and Wool, I. G.,** The phosphorylation of liver ribosomal proteins *in vivo, J. Biol. Chem.,* 249, 6917, 1974.
277. **Rosen, H., Di Segni, G., and Kaempfer, R.,** Translational control by messenger RNA competition for eukarytic initiation factor 2, *J. Biol. Chem.,* 257, 946, 1982.
278. **Lodish, H. F.,** Alpha and beta globin mRNA: different amounts and rates of initiation of translation, *J. Biol. Chem.,* 246, 7131, 1971.
279. **Esteban, M. and Paez, E.,** unpublished data, 1983.
280. **Bablanian, R. and Esteban, M.,** unpublished data, 1984.
281. **Van Steeg, M., Van Grinsven, M., Van Mansfeld, F., Voorma, H. O., and Benne, R.,** Initiation of protein synthesis in neuroblastoma cells infected by Semliki Forest virus, *FEBS Lett.,* 129, 62, 1981.
282. **Daniels-McQueen. S., Detjen, B., Grifo, J. A., Merrick, W. C., and Thach, R. E.,** Unusual requirements for optimum translation of polio viral RNA *in vitro, J. Biol. Chem.,* 258, 7195, 1983.
283. **Sonenberg, N.,** ATP/Mg^{2+}-dependent cross-linking of cap binding proteins to the 5′ end of eukaryotic mRNA, *Nucl. Acids. Res.,* 9, 1643, 1981.
284. **Kaempfer, R., Hollender, R., Abrams, W. R., and Israeli, R.,** Specific binding of messenger RNA and methionyl-$tRNA_f^{Met}$ by the same initiation factor for eukaryotic protein synthesis, *Proc. Natl. Acad. Sci. U.S.A.,* 75, 209, 1978.
285. **Lerner, W. C., Andrews, N. C., Miller, G., and Steitz, J. A.,** Two small RNAs encoded by Epstein-Barr virus and complexed with proteins are precipitated by antibodies from patients with systemic lupus erythematosus, *Proc. Natl. Acad. Sci. U.S.A.,* 78, 805, 1981.
286. **Katze, M. G., Chen, Y. T., and Krug, R. M.,** Nuclear-cytoplasmic transport and VAI RNA-independent translation of influenza viral messenger RNAs in late adenovirus-infected cells, *Cell,* 37, 483, 1984.
287. **Hunter, T., Hunt, T., Jackson, R. J., and Robertson, H. D.,** The characteristics of inhibition of protein synthesis by double-stranded ribonucleic acid in reticulocyte lysates, *J. Biol. Chem.,* 250, 409, 1975.
288. **Bhat, R. and Thimmappaya, B.,** Two small RNAs encoded by Epstein-Barr virus can functionally substitute for the virus associated RNAs in the lytic growth of adenovirus 5, *Proc. Natl. Acad. Sci. U.S.A.,* 80, 4789, 1983.
289. **Bablanian, R. and Banerjee, A.,** Polyriboadenylic acid preferentially inhibits *in vitro* translation of cellular compared to vaccinia virus mRNAs: possible role in vaccinia virus cytopathology, *Proc. Natl. Acad. Sci. U.S.A.,* 83, 1290, 1986.
290. **Simons, R. W. and Kleckner, N.,** Translational control of IS 10 transposition, *Cell,* 34, 683, 1983.
291. **Mizuno, T., Chou, M.-Y., and Inouye, M.,** A unique mechanism regulating gene expression: translational inhibition by a complementary RNA transcript (micRNA), *Proc. Natl. Acad. Sci. U.S.A.,* 81, 1966, 1984.

292. **Izant, J. G. and Weintraub, H.,** Inhibition of thymidine kinase gene expression by anti-sense RNA: a molecular approach to genetic analysis, *Cell,* 36, 1007, 1984.
293. **Kim, S. K. and Wold, B.,** Stable reduction of thymidine kinase activity in cells expressing high levels of anti-sense RNA, *Cell,* 42, 129, 1985.
294. **Pestka, S., Daugherty, B. L., Jung, V., Hotta, K., and Pestka, R. K.,** Anti-mRNA: specific inhibition of translation of single mRNA molecules, *Proc. Natl. Acad. Sci. U.S.A.,* 81, 7525, 1984.
295. **Melton, D. A.,** Injected anti-sense RNAs specifically block messenger RNA translation *in vivo, Proc. Natl. Acad. Sci. U.S.A.,* 82, 144, 1985.
296. **Izant, J. G. and Weintraub, H.,** Constitutive and conditional suppression of exogenous and endogenous genes by anti-sense RNA, *Science,* 229, 345, 1985.
297. **Coleman, J., Hirashima, A., Inokuchi, Y., Green, P. J., and Inouye, M.,** A novel immune system against bacteriophage infection using complementary RNA (micRNA), *Nature (London),* 315, 601, 1985.

Chapter 5

CELL KILLING BY ENVELOPED RNA VIRUSES

E. T. Ulug, R. F. Garry, and H. R. Bose, Jr.

TABLE OF CONTENTS

I. INTRODUCTION: MECHANISMS OF CELL KILLING BY RNA VIRUSES

The ability of a virus to kill the host cell or to inhibit normal cellular functions underlies the pathological basis for many viral diseases. The death of cells infected with various lytic animal viruses probably represents the culmination of many structural, metabolic, and biosynthetic lesions within the infected cells. Mechanisms for the induction of cytopathology by viruses fall into two broad categories: inhibition of host macromolecular synthesis and cell injury. In this review, a comparison of the mechanisms by which widely diverse enveloped RNA animal viruses induce cell death will be discussed.

A. Inhibition of Host Macromolecular Synthesis

Following infection of cultured cells with many lytic viruses, host macromolecular synthesis is rapidly inhibited. Though required for cell proliferation, DNA synthesis is not essential for cell survival. Most cells in differentiated tissue, for example, do not divide. Host RNA and protein synthesis, by contrast, are essential for cell survival. Lytic viruses frequently terminate host RNA synthesis concurrent with or after protein synthesis has been arrested. Therefore, the termination of host protein synthesis has been most extensively studied. Cells infected with a variety of lytic viruses exhibit two effects on host protein synthesis. The first is characterized by a reduction in the overall rate of protein synthesis. The nature of this effect may be nonselective, and it may affect both host and viral gene expression. The second effect is the selective inhibition of host translation and a switch from cell- to virus-specified polypeptide synthesis. Both the decline in the overall rate and the selective termination of host translation involve the initiation steps of protein synthesis. Several mechanisms have been proposed to account for the selective inhibition of host protein synthesis by lytic RNA viruses. The models outlined here will be elaborated upon in the sections dealing with the specific virus groups.

The competition model[1,2] proposes that the amplification of viral RNA during infection results in a large excess of viral mRNAs that compete with host mRNA for a limited number of ribosomes or other components involved in translation. Host mRNAs are not actively degraded in the lytic virus-infected cells and are functional in in vitro protein synthesis systems.[1,3,4] In the strictest sense, the competition model assumes that viral and cellular mRNAs bind components of the initiation complex or ribosomes with equivalent efficiencies, and that ribosomes initiate and translate the RNAs with equivalent rates.

Other models have been proposed in which viral mRNAs possess a competitive advantage over host mRNAs for initiation under conditions imposed upon the cell by the viral infection. One assumption of these models is that viral mRNAs bind to some component(s) of the initiation complex more efficiently than do most host mRNAs. Thus, a relatively small proportion of viral mRNA would be able to successfully compete with an excess of cellular mRNAs. A modest reduction in the overall rate of initiation could further shift the balance in favor of mRNAs with a greater propensity to be translated. The initiation frequency of viral and cellular mRNAs may be differentially modulated under certain experimental conditions. For example, if infected cells are incubated in medium containing increased (or decreased) concentrations of NaCl or exposed to membrane active drugs, the initiation of translation of the majority of cellular mRNAs may be inhibited to a greater extent than most viral mRNAs. This observation has been documented for picornaviruses,[5] alphaviruses,[6] rhabdoviruses,[7,8] cytopathic retroviruses,[9] and several DNA viruses including herpes viruses.[10] A few cellular mRNAs may be translated under these conditions. These include the mRNAs for β-interferon and certain stress proteins.[11] These results suggested that many viral mRNAs and a limited subclass of cellular mRNAs may contain a feature(s) that permits translation under conditions where the initiation step of cellular protein synthesis is restricted. Studies with hypertonic media provided the basis for the theory first proposed by Carrasco and Smith[4] that the inhibition of protein synthesis by lytic viruses may be due to a virus-induced increase in the intracellular concentration of Na^+. This would favor the translation of lytic virus mRNAs that are able to initiate translation in the presence of high intracellular concentrations of Na^+.

An additional means by which lytic viruses may subordinate host protein synthesis involves the modification or inhibition of ribosomal components or initiation factors involved in host translation. For example, modification of the cap binding protein by poliovirus (see Chapter 6) may enable the virus to selectively inhibit the translation of host mRNA. Picornavirus RNA is not capped and, therefore, does not require this factor for translation. Likewise, inhibition of initiation factors by vesicular stomatitis virus and Semliki Forest virus may result in the selective translation of viral RNAs.

B. Virus-Induced Cell Injury

The inhibition of host macromolecular synthesis and the induction of cell injury are not necessarily independent mechanisms that lead to cell death. The viral function that inhibits cell protein synthesis may also contribute to cell killing by certain lytic RNA viruses.[12] For example, displacement of polysomes from the cytoskeletal network following picornavirus infection has been proposed to correlate with the inhibition of host protein synthesis.[13] Likewise, changes in plasma membrane permeability or inhibition of plasma membrane functions may influence the rate of protein synthesis at both the translational and transcriptional level.[14-16]

Inhibition of host macromolecular synthesis cannot by itself account for the rapid lysis and death of lytic virus-infected cells. Treatment of cells with inhibitors of protein, RNA, or DNA synthesis does not result in the induction of a cytopathic effect or in the loss of metabolic functions with the same rapid kinetics as productive virus infection.[10,17,18] Cell lysis induced by lytic viruses, therefore, is unlikely to result solely from inhibition of macromolecular synthesis.

The accumulation of lesions in virus-infected cells often results in morphological changes. These cytopathic effects (CPE) induced by various lytic viruses may be quite diverse and, in the case of enveloped lytic viruses, may not result in cell lysis until relatively late in the infection cycle. Many of the cytopathic lesions observed in lytic virus-infected cells are due to virus-induced changes in the plasma membrane. One characteristic feature of virus-induced injury to the plasma membrane is the depolarization of ion gradients.[19] The maintainance of ion gradients across the membrane is necessary for the regulation of the cell cycle, the induction of cell proliferation by a variety of growth factors, cell differentiation, cell-cell coupling, and the uptake of various metabolites.[20,21] The regulation of ion transport is also required in a number of specialized cell functions, including impulse transduction in nerve and muscle, and salt secretion and absorption by epithelium. Alterations in plasma membrane function following infection by lytic viruses, therefore, may contribute to a loss of many activities required for cellular proliferation and survival.

Virus-induced damage to the cell membrane may occur as a consequence of envelope glycoprotein insertion and virus budding. Two general mechanisms may account for plasma membrane injury by lytic enveloped viruses. In the first, lytic virus infection results in pore formation in the plasma membrane that results in nonspecific permeabilization of the cell membrane. Carrasco[22] has shown that virus infection renders the cell membrane permeable to toxins and phosphorylated intermediates of metabolism. Impraim and co-workers[23] have reported that Sendai infection of Letree cells results in the formation of pores that support the gradient-dependent flux of ions and metabolites ($M_r \leq 1000$) but not larger compounds. Alternatively, inactivation of cellular enzymes that maintain cellular homeostasis may contribute to plasma membrane injury. In Sindbis virus-infected chick cells, inhibition of Na^+K^+ATPase activity results in depolarization of cellular Na^+ and K^+ gradients.[24] In rabies virus-infected neuroblastoma cultures, inactivation of the acetylcholine receptor may correlate with the killing of neurons in vivo.[25]

The cytoskeletal system of eukaryotic cells is of major importance in the maintenance of cell shape and morphology. Cytoskeletal disruption by chemical agents has been shown to induce morphological changes that may resemble the CPE of many lytic virus infections.[26] Since many cytoskeletal elements are anchored in or near specialized regions of the plasma membrane, cytoskeletal changes may also be a consequence of plasma membrane damage. The induction of cytopathology, therefore, may result from both structural and metabolic damage induced by lytic virus infections.

II. CELL KILLING BY ALPHAVIRUSES

A. Virion Structure and Strategy of Replication

Sindbis virus and Semliki Forest virus, prototypes of the alphavirus group of the togavirus family, are small, enveloped plus-stranded RNA viruses.[27,28] Electron microscopy of alphaviruses reveals spherical particles with an envelope containing surface projections and an electron-dense core. The envelope of Sindbis virus consists of two glycoproteins and host-derived lipid. Semliki Forest virus contains a third glycoprotein. The nucleocapsid of alphaviruses is icosahedral and contains multiple copies of the capsid protein that surround a 49S plus-stranded RNA genome (11,703 nucleotides in length). The genomic RNA encodes at least four nonstructural proteins that participate in viral RNA synthesis. A genome length negative-stranded RNA template is synthesized, from which plus-stranded transcripts are copied. During infection a subgenomic (26S) RNA species (4200 bases in length) is also produced in large amounts. The 26S RNA contains sequences corresponding to the 3′ third of the virus genome and encodes the viral structural proteins.

The induction of cell death by alphaviruses contributes to the development of disease in infected animals. Attenuated mutants of Semliki Forest virus have been utilized to study cell tropism and the molecular components involved in neurovirulence.[29] Unlike the virulent strains, the mutants killed neither neuroblastoma cells grown in culture nor neurons in the brains of infected mice. The lethal event in virulent infections, therefore, may be due to the selective killing of neurons by the virus. Cultured oligodendrocytes infected with these mutants, however, exhibited an extensive cytopathic effect that correlated with the capacity of the mutants to induce demyelating disease in the infected animals.[29,30] Other attenuated mutants of Sindbis virus exhibited altered glycoprotein conformations which may account for their loss of virulence in vivo.[31]

B. Inhibition of Protein Synthesis by Alphaviruses

The subordination of host macromolecular synthesis by alphaviruses is characterized by a rapid inhibition of host protein, DNA, and RNA synthesis.[32-34] Vertebrate cells infected with alphaviruses exhibit a reduced overall rate of protein synthesis, and virus-specified polypeptides account for over 95% of the proteins synthesized late in the infectious cycle.

The transcription of viral RNA may be essential for the inhibition of host protein synthesis by alphaviruses. Only alphaviruses mutants capable of viral RNA synthesis (RNA^+ ts mutants) were able to inhibit host protein synthesis.[35,36] Likewise, alphavirus-directed transcription temporally correlated with the inhibition of host protein synthesis.[32] The subgenomic 26S RNA is the predominant viral RNA associated with the polyribosomes following alphavirus infection.[37] One plausible mechanism, therefore, which contributes to the induction of cell death by alphaviruses, is the direct competition of host viral and 26S RNA for the translational machinery in the infected cell.

Additional evidence for the role of viral RNA in the inhibition of host protein synthesis was obtained with the use of defective interfering (DI) particles. Certain undiluted DI stocks of Semliki Forest virus abrogated the inhibitory effect of infectious virus on host protein synthesis.[38] In the absence of infectious virus, these DI virus preparations did not affect host protein synthesis. Coinfection of cells with the infectious virus and dilutions of the DI stock, however, resulted in the inhibition of both host and viral protein synthesis. A limited level of viral RNA synthesis, therefore, may be sufficient to shut off host protein synthesis in the absence of detectable virus protein synthesis. It should be worthwhile to determine the nature of the viral RNA that induces this effect. DI RNAs are heterogeneous, and the 5′ end of certain DI RNAs have been shown to contain asparagine-tRNA sequences that could conceivably affect viral or host protein synthesis.[39]

Although alphavirus and cellular mRNAs are present in nearly equal levels late in infection,[40] most cellular mRNA is replaced by viral RNA in the polysomes. The selective translation of viral RNA, therefore, cannot result solely from simple mRNA competition. The selective initiation of alphavirus translation may reflect a distinguishing feature of the 26S RNA. Though capped at the 5′ end with 7-methyl guanosine,[41] the alphavirus 26S RNA may contain sequences at the 5′ end that bind efficiently to components of the initiation complex.

The selective translation of alphaviruses mRNA may also involve alterations in component(s) of the initiation complex. Ribosomal washes of Semliki Forest virus-infected neuroblastoma cells inhibited the translation of 49S viral and cellular mRNA, but not translation of 26S viral RNA.[42,43] The inhibitory component in these preparations was identified as capsid protein. Exogenously added capsid protein selectively inhibited translation of host and 49S RNA, but did not affect 26S RNA-directed protein synthesis in an in vitro system. The inhibitory component interfered with 80S complex formation by host mRNA. Initiation was restored by the addition of purified eukaryotic initiation factor-4B (eIF-4B) to the reactions.[43,44] Topological modifications of ribosomes isolated from alphavirus-infected cells have also been reported.[45] These preparations contain a substantial amount of capsid protein bound to ribosomes.

Inconsistent with the notion that capsid protein selectively restricts host translation are studies in which tunicamycin was shown to partially prevent the selective termination of host protein synthesis by Sindbis virus.[18] Although the level of capsid protein synthesis was not affected, actin synthesis was elevated two- to threefold as a consequence of tunicamycin treatment. Moreover, expression of Semliki Forest virus capsid protein by SV40 recombinants did not result in cell killing.[46]

The selective translation of viral RNA may also be facilitated by viral functions that affect the metabolic and ionic status of the infected cells. Alterations in the levels of intracellular Na^+ and K^+ have been shown to significantly alter the rate of protein synthesis in cultured cells.[47] Moreover, the replication of several viruses, including Sindbis virus, was resistant to treatments that alter intracellular salt levels.[48] In alphavirus-infected cells, protein synthesis directed by 26S RNA, but not 49S or host RNA, was resistant to alterations of the Na^+ concentration in the growth medium, or to agents that alter the intracellular Na^+ and K^+ levels.[6,49] In cells infected with Sindbis virus, an inhibition of host mRNA translation temporally correlated with alterations in the intracellular concentrations of Na^+ and K^+.[6,24] The inhibition of host cell protein synthesis by Sindbis virus is, however, dependent upon the multiplicity of infection (MOI).[18] At elevated multiplicities, virus protein synthesis predominated in the absence of functional plasma membrane modification. Under these conditions, viral RNA was synthetized to increased levels and presumably maintained a competitive advantage for translation. At reduced multiplicites of infection, however, membrane-mediated alterations in ion levels may account for the decreased amount of host-specified protein synthesis in the infected cells. Alterations in the intracellular monovalent cation levels were due, at least in part, to a selective inhibition of Na^+K^+ATPase activity by Sindbis virus.[24] Disruption of plasma membrane enzyme function, therefore, may contribute to the inhibition of host protein synthesis by Sindbis virus.

C. Induction of Cell Injury by Alphaviruses

Shut-off of host protein synthesis in the absence of functional membrane modification does not induce a rapid development of the cytopathic effect in alphavirus-infected cells.[18] Plasma membrane modification by Sindbis virus is characterized by the insertion of virus glycoproteins and the assembly and release of virus particles by budding. Plasma membrane modification in alphavirus-infected cells is accompanied by alterations in lipid composition, fluidity, enzyme function, and membrane permeability.

Sindbis virus induces a cytopathic effect characterized by the shrinking and rounding of infected avian cells.[50] The development of this cytopathic effect may be a consequence, in part, of the assembly and release of virus particles at the cell surface. Tunicamycin, which inhibits virus particle assembly and release, prevented the development of a typical cytopathic effect in avian cultures infected with Sindbis virus.[18] Inhibition of Na^+K^+ATPase activity by Sindbis virus was also prevented by treatment of the infected cultures with tunicamycin. Perturbation of plasma membrane integrity, which may occur as a consequence of terminal events in virus assembly and release, therefore correlates with the development of a cytopathic effect.

The insertion of alphavirus envelope proteins into lipid bilayers resulted in an increase in microviscosity.[51,52] Removal of the accessible protein by proteolysis or substitution of plasma membrane proteins for viral proteins resulted in increases in the fluidity of the membrane. The insertion of unglycosylated envelope proteins into the membranes of tunicamycin-treated, Sindbis virus-infected cells, however, did not result in alterations in Na^+K^+ATPase activity.[18] It is unclear, therefore, whether changes in membrane fluidity per se contribute to the induction of the cytopathic effect. One peculiar aspect of the lipids present in alphavirus-infected cells is a specific reduction in the saturation of 18 carbon fatty acids.[53] The basis for this alteration has not been examined.

Carrasco[16] proposed that enhanced nonspecific permeability of the plasma membrane occurs as a consequence of viral infection. Indeed, alphavirus-infected cells release ^{51}Cr at the time of maximal virus budding.[53] Alphavirus-infected cells, however, were not as permeable to toxins as cells infected by picornaviruses or rhabdoviruses.[54] Furthermore, the uptake of other impermeable compounds, i.e., phosphorylated sugars, was not enhanced in alphavirus-infected cells;[55] nor could evidence for drug-insensitive K^+ leakage be detected in Sindbis virus-infected avian cells.[24] Therefore, gross changes in membrane permeability do not contribute to rapid development of alphavirus-induced cytopathology.

D. Effect of Alphaviruses on Host Cell Metabolism

Alterations in host cell metabolism may also contribute to the death of alphavirus-infected cells. Cells infected with alphaviruses exhibited an enhanced rate of glucose catabolism, and the activities of several enzymes required for various stages in glucose catabolism were enhanced.[56] These enzymes all exhibited a requirement for nicotinamide adenine dinucleotide as a cofactor. The enhanced activity of these constitutively controlled enzymes was prevented by actinomycin D treatment, suggesting that synthesis of these proteins was induced as a consequence of alphavirus infection.[56] It is not known whether these enzymes correspond to the classes of proteins induced after stress treatment of alphavirus-infected cells.[57]

The uptake of 2-deoxy-D-glucose is enhanced in alphavirus-infected cells.[58] The stimulation in glucose uptake has been proposed to result from plasma membrane modification.[58] In light of the enhanced rate of glucose catabolism in alphavirus-infected cells, stimulation of glucose uptake may also result from a decrease in the intracellular level of glucose. This latter view is supported by studies in which tunicamycin was shown to exhibit no selective effect on the stimulation of glucose uptake in alphavirus-infected cells.[59] Enhanced glucose uptake may also result from enhanced Na^+ flux into alphavirus-infected cells.

Reductions in the intracellular concentrations of nucleoside triphosphates are other characteristics of alphavirus infection that may contribute to the induction of cell death. The adenylate and guanylate energy of cells infected with alphaviruses were reduced due to a selective reduction in the amount of ATP and GTP present in the cells.[56,60] ATP and GTP are required for the initiation steps of protein synthesis and may contribute to the overall reduction in the rate of protein synthesis in alphavirus-infected cells. The levels of nucleoside triphosphates may be specifically reduced by a nucleoside triphosphate phosphohydrolase (NTPase) in cells infected with alphaviruses.[61,62] NTPase activity was correlated with the inhibition of DNA polymerase activity in vitro. The inhibition of NTPase activity may, therefore, be responsible for the inhibition of cellular DNA synthesis following alphavirus infection. This enzyme activity was actinomycin D resistant and copurified with the 82 kdalton nonstructural protein of western equine encephalitis virus.[63]

III. CELL KILLING BY RHABDOVIRUSES

A. Virion Structure and Strategy of Replication

Rhabdoviruses, which include vesicular stomatitis virus (VSV) and rabies virus, are cylindrical particles (70 by 170 nm) surrounded by a lipid-containing envelope.[64,65] Embedded in the envelope are prominent spikes about 10 nm long composed of the viral glycoprotein (G). Internally, the virion contains a tightly coiled ribonucleoprotein (RNP) helix of genomic RNA and a nucleocapsid protein (N). Associated with the RNA are several molecules of the "large" protein (L) and the nonstructural protein (NS, a misnomer since the protein is virion-associated). Together the L and NS molecules serve as the transcriptase for the virus. The N protein is also required for transcription. A fifth virus-encoded protein, M, variously named the membrane or matrix protein, is present in the viral envelope and forms a complex with the viral glycoprotein and N molecules of the RNP structure.

The genome of rhabdoviruses is a negative-polarity (antisense) RNA molecule of approximately 11,200 nucleotides.[64,66] The gene order is 3′-1-N-NS-M-L-5′, with 1 representing the sequence encoding the plus-leader, a small species of RNA that is actively transcribed. Base sequence analysis reveals two regions at the 3′ end of the genomic RNA that may serve as transcriptase binding sites.[67] Initiation of transcription at the site at the extreme 3′ end may result in the formation of either leader RNA or full length positive-stranded RNA molecules. Initiation at the second site at the 3′ end of the N gene results in synthesis of mRNAs. Alternatively, it has been suggested that transcription is initiated exclusively at the extreme 3′ site.[68] Transcription of the mRNAs exhibits polarity: genes at the 3′ end are transcribed at a higher frequency than those at the 5′ end.[69,70] The N and NS genes at the 3′ end of the genome are the least sensitive to UV-irradiation, while the L gene at the distal end is the most sensitive. These results suggest that transcription of viral mRNA proceeds sequentially through the structural genes from the 3′ to the 5′ end of the genome.

B. Inhibition of Protein Synthesis by Rhabdoviruses

In susceptible vertebrate cells lytically infected with a low multiplicity of VSV (less than 10 PFU/cell), the overall rate of protein synthesis begins to decline early in infection.[33,71-73] Selective inhibition of host protein synthesis begins later, coincident with the appearance of viral proteins. Increases in the MOI may accelerate the appearance of viral proteins, inhibition of the overall rate of protein synthesis, and development of the CPE.[74-76] Furthermore, both the virus strain and the cell type used may influence the degree and kinetics of the inhibition of protein synthesis.[2,77]

Termination of host protein synthesis by VSV at low MOI requires the transcription of virion RNA.[12,78-84] The selective inhibition of host protein synthesis may therefore result from competition between cellular and viral mRNA.[1,2] Support for this model was initially provided by the use of viral mutants. Inhibition of host protein synthesis did not take place in cells infected with transcription-negative ts mutants cultured at the nonpermissive temperature.[78,79] Likewise, the R1 mutant, a revertant of ts1026, is delayed in inhibition of protein synthesis (a phenotype referred to a P minus).[85] Cells infected with R1 mutants accumulate mRNA more slowly than wild-type.[2] Moreover, the rt-mutants of VSV, which overproduce genomic RNA and were deficient in the accumulation of plus leader and mRNA species, were delayed in inhibition of cellular protein synthesis.[86]

Further support for the competition model was provided by the analysis of polyribosome structures in VSV-infected cells. In uninfected cells, the mRNAs that encode various cellular proteins were found on polyribosomes of distinct size.[1] For example, the mRNAs that encode actin were generally associated with 12 ribosomes. The N mRNA and the actin mRNA both encode proteins of approximately 43 kdalton and were found associated with six ribosomes in cells infected with the Glascow strain of VSV. These results suggested that the overall rate of initiation is reduced and/or that an increased number of mRNAs was associated with a fixed number of active ribosomes. Furthermore, the rate of initiation of viral and host mRNAs was approximately equal.

The simple competition model is insufficient to account totally for inhibition of protein synthesis by all strains of VSV. In cells infected with the San Juan or other virulent strains of VSV, selective inhibition of host protein synthesis occurred at a time in infection when the levels of total mRNA have increased only two- to threefold.[1,2,87,88] Viral proteins should, therefore, have accounted for 50 to 65% of the total protein synthesis in infected cells. However, major cellular proteins such as actin were synthesized at less than 20% of the rate in uninfected cells. The synthesis of other cellular proteins was reduced to a significantly greater extent, resulting in more than 95% inhibition of cellular protein synthesis. The competition model also fails to account for the mechanism by which VSV reduces the overall rate of protein synthesis. Even in the case of "weakly-inhibiting" strains, such as VSV-Glascow, the overall rate of protein synthesis decreased at least 33%.[1,2] In competition among mRNAs with roughly equal initiation frequencies, the overall rate of protein synthesis should remain constant irrespective of the proportion of viral and host mRNAs present. Furthermore, the S2 small plaque mutant of VSV produces a four- to fivefold excess of viral mRNA that can be translated in vitro with the same efficiency as wild-type mRNA.[89] Cells infected with this mutant, however, produce less viral protein than cells infected with wild-type VSV. Simple competition would predict that cells infected with S2 mutants would synthesize more viral protein than wild-type VSV-infected cells. Clearly, certain strains of VSV encode functions that favor the translation of viral mRNAs.

Further evidence against the role of mRNA competition was provided by UV-inactivation studies, which demonstrated that cell killing could be mediated by products of primary transcription.[73,84] Although VSV induces an initial multiplicity-dependent, UV-resistant inhibition in the overall rate of protein synthesis, the selective inhibition of host protein synthesis was UV-sensitive.[73] The selective inhibition of cellular protein synthesis and cell killing by UV-irradiated VSV required transcription of N and perhaps NS genes.[84] Therefore, inhibition of protein synthesis does not require amplification of viral mRNA that would occur as a consequence of secondary rounds of transcription.

The role of primary transcription in the inhibition of protein synthesis by VSV may be more complex than previously appreciated. Dunigan and Lucas-Lenard[90] showed that the inhibition of protein synthesis by UV-inactivated VSV was biphasic. This suggested that the transcription of more than one region of the viral genome is involved in inhibition of protein synthesis. Low doses of UV-irradiation (550 ergs/mm^2) inactivated transcription of a 373 nucleotide product, which could correspond to a partial transcript of the N gene. At higher doses (4000 ergs/mm^2) a smaller transcript corresponding to about 42 nucleotides was not synthesized. The identity of this smaller transcript is uncertain. It may correspond to the plus-leader. Alternatively, the small RNA product may correspond to several small

RNA species that arise from initiation at the 3′ end of viral genes. These small RNAs, which possess ribosome binding sites, could block the initiation of mRNA translation. At doses of greater than 18,000 ergs/mm^2, VSV was no longer capable of inhibiting protein synthesis. The ability of VSV to inhibit protein synthesis was also abolished by heat treatment that inactivated the virion-associated transcriptase. They inferred from these results, as did Marcus, Marvaldi, and co-workers, that limited viral transcription is necessary to inhibit protein synthesis;[12,80-84] however, high doses of UV-irradiation or heat could denature other virion proteins. Furthermore, a reduction in the overall rate of protein synthesis may occur within the first hour after infection. In addition, DI particles that produce little mRNA reduce the overall rate of protein synthesis.[88] Collectively, these observations suggest that virion components may contribute to the inhibition of protein synthesis by VSV.

The selective translation of VSV mRNA may reflect unique features at the 5′ end of the viral message, which efficiently bind to components involved in initiation. The initiation of rhabdovirus translation occurs over a broader range of monovalent ion concentrations than does cellular mRNA translation.[7,8,91,92] Resistance of VSV mRNA translation to hypertonic medium correlates with the distance between the cap and initiation codon in the mRNA.[93] Changes in Na^+ and K^+ levels in VSV-infected cells, therefore, could contribute to the selective inhibition of host protein synthesis. In some cases the changes in intracellular ion levels were shown to occur early in infection, concomitant with the selective inhibition of host protein synthesis (2 to 4 hr).[11,94,95] In rt-mutants which are delayed in the inhibition of host protein synthesis, a correlation also exists between the onset of host protein synthesis inhibition and the inhibition of $^{86}Rb^+$ (a K^+ tracer) uptake.[95] Other investigators, however, have reported that significant changes in intracellular Na^+ and K^+ concentrations could not be observed until late after VSV infection, suggesting that selective termination of host protein synthesis is not mediated by changes in intracellular ion concentration.[58,96]

An additional approach to define the mechanisms by which viruses inhibit initiation of protein synthesis has been the reconstitution of in vitro protein synthesis systems using purified components from both infected and uninfected cells. Lysates of VSV-infected cells show a reduction in the ability to translate endogenous host and viral mRNAs.[97-102] The inhibition occurs primarily, if not exclusively, at the initiation step in protein synthesis. Purified eIF-2, eIF-3, and eIF-4B have been shown to stimulate protein synthesis in lysates of VSV-infected cells.[87,97,100,102] Synthesis directed by endogenous cellular mRNAs was stimulated about twofold, while synthesis directed by viral mRNAs was stimulated to a greater extent. eIF-2 is involved in activation of initiator tRNA, eIF-3 is involved in recycling ribosomes, and eIF-4B participates in the recognition of capped RNAs.[104] The addition of purified initiation factors to lysates from virus infected cells has not been shown to selectively stimulate the translation of host mRNA. The inactivation of these factors in lysates of VSV-infected cells, therefore, may account for the overall reduction in protein synthesis.

The product responsible for the decreased protein synthesis capacity of lysates from VSV-infected cells has not been identified. However, double stranded (ds) RNA present in mRNA preparations from VSV-infected cells has been found to inhibit the initiation step in protein synthesis.[101] Double stranded RNA may activate a ds RNA-dependent protein kinase that reversibly inhibits the activity of eIF-2.[103,104] Marvaldi and Lucas-Lenard[105] reported that VSV stimulated the phosphorylation of five ribosome-associated proteins in L cells. The relationship of these modifications to inhibition of protein synthesis by VSV remains to be determined.

Another mechanism proposed to account for the inhibition of protein synthesis by VSV involves the formation of messenger RNP particles (mRNP).[1,87] These mRNP complexes that accumulate during lytic infection contain all five species of VSV mRNA and N protein. The viral mRNP inhibits in vitro protein synthesis by decreasing ternary complex formation (eIF-2-GTP-Met-tRNA).[87] This inhibition was partially reversed by purified eIF-2. The requirement for N protein in inhibition of protein synthesis by VSV may be related to its structural role in formation of the mRNP complex.

C. Induction of Cell Injury by Rhabdoviruses

The induction of cell injury by rhabdoviruses is primarily due to the inherent toxicity of virion components. Two forms of rhabdovirus cytopathogenicity can be distinguished on the basis of MOI.[75,76] Following high multiplicity infections with VSV, cultured cells rapidly exhibit cytopathic changes characterized by significant cell rounding and detachment from the substratum. This killing "from without" is also induced by DI particles and is, therefore, not dependent upon virus gene expression. By contrast, the development of CPE in cells infected with lower multiplicities occurs later in the course of infection and requires virus gene expression. Although limited viral transcription ultimately results in cell death,[12,80] viral protein synthesis and/or secondary transcription accounts for the rapid development of VSV-induced cytopathology during productive virus infections. Treatment of VSV-infected cells with cycloheximide, for example, does not effect primary transcription, yet prevents the rapid development of CPE. Likewise, ts mutant studies indicate that the induction of the cytopathic effect by VSV is largely due to the activity of the G protein.[26]

The morphological changes detected in rhabdovirus-infected cells may be due, in part, to disruption of the cytoskeleton. The development of CPE during productive VSV infections resembles many of the characteristics of cells treated with cytochalasin B and other agents that disrupt the cytoskeleton.[26] The relative insensitivity of VSV replication to these drugs suggests that an intact cytoskeleton is not required. Immunofluorescence studies have demonstrated that the disruption of the cytoskeleton by VSV occurs, at least in part, because of depolymerization of actin bundles.[106] Many cytoskeletal structures are anchored in specialized regions of the plasma membrane known as adhesion plaques. Disruption of the cytoskeleton by rhabdoviruses, therefore, may result from insertion of viral envelope proteins.

One consequence of cytoskeleton disruption may be the inhibition of fluid phase pinocytosis, which rapidly occurs after VSV infection of BHK cells.[107] Pinocytosis, which is required for the uptake of many nutrients and growth factors, requires a structurally intact cytoskeletal system. Inhibition of pinocytosis by VSV required gene expression; little or no inhibition was detected in cells infected with UV-irradiated virus or in virus-infected cells treated with cycloheximide. Inhibition of pinocytosis required expression of the viral glycoprotein.

As previously indicated, perturbation of plasma membrane activity may also influence macromolecular synthesis in lytic virus-infected cells. Genty[108] has shown that a reduction in uridine transport may account, in part, for the inhibition of RNA synthesis in chick cells infected with VSV. Reduction in the capacity to transport uridine occurred rapidly after high MOI, even in the presence of cycloheximide, suggesting that virion components were cytotoxic. By contrast, following low multiplicity infections, inhibition of uridine transport was delayed. Cycloheximide prevented the inhibition of transport following low multiplicity infections, suggesting that viral gene expression was required. Mutants in the M protein exhibited a delayed response, indicating that membrane modification was, in part, responsible for this loss in cell membrane function.

Changes in intracellular monovalent cation contents have also been observed in cells infected with VSV.[11,95,96,109] The degree to which the intracellular ion concentrations were affected in VSV infected cells, however, was variable. This may be due, in part, to the cell types examined. Renal, epithelial, and neural cells utilize different transport mechanisms. In BHK cells infected with VSV, for example, changes in intracellular Na^+ or K^+ levels were not detected until late after infection.[58] By contrast, infection of MDCK cells with VSV initially stimulated Na^+K^+ATPase activity.[110] At later times p.i. (postinfection), the activity of the Na^+K^+ATPase returned to that detected in uninfected cells. Passive ion transport, however, was inhibited late p.i. These latter changes coincided with the appearance of viral envelope proteins on the cell surface and virus release.

Infection of cells by VSV may also result in an increase in nonspecific permeability of the plasma membrane. VSV infection renders L cells permeable to hygromycin, an antibiotic that inhibits protein synthesis.[54] Sensitivity of VSV infected cells to hygromycin was multiplicity-dependent and occurred rapidly after infection. By contrast, Semliki Forest virus-infected cells exhibited little sensitivity to hygromycin, even after high multiplicity infections.

Alterations in ion permeability and membrane fluidity following VSV infection have been correlated with conductance changes in the plasma membrane. Exposure of planar lipid bilayers to VSV results in an increase in membrane conductance at pH 5.0 to 5.2, the pH range which activates the glycoprotein fusion activity.[111] Likewise, Lopez-Vancell and co-workers[110] detected an increase in the transepithelial electrical conductance in MDCK monolayers after infection with VSV. The reduction in resistance may, however, be due to changes in the intercellular junctions that takes place during the rounding and detachment of cells from the substratum.

Many of the toxic effects of VSV on the host cell membrane are due to activities associated with the viral glycoprotein. Treatment of cells with highly purified viral glycoproteins results in a toxic effect on cellular DNA and RNA synthesis.[112] The glycoprotein of VSV is responsible for the fusion activity of the virus. The expression of a molecularly cloned VSV glycoprotein gene results in cell fusion.[113] Likewise, manipulation of cellular conditions that results in the overproduction of G, relative to M, induces cell fusion.[114] Under normal conditions, the fusion activity of the VSV glycoprotein is activated after exposure to low pH (pH 5.0 to 5.3). It has been proposed that these conditions resemble those found in the lysosomal microenvironment, where viral fusion activity acts to uncoat the infecting particles (see Chapter 2).

Many of the biological properties of the VSV glycoprotein are contained in a 25 amino acid peptide.[115] This peptide, which corresponds to the amino terminus of the protein, is a pH-dependent hemolysin. When applied to mammalian cells, the peptide is cytotoxic and affects gross changes in cell morphology and permeability. The amino termini of the glycoproteins of VSV and rabies virus exhibit extensive sequence homology,[116] suggesting that this domain has an important role in the function of rhabdovirus glycoproteins. Another domain (between residues 189 and 214) of the rabies virus glycoprotein may play an important role in cell killing and/or neurotropism.[25,117] This region shares considerable sequence homology with the "toxic loop" of curaremimetic neurotoxins. These toxins interact with the binding site of the acetylcholine receptor. This region of the viral glycoprotein, therefore, may recognize the acetylcholine receptor and contribute to the neurovirulence of this virus.

D. Development of Persistence: Infection of Insect Cells by Alphaviruses and Rhabdoviruses

Many togaviruses and rhabdoviruses alternate in nature between insect and vertebrate hosts and can replicate in both insect and vertebrate cells.[118,119] Although Sindbis virus, Semliki Forest virus, and VSV replicate to high titers in mosquito cells, the majority of cells survives the initial infection and continues to proliferate to form a persistently infected culture. Once the persistent infection is established, host macromolecular synthesis is not inhibited. The mechanism by which viruses establish persistent infections has not been determined. The establishment of a persistent infection may require interferon (or interferon-like substances) or the evolution of virus variants such as ts mutants or DI particles. In arthropod cell cultures, a soluble factor released from infected cells has been reported to induce the persistent state.[120]

Failure of viruses to establish a lytic infection in mosquito cells may reflect an intrinsically greater resistance of mosquito cells to altered intracellular levels of Na^+ and K^+. Mitshuhashi and Maramorosch medium, used in the isolation and establishment of mosquito cells, is over 300 m*M* in NaCl. Infection of mosquito cells grown in Mitshuhashi and Maramorosch medium results in virus replication (over 10^8 PFU/mℓ), without the induction of a CPE. Mosquito cells that have been adapted to grow in Eagle's minimal essential medium (MEM), which is approximately 150 m*M* NaCl, however, exhibit a heterologous response to alphavirus or VSV infection.[121-123] While many clones fail to develop a CPE (CPE−), other clones exhibited extensive CPE (CPE+) after exposure to Sindbis virus or VSV when the cultures were incubated at 34°C in the presence of a serum.[122] The normal incubation temperature for cells from poikilothermic organisms is 28°C. Both viral and host protein synthesis is decreased following infection of CPE+ cells by Sindbis virus and VSV at 34°C. In contrast, the development of CPE in Semliki Forest virus-infected cells occurred at 28°C and there was a selective inhibition of host protein synthesis.[123] Uninfected mosquito cells incubated at 37°C in the presence of serum exhibited a CPE.[95] The incubation of CPE+ mosquito cells at 34°C, therefore, may induce alterations in cell metabolism that play a synergistic role in inhibition of protein synthesis by Sindbis virus and VSV.

Mosquito cells may fail to develop a CPE because the viral mRNAs do not enjoy a competitive advantage over the host mRNAs. Adaptation of mosquito cells from Mitshuhashi and Maramorosch medium to MEM may alter the ability of certain cells to respond homeostatically to changes in the intracellular ionic conditions. Alternatively, it is possible that changes in intracellular ion concentrations occur in CPE+ cells grown in MEM, but fail to occur in CPE− clones. A reduction in $^{86}Rb^+$ transport was observed following Sindbis virus or VSV infection of CPE+ mosquito cells grown in MEM at 34°C in the presence of serum.[95] Reduced $^{86}Rb+$ transport was not, however, observed in the parental CPE− cell line.[95] Phosphorylation of specific ribosomal proteins may also result in inhibition of protein synthesis in VSV infected (CPE+) mosquito cells. Gillies and Stollar[124] determined that a reduction in the phosphorylation of the ribosomal protein P1 in cells correlated with the decreased capacity of cell extracts to translate endogenous or exogenously added viral and cellular mRNAs. The addition of an S-100 fraction from uninfected cells restored protein synthesis capacity, stimulated formation of 80S complexes, and increased P1 phosphorylation.

IV. CELL KILLING BY PARAMYXOVIRUSES

A. Virion Structure and Strategy of Replication

Newcastle disease virus (NDV) and Sendai virus, prototypes of the paramyxovirus family, are pleomorphic enveloped viruses that contain a single-stranded genome of negative polarity (5.1 to 5.7 × 10^6 daltons).[125,126] The paramyxoviruses comprise an important group of human and animal pathogens including measles, mumps, parainfluenza, respiratory syncytial, distemper, rinderpest, and Newcastle disease virus. Paramyxoviruses are irregular particles of both spherical and filamentous morphology. The virions contain a large virus-encoded protein (L) and a nucleocapsid protein (NP), which together constitute the transcriptase. Paramyxovirus particles also contain a membrane protein (M) and a phosphoprotein (P). Certain paramyxoviruses may also contain one or two small virion proteins. Unlike rhabdoviruses, however, the envelope of paramyxoviruses contains two biologically active glycoprotein spikes. One population of spikes (designated HN) contains hemagglutination and neuraminidase activities and is involved in the attachment of virions to susceptible cell surfaces.[127] The second form, which contains two disulfide-linked glycoproteins (designated F1 and F2), confers fusion and hemolytic activities on the virion and is involved in virion penetration. Avirulent or "early harvest" paramyxovirus particles may contain larger precursors to the F glycoproteins (F_o). These stocks are not infectious when introduced into cell culture unless exposed to mild trypsinization that cleaves the fusion protein precursors. Protease-deficient cell types also release noninfectious wild-type virus, indicating that F_o cleavage utilizes a cellular protease.[128]

The infection cycle of paramyxoviruses resembles that previously described for rhabdoviruses. Transcription of the negative-stranded paramyxovirus genome results in six or seven unique RNA species that encode the virus-specified polypeptides. A comparison of the UV target sizes with the corresponding gene sizes suggests that transcription of NDV initiates at a single promoter and proceeds in the order 3′ NP-P-(F_o,M)-HN-L 5′.[129,130] A slightly different gene order has been reported for Sendai virus: 3′ NP-F_o-M-P-HN-L 5′.[131] As described for rhabdoviruses, a small leader RNA has been detected in NDV-infected cells.[129]

Insight into the mechanism of cell killing by NDV has been provided by the isolation of noncytopathic (NC) mutants of the Australia-Victoria wild-type strain. The NC mutants formed hemeadsorption centers, but did not form plaques under defined experimental conditions.[132,133] These mutants were less cytopathic in cell culture and extended the mean embryo death time, yet produced amounts of progeny virus equivalent to the wild-type virus. Analysis of plaque forming revertants of the NC mutants indicated that the capacity to synthesize viral RNA correlated with the ability of the virus to kill cells and form plaques. Viral RNA synthesis and cytopathology may, therefore, be causally related.[132,133]

B. Inhibition of Cellular Protein Synthesis by Paramyxoviruses

The inhibition of cellular protein synthesis by virulent strains of paramyxoviruses occurs much slower than by alphaviruses, rhabdoviruses, or picornaviruses, and has been studied in less detail. By 6 hr p.i., of chick cells with NDV, the overall rate of protein synthesis is reduced approximately 50% relative to uninfected cells. Concomitant with the reduction in the overall rate of protein synthesis is a gradual transition from host-specified to virus-specified polypeptide synthesis.[134,135] These results cannot be explained by a reduction in the rate of amino acid uptake or by enhanced protein degradation. Correlated with the reduction in the rate of host protein synthesis was a reduction in the number of actively translating ribosomes.[136]

The inhibition of host protein synthesis by NDV requires viral RNA synthesis. Chicken embryo cells infected with noncytopathic strains of NDV accumulated less virus-specified RNA and failed to inhibit host protein synthesis.[133] Furthermore, certain ts mutants of NDV did not inhibit host protein synthesis at the nonpermissive temperature. After shift to the nonpermissive temperature, virus-specified proteins were no longer synthesized and host protein synthesis was reinitiated.[137] The recovery of host protein synthesis was inhibited by actinomycin D, but was not affected when cycloheximide was present during shift to the nonpermissive temperature.[137] The recovery of host protein synthesis, therefore, requires cellular mRNA synthesis, but utilizes transcription and translation factors synthesized prior to protein synthesis inhibition. Thus, NDV-induced termination of host protein synthesis may result from inhibition of host mRNA synthesis or host mRNA destabilization.

Virion components of paramyxoviruses may also contribute to the inhibition of host protein synthesis and cell killing. Temperature-sensitive mutants of Sendai virus that contain lesions in the HN glycoprotein do not inhibit host protein synthesis at nonpermissive temperature.[138] Modification of the plasma membrane by the viral glycoprotein(s), therefore, may be an important determinant in cell killing by paramyxoviruses. However, in contrast to other enveloped RNA viruses, protein synthesis by paramyxoviruses is sensitive to ouabain, which alters the intracellular Na^+ and K^+ concentrations by inhibiting the Na^+K^+ATPase.[48] It is unlikely, therefore, that alterations in intracellular salt concentrations account for the selective translation of paramyxovirus mRNA.

C. Induction of Cell Injury by Paramyxoviruses

The pathogenicity of paramyxoviruses is conferred, in part, by the nature of the virus envelope. Pathogenicity of NDV strains in chickens differs with respect to the cleavage of the glycoproteins.[139] Nonpathogenic NDV strains exhibited reduced levels of the cleaved glycoproteins in comparison with the virulent strains. Coinfection experiments using both virulent and avirulent strains suggested that the failure to cleave the envelope precursors was an inherent property of the viral glycoproteins and not a reduction in the level of cellular proteases. As expected, viruses lacking cleaved HN or F proteins failed to successfully initiate infection. This was due in part to a reduced rate of adsorption or penetration.[139,140] If, however, the virions were treated with trypsin, elevated levels of infectivity and virus multiplication were observed.

Polykaryocytosis is the principal form of cytopathic effect produced by virulent paramyxoviruses grown in cell culture. A good correlation between the ability of paramyxovirus strains to induce polykaryocytosis in culture and to induce pathological effects in animals has been established.[140,141] Two forms of polykaryocytosis have been described.[142] The first, called ''fusion from without'', occurs in the absence of virus replication. Fusion from without can be induced by noninfectious virus and occurred within 2 to 3 hr p.i. at high multiplicities. The second form of polykaryocytosis, called ''fusion from within'', was detected in cells infected at low multiplicities and required virus gene expression.

Paramyxoviruses have been useful in elucidating the mechanisms involved in virus-induced membrane damage because virus-induced cell changes (e.g., fusion from without) can be monitored quickly after virus infection. The initial event after the addition of Sendai virus to cells is the rapid depolarization of the cell membrane.[143] This is sequentially followed by changes in the intracellular ion concentrations that result in cell swelling.[144] As progressively larger lesions develop in the cell membrane, the cells exhibit enhanced permeability to phosphorylated intermediates of metabolism and proteins.[19,144] These immediate permeability changes induced by Sendai virus required the fusion of the virus and cellular membranes[145] and resemble many of the characteristics of membrane injury induced by toxins and activated complement.[146] These changes, however, are transient and do not necessarily result in cell lysis.[23] Sendai virus-induced permeability changes may be prevented by calcium,[146,147] though calcium does not prevent the fusion of cellular and virus membranes.[147]

The induction of cell permeability changes by Sendai virus appears to require the fusion activity of the infecting virus particles. The leakage of nutrients and ions from cells is a secondary effect. Calcium can be used to protect cells from becoming permeabilized if added following the onset of fusion.[148] This may represent a common pathway in the onset of permeability changes by other viruses. If influenza virus is bound to cells at pH 5.3 where fusion occurs, the permeability changes can be blocked if calcium is added to the culture medium before the onset of the permeability changes.[148] It may be concluded that the fusion event is related to the induction of permeability changes through the attainment of a threshold level of membrane damage.

Cell fusion, however, is not the only factor involved in cell killing by paramyxoviruses. Certain plaque forming revertants of the NC mutants of NDV (defective in the P protein) lost the capacity to fuse cells, yet were virulent in chick embryos. The accumulation of lesions in the P protein affects the ability of the virus to spread.[132] Mutations in the HN also resulted in a reduction in the cytopathology and inhibition of cell protein synthesis by Sendai virus. This indicates that the native HN glycoprotein is an important factor in cell killing by paramyxoviruses.[138] Several virion proteins may, therefore, contribute to the cell killing by paramyxoviruses.

V. CELL KILLING BY RETROVIRUSES

A. Virion Structure and Strategy of Replication

Retroviruses are a diverse group of pathogens that have been found in nearly every vertebrate species, including man.[149] Retroviruses are composed of an icosohedral nucleocapsid consisting of four to five proteins that contain the group (species) specific antigens.

The virions are enveloped and contain at least two glycoproteins that provide the type specificities. A number of enzymatic activities is associated with the virions: a protease activity involved in the processing of virion proteins and an RNA-dependent DNA polymerase (reverse transcriptase) and endonuclease activity.

The genome of retroviruses is a single-stranded RNA of positive polarity that is capped and polyadenylated.[150] Virions contain two copies of the RNA genome joined by sequence homologies at the 5′ ends. The genomes of replication-competent retroviruses contain at least three genes: *gag* — which encodes the group-specific nucleocapsid proteins; *pol* — which encodes the virion-associated polymerase; and *env* — which encodes the envelope proteins. Upon infection, the virion RNA is transcribed into a linear double-stranded DNA molecule (provirus) in the cytoplasm of the host cell.[151] The linear DNAs are circularized and transported into the nucleus where they are integrated into the host chromosomal DNA. During reverse transcription and circularization, unique and repeated sequences from both the 5′ and 3′ regions of the genomic RNA are joined and duplicated to generate long terminal repeats (LTRs). LTRs are found at both ends of the integrated provirus. The LTRs of many retroviruses contain strong promoters and enhancer sequences. Thus, the integrated proviruses may be transcribed using host RNA polymerase. The *gag* and *pol* genes are transcribed into a single mRNA species whose size corresponds to a genomic subunit, and the *env* gene is transcribed into a subgenomic mRNA. Most of the acutely transforming retroviruses are replication-defective. Replication-defective retroviruses have deleted a portion of the genes involved in replication and contain a cell-derived sequence.[150] These host-derived sequences, referred to as viral oncogenes, permit these viruses to transform cells or cause tumors in animals.[152] The expression of certain viral oncogenes (*abl, mos, src*) may be cytotoxic.[153-155] However, the mechanisms, by which the expression of various viral oncogenes contributes to cell death, have not been investigated.

B. Mechanisms of Cell Killing by Retroviruses

The replication strategy of retroviruses is ideally suited for formation of persistent infections. In fact, all vertebrate species are known to harbor retrovirus-like elements in their germline DNA. Nevertheless, infections by some replication-competent retroviruses lead to the development of extensive CPE and cell killing both in cultured cells and in vivo. The mechanisms employed by replication-competent retroviruses to kill cells have been most extensively studied in the avian reticuloendotheliosis virus (RE) system and the avian leukosis virus (ALV) system. The replication-competent members of the reticuloendotheliosis virus group replicate in and establish persistent infections in avian fibroblast cultures.[156,157] The initial (acute) phase of infection by these nontransforming RE viruses is characterized by the development of an extensive CPE.[157,158] The development of CPE in cultured fibroblasts during the acute phase may take place in two temporally displaced rounds. It has been suggested that the second round of cell killing is due to the spread of progeny virus.[159] The cells that survive both rounds of the acute phase proliferate and form a persistently infected culture that is morphologically indistinguishable from uninfected fibroblast cultures.[156,157] The acute and chronic phases of infection by RE viruses are not restricted to avian fibroblasts since lymphoid cells are also sensitive to cell killing.[160]

Cells infected with replication-competent members of the ALV group also form persistently infected cultures. Development of the acute phase, characterized by extensive cytopathic effects, is associated only with members of the subgroups B, D, and F of the ALV group.[161,162] As noted, subgroup specificies of retroviruses reside primarily in the envelope determinants, suggesting that virus-specified envelope glycoproteins are involved in cell killing by ALV.

During the acute phase, RE provirus DNA sequences are integrated at multiple sites in chromosomal DNA, and large amounts of unintegrated (mostly linear) proviral DNA transiently accumulate.[163,164] Cells chronically infected with these viruses contain an average of one copy of infectious integrated viral DNA, and the level of unintegrated linear proviral DNA is 50- to 100-fold lower than that detected in acutely infected cells.[163-165] Based on these observations, Keshet and Temin[164] suggested that cell death during the acute phase is a consequence of the toxic effect of the large amounts of unintegrated proviral DNA.

Weller et al.[159] have also found that detached, but not yet lysed, cells infected with Rous associated virus 2 (RAV-2, ALV subgroup B) contained substantially more unintegrated linear proviral DNA than the attached cells that survive to form the persistently-infected culture. Furthermore, during acute infection, there was approximately 40-fold more unintegrated linear DNA in RAV-2-infected cells than in cells infected with RAV-1, a virus which fails to induce a CPE (ALV, subgroup A). These results are consistent with the hypothesis that the transient accumulation of linear proviral DNA during infection by lytic retroviruses is responsible for cell killing.

The accumulation of linear DNA may not be the only mechanism of cell killing by retroviruses. After infection of CEF cultures with replication-competent RE viruses, a CPE develops within the first 24 hr.[157,160] Substantial accumulation of linear proviral DNA does not occur until 48 to 72 hr p.i. Furthermore, UV-inactivated RE viruses that were incapable of transcribing viral RNA into proviral DNA could induce an acute cytopathic effect and inhibit protein synthesis as early as 1 hr after high multiplicity infection.[160] The induction of CPE by RE viruses is also multiplicity-dependent.[157] Therefore, a virion component may participate in the development of the acute phase of infection.

Replication-competent retroviruses induce a number of symptoms in infected animals that may be related to their ability to kill cells. Of great contemporary interest is the ability of the human replication-competent retroviruses (designated human T-cell lymphotrophic virus III/lymphadnopathy associated virus [HTLV-III/LAV]) to induce extensive CPE.[166] Killing of helper T-lymphocytes by HTLV-III/LAV may underlie the ability of this virus to induce the acquired immune deficiency syndrome (AIDS). As in the ALV and RE virus systems, the ability to kill cells correlates with the ability to induce immunosuppression.[167,168]

VI. CONCLUSIONS

The subordination of host macromolecular synthesis rapidly follows infection of cells with lytic RNA viruses and is a key determinant in cell killing. Inhibition of cellular protein synthesis may require limited viral transcription. Viral mRNAs may, therefore, directly or indirectly compete with cellular mRNAs for protein synthesis initiation. Specific viral functions (''P functions'') have been proposed to facilitate the efficient translation of viral mRNA. One possible target of these P products is initiation factors, some of which may be required for the synthesis of host but not viral proteins. Viral mRNAs also possess features that ensure their efficient translation under conditions where cellular translation is inhibited. Damage to the cell membrane and cytoskeleton has been proposed as a factor that inhibit cellular translation. The accumulation of virion polypeptides may damage these cellular structures and, therefore, inhibits host translation. Lytic viruses may, therefore, utilize multiple mechanisms to inhibit cellular macromolecular synthesis.

Although limited viral transcription inhibits host macromolecular synthesis and eventually results in cell death, the rapid killing of cells during permissive virus infections probably results from the accumulation of virus structural components. Nonreplicating and DI virus particles are also cytotoxic at high multiplicities, indicating that virus structural polypeptides are inherently toxic. Viral membrane and nucleocapsid proteins contain many activities required for cell penetration and/or virus assembly, and in high doses may affect the activities of multiple cellular components. A single viral gene product (e.g., the rabies G protein) may inhibit several cellular processes required for cell survival.

It is likely that several mechanisms account for cell killing, even within the same family of enveloped viruses. It is noteworthy that cell killing by lytic viruses occurs at the time of or after the release of the majority of the progeny virions. By contrast, persistently infected cells induced by defective or attenuated viruses that may harbor several genetic alterations are generally not killed as a consequence of infection.

The advantage of cell killing in the replication strategy of lytic RNA viruses is not understood. Undoubtedly, competition of the virus for host components necessary for its replication ensures the production of large levels of progeny virions within a relatively short period of time. Rapid virus multiplication may be a prerequisite for escaping the immune surveillance mechanisms of the host organism.

ACKNOWLEDGMENTS

The work was supported by Public Health Service grants CA 26169 and CA 33192 from the National Cancer Institute to H. R. Bose, Jr., and by a grant from the Edward G. Schlieder Educational Foundation to R. F. Garry. E. T. Ulug is a postdoctoral fellow of the Robert A. Welch Foundation (F-849).

REFERENCES

1. **Lodish, H. F. and Porter, M.,** Translational control of protein synthesis after infection by vesicular stomatitis virus, *J. Virol.*, 36, 719, 1980.
2. **Lodish, H. F. and Porter, M.,** Vesicular stomatitis virus mRNA and inhibition of translation of cellular mRNA — Is there a P function in vesicular stomatitis virus?, *J. Virol.*, 38, 504, 1981.
3. **Jaye, M. C., Godchaux, W., III, and Lucas-Lenard, J.,** Further studies on the inhibition of cellular protein synthesis by vesicular stomatitis virus, *Virology*, 116, 148, 1982.
4. **Carrasco, L. and Smith, A. E.,** Sodium ions and the shut-off of host cell protein synthesis by picornavirus, *Nature (London)*, 264, 807, 1976.
5. **Ramabhadran, T. V. and Thach, R. E.,** Translational elongation rate changes in encephalomyocarditis virus-infected and interferon-treated cells, *J. Virol.*, 39, 573, 1981.
6. **Garry, R. F., Bishop, J. M., Parker, S., Westbrook, K., Lewis, G., and Waite, M. R. F.,** Na^+ and K^+ concentrations and the regulation of protein synthesis in Sindbis virus-infected chick cells, *Virology*, 96, 108, 1979.
7. **Nuss, D. L., Oppermann, H. O., and Koch, G.,** Selective blockage of initiation of host protein synthesis in RNA-virus infected cells, *Proc. Natl. Acad. Sci. U.S.A.*, 72, 1258, 1975.
8. **Nuss, D. L. and Koch, G.,** Differential inhibition of vesicular stomatitis virus polypeptide synthesis by hypertonic initiation block, *J. Virol.*, 17, 283, 1976.
9. **Bishop, J. M., Maldonado, R. L., Garry, R. F., Allen, P. A., Bose, H. R., and Waite, M. R. F.,** Effect of medium of lowered NaCl concentration on virus release and protein synthesis in cells infected with reticuloendotheliosis virus, *J. Virol.*, 17, 446, 1976.
10. **Hackstadt, T. and Mallavia, L. P.,** Na^+ and K^+ transport in herpes simplex virus-infected cells, *J. Gen. Virol.*, 60, 199, 1982.
11. **Garry, R. F., Ulug, E. T., and Bose, H. R., Jr.,** Membrane-mediated alterations of intracellular Na^+ and K^+ on lytic virus-infected and retrovirus-transformed cells, *Biosci. Rep.*, 2, 617, 1982.
12. **Marvaldi, J. L., Lucas-Lenard, J., Sekellick, M. J., and Marcus, P. I.,** Cell killing by viruses *IV:* cell killing and protein synthesis inhibition by vesicular stomatitis virus require the same gene functions. *Virology*, 79, 267, 1977.
13. **Cervera, M., Dreyfuss, G., and Penman, S.,** Messenger RNA is translated when associated with the cytoskeletal framework in normal and VSV-infected HeLa cells, *Cell*, 23, 113, 1981.
14. **Saborio, J. L., Pong, S. S., and Koch, G.,** Selective and reversible inhibition of initiation of protein synthesis in mammalian cells, *J. Mol. Biol.*, 85, 195, 1974.
15. **Groudine, M. and Weintraub, H.,** Propagation of globin DNAase I-hypersensitive sites in absence of factors required for induction: a possible mechanism for determination, *Cell*, 30, 131, 1982.
16. **Carrasco, L.,** The inhibition of cell functions after viral infection. A proposed general mechanism, *FEBS Lett.*, 75, 11, 1977.

17. **Garry, R. F.,** Na^+ and K^+ Concentrations and the Regulation of Macromolecular Synthesis in Lytic Virus Infected and Retrovirus Transformed Cells, Dissertation, The University of Texas at Austin, Austin, 1978.
18. **Ulug, E. T. and Bose, H. R., Jr.,** Effect of tunicamycin on the development of the cytopathic effect in Sindbis virus-infected avian fibroblasts, *Virology,* 143, 546, 1985.
19. **Pasternak, C. A. and Micklem, K. J.,** Virally induced alterations in cellular permeability: a basis of cellular and physiological damage?, *Biosci. Rep.,* 1, 431, 1981.
20. **Glynn, I. M. and Karlish, S. J. D.,** The sodium pump, *Annu. Rev. Physiol.,* 37, 13, 1975.
21. **Kaplan, J. G.,** Membrane cation transport and the control of proliferation of mammalian cells, *Annu. Rev. Physiol.,* 40, 19, 1978.
22. **Carrasco, L.,** Membrane leakiness after viral infection and a new approach to the development of antiviral agents, *Nature (London),* 272, 694, 1978.
23. **Impraim, C. C., Foster, K. A., Micklem, K. J., and Pasternak, C. A.,** Nature of virally mediated changes in membrane permeability to small molecules, *Biochem. J.,* 186, 847, 1980.
24. **Ulug, E. T., Garry, R. F., Waite, M. R. F., and Bose, H. R., Jr.,** Alterations in monovalent cation transport in Sindbis virus-infected chick cells, *Virology,* 132, 118, 1984.
25. **Lentz, T. L., Wilson, P. T., Hawrot, E., and Apeicher, D. W.,** Amino acid sequence similarity between rabies virus glycoprotein and snake venom curaremimetic neurotoxins, *Science,* 226, 847, 1984.
26. **Genty, N. and Bussereau, F.,** Is cytoskeleton involved in vesicular stomatitis virus reproduction?, *J. Virol.,* 34, 777, 1980.
27. **Pfefferkorn, E. R. and Shapiro, D.,** Reproduction of togaviruses, in *Comprehensive Virology,* Vol. 2, Fraenkel-Conrat, H. and Wagner, R. R., Eds., Plenum Press, New York, 1975, 171.
28. **Strauss, E. G., Rice, C., and Strauss, J. H.,** Complete sequence of the genomic RNA of Sindbis virus, *Virology,* 133, 92, 1984.
29. **Atkins, G. J., Sheahan, B. J., and Dimmock, N. J.,** Semliki Forest virus infection of mice: a model for genetic and molecular analysis of viral pathogenicity, *J. Gen. Virol.,* 66, 395, 1985.
30. **Atkins, G. J. and Sheahan, B. J.,** Semliki Forest Virus neurovirulence mutants have altered cytopathogenicity for central nervous system cells, *Infect. Immunol.,* 36, 333, 1982.
31. **Olmsted, R. A., Baric, R. S., Sawyer, B. A., and Johnston, R. E.,** Sindbis virus mutants selected for rapid growth in cell culture display attenuated virulence in animals, *Science,* 225, 424, 1984.
32. **Wengler, G.,** Effects of alphaviruses on host cell macromolecular synthesis, in *The Togaviruses,* Schlesinger, R. W., Ed., Academic Press, New York, 1980, 459.
33. **Kaariainen, L. and Rank, M.,** Inhibition of cell functions by RNA-virus infections, *Am. Rev. Microbiol.,* 38, 91, 1984.
34. **Simizu, B.,** Inhibition of host cell macromolecular synthesis following togavirus infection, in *Comprehensive Virology,* Vol. 19, Fraenkel-Conrat, H. and Wagner, R. R., Eds., Plenum Press, 1984, 465.
35. **Keranen, S. and Kaariainen, L.,** Proteins synthesized by Semliki Forest virus and its 16 temperature-sensitive mutants, *J. Virol.,* 16, 388, 1975.
36. **Atkins, G. J.,** Effect of infection with Sindbis virus and its temperature-sensitive mutants on cellular protein and DNA synthesis, *Virology,* 71, 593, 1976.
37. **Wengler, G. and Wengler, G.,** Protein synthesis in BHK-21 cells infected with Semliki Forest virus, *J. Virol.,* 17, 10, 1976.
38. **Barrett, A. D. T. and Dimmock, N. J.,** Differential effects of defective interfering Semliki Forest virus on cellular and virus polypeptide synthesis, *Virology,* 142, 59, 1985.
39. **Monroe, S. S. and Schlesinger, S.,** RNAs from two independently isolated DI particles of Sindbis virus contain a cellular tRNA sequence at their 5′ ends, *Proc. Natl. Acad. Sci. U.S.A.,* 80, 3279, 1983.
40. **Tuomi, K., Kaariainen, L., and Soderlund, H.,** Quantitation of Semliki Forest virus RNAs in infected cells using ^{32}P equilibrium labelling, *Nucl. Acids Res.,* 2, 555, 1975.
41. **Dubin, D. T., Stollar, V., Hsuchen, C. C., Timko, K., and Guild, G. M.,** Sindbis virus mRNA: the 5′ termini and methylation residues of 26 and 42S RNA, *Virology,* 77, 457, 1977.
42. **Van Steeg, H., Thomas, A., Verbeek, S., Kasperiatis, M., Voorma, H. O., and Beene, R.,** Shutoff of neuroblastoma cell protein synthesis by Semliki Forest virus: loss of ability of crude initiation factors to recognize early Semliki Forest Virus and host mRNAs, *J. Virol.,* 38, 728, 1981.
43. **Van Steeg, H., Kasperaitis, M., Voorma, H. O., and Beene, R.,** Infection of neuroblastoma cells by Semliki Forest Virus: the interference of viral capsid protein with the binding of host messenger RNAs into infection complexes is the cause of shut-off of host protein synthesis, *Eur. J. Biochem.,* 138, 473, 1984.
44. **Saliman, M. M. M., Van Hengten, H. A. A., Van Steeg, H., and Voorma, H. O.,** Effect of serum deprivation on the inhibition of initiation of protein synthesis in mouse neuroblastoma cells, *Biochim. Biophys. Acta,* 824, 16, 1985.
45. **McGuire, R. P., Eaton, B., and Kisilevsky, R.,** Ribosome topography in baby hamster kidney cells infected with Sindbis virus and vesicular stomatitis virus, *Biochim. Biophys. Acta,* 741, 258, 1983.
46. **Jalanko, A.,** Expression of Semliki Forest virus capsid protein from SV40 recombinant virus, *FEBS Lett.,* 186, 59, 1985.

47. **Lubin, M.,** Intracellular potassium and macromolecular synthesis in mammalian cells, *Nature (London),* 213, 451, 1967.
48. **Nagai, Y., Maeno, K., Iinuma, M., Yoshida, T., and Matsumoto, T.,** Inhibition of virus growth by ouabain: effect of ouabain on the growth of HVJ in chick embryo cells, *J. Virol.,* 9, 234, 1972.
49. **Garry, R. F., Westbrook, K., and Waite, M. R. F.,** Differential effects of ouabain on host and Sindbis virus-specific protein synthesis, *Virology,* 99, 179, 1979.
50. **Frothingham, T. E.,** Tissue culture applied to the study of Sindbis virus, *Am. J. Trop. Med. Hyg.,* 4, 863, 1955.
51. **Sefton, B. M. and Gaffney, B. J.,** Effect of the viral proteins on the fluidity of the membrane lipids of Sindbis virus, *J. Mol. Biol.,* 90, 343, 1974.
52. **Moore, N. R., Barenholz, Y., and Wagner, R. R.,** Microviscosity of togavirus membranes studied by fluorescence depolarization: influence of envelope proteins and the host cell, *J. Virol.,* 19, 126, 1976.
53. **Nozawa, C. M. and Apostolov, K.,** Association of the cytopathic effect of Sindbis virus with increased fatty acid saturation, *J. Gen. Virol.,* 59, 219, 1982.
54. **Carrasco, L.,** Modification of membrane permeability induced by animal viruses early in infection, *Virology,* 113, 623, 1981.
55. **Gray, M. A., Austin, S. A., Clemens, M. J., Rodrigues, L., and Pasternak, C. A.,** Protein synthesis in Semliki Forest virus-infected cells is not controlled by permeability changes, *J. Gen. Virol.,* 64, 2631, 1983.
56. **Cassels, A. C. and Burke, D. C.,** Changes in the constitutive enzymes of chick cells following infection with SFV, *J. Gen. Virol.,* 18, 135, 1973.
57. **Garry, R. F., Ulug, E. T., and Bose, H. R.,** Induction of stress proteins in Sindbis virus and vesicular stomatitis virus-infected cells, *Virology,* 129, 319, 1983.
58. **Gray, M. A., Micklem, K. J., Brown, F., and Pasternak, C. A.,** Effect of vesicular stomatitis virus and Semliki Forest virus on uptake of nutrients and intracellular ion concentration, *J. Gen. Virol.,* 64, 1449, 1983.
59. **Ulug, E. T.,** unpublished data, 1985.
60. **Whitehead, F. W., Trip, E., and Vance, D. E.,** Semliki Forest virus does not inhibit phosphatidycholine biosynthesis in BHK-21 cells, *Can. J. Biochem.,* 59, 38, 1981.
61. **Koizumi, S., Simizu, B., Hashimoto, K., Oya, A., and Yamada, M.,** Inhibition of DNA synthesis in BHK cells infected with western equine encephalitis virus. I. Isolation of an inhibitory factor of cellular DNA polymerase activity, *Virology,* 94, 314, 1979.
62. **Koizumi, S., Simizu, B., Ishida, I., Oya, A., and Yamada, M.,** Inhibition of DNA synthesis in BHK cells infected with western equine encephalitis virus. II. Properties of the inhibitory factor of DNA polymerase induced in infected cells, *Virology,* 98, 439, 1979.
63. **Ishida, I., Simizu, B., Koizumi, A., Oya, A., and Yamada, M.,** Nucleotide triphosphate phosphohydrolase produced in BHK cells infected with western equine encephalitis virus is probably associated with 82K dalton nonstructural polypeptide, *Virology,* 108, 13, 1981.
64. **Wagner, R. R.,** Reproduction of rhabdoviruses, in *Comprehensive Virology,* Vol. 2, Fraenkel-Conrat, H. and Wagner, R. R., Eds., Plenum Press, New York, 1975, 1.
65. **Wagner, R. R., Snyder, R. M., and Yamazaki, S.,** Proteins of vesicular stomatitis virus: kinetics and cellular sites of synthesis, *J. Virol.,* 5, 548, 1970.
66. **Ball, L. A. and Wertz, G. W.,** VSV RNA synthesis: how can you be positive?, *Cell,* 26, 143, 1981.
67. **McGeach, F. J.,** Structure of the gene N: gene NS intercistronic junction in the genome of vesicular stomatitis virus, *Cell,* 17, 673, 1979.
68. **Emerson, S. U.,** Reconstitution studies detect a single polymerase entry site on the vesicular stomatitis virus genome, *Cell,* 31, 635, 1982.
69. **Abraham, G. and Banerjee, A. K.,** Sequential translation of the genes of vesicular stomatitis virus, *Proc. Natl. Acad. Sci. U.S.A.,* 73, 1504, 1976.
70. **Ball, L. A. and White, C. N.,** Order of transcription of genes of vesicular stomatitis virus, *Proc. Natl. Acad. Sci. U.S.A.,* 73, 442, 1976.
71. **Wagner, R. R., Thomas, J. R., and McGowan, J. J.,** Rhabdovirus cytopathology: effects on cellular macromolecular synthesis, in *Comprehensive Virology,* Vol. 19, Fraenkel-Conrat, H. and Wagner, R. R., Eds., Plenum Press, New York, 1984, 223.
72. **Wertz, G. W. and Youngner, J. S.,** Interferon production and inhibition of host synthesis in cells infected with vesicular stomatitis virus, *J. Virol.,* 6, 476, 1970.
73. **Wertz, G. W. and Youngner, J. S.,** Inhibition of protein synthesis in L cells infected with vesicular stomatitis virus, *J. Virol.,* 9, 85, 1972.
74. **Bablanian, R.,** Structural and functional alterations in cultured cells infected with cytocidal viruses, *Prog. Med. Virol.,* 19, 40, 1975.
75. **Baxt, B. and Bablanian, R.,** Mechanisms of vesicular stomatitis virus-induced cytopathic effects. I. Early morphologic changes induced by infectious and defective-interfering particles, *Virology,* 72, 370, 1976.

76. **Baxt, B. and Bablanian, R.,** Mechanisms of vesicular stomatitis virus-induced cytopathic effects. II. Inhibition of macromolecular synthesis induced by infectious and defective-interfering particles, *Virology,* 72, 383, 1976.
77. **Otto, M. J. and Lucas-Lenard, J.,** The influence of the host cell on the inhibition of virus protein synthesis in cells infected with vesicular stomatitis virus and mengovirus, *J. Gen. Virol.,* 50, 29, 1980.
78. **McAllister, P. E. and Wagner, R. R.,** Differential inhibition of host protein synthesis in cells infected with RNA-temperature-sensitive mutants of vesicular stomatitis virus, *J. Virol.,* 18, 550, 1976.
79. **Obijeski, J. F. and Simpson, R. W.,** Conditional lethal mutants of vesicular stomatitis virus. II. Synthesis of virus-specific polypeptides in nonpermissive cells infected with RNA-host-restricted mutants, *Virology,* 57, 369, 1974.
80. **Marcus, P. I. and Sekellick, M. J.,** Cell killing by viruses. I. Comparison of cell-killing, plaque-forming, and defective-interfering particles of vesicular stomatitis virus, *Virology,* 57, 321, 1974.
81. **Marcus, P. I. and Sekellick, M. J.,** Cell killing by viruses. II. Cell killing by vesicular stomatitis virus: a requirement for virion-derived transcription, *Virology,* 63, 176, 1975.
82. **Marcus, P. I. and Sekellick, M. J.,** Cell killing by viruses. III. The interferon system and inhibition of cell-killing by vesicular stomatitis virus, *Virology,* 63, 378, 1976.
83. **Marcus, P. I., Sekellick, M. J., Johnson, L. D., and Lazzarini, R. A.,** Cell killing by viruses. V. Transcribing defective-interfering particles of vesicular stomatitis virus function as cell-killing particles, *Virology,* 82, 242, 1977.
84. **Marvaldi, J., Sekellick, M. J., Marcus, P. I., and Lucas-Lenard, J.,** Inhibition of mouse cell protein synthesis by ultraviolet-irradiated vesicular stomatitis virus requires viral transcription, *Virology,* 84, 127, 1978.
85. **Stanners, C. P., Francoeur, A. M., and Lam, T.,** Analysis of VSV mutant with attenuated cytopathogenicity in viral function, P, for inhibition of protein synthesis, *Cell,* 11, 272, 1977.
86. **Frey, T. K. and Youngner, J. S.,** Further studies on the RNA synthesis phenotype selected during persistent infection with vesicular stomatitis virus, *Virology,* 136, 211, 1984.
87. **Rosen, C. A., Ennis, H. L., and Cohen, P. S.,** Translational control of vesicular stomatitis virus protein synthesis: isolation of an mRNA-sequestering particle, *J. Virol.,* 44, 932, 1982.
88. **Schnitzlein, W. M., O'Banion, M. K., Poirot, M. K., and Reichmann, M. E.,** Effect of intracellular vesicular stomatitis virus mRNA concentration on the inhibition of host cell protein synthesis, *J. Virol.,* 45, 206, 1983.
89. **Davis, N. L. and Wertz, G. W.,** A VSV mutant synthesizes a large excess of functional mRNA but produces less viral protein than its wild-type parent, *Virology,* 103, 21, 1980.
90. **Dunigan, D. D. and Lucas-Lenard, J. M.,** Two transcription products of the vesicular stomatitis virus genome may control L-cell protein synthesis, *J. Virol.,* 45, 618, 1983.
91. **Nuss, D. L. and Koch, G.,** Translation of individual host mRNA's in MPC-11 cells is differentially suppressed after infection by vesicular stomatitis virus, *J. Virol.,* 19, 572, 1976.
92. **Madore, H. P. and England, J. M.,** Selective suppression of cellular protein synthesis in baby hamster kidney (BHK-21) cells infected with rabies virus, *J. Virol.,* 16, 1351, 1975.
93. **Rose, J. K.,** Complete intergenic and flanking gene sequences from the genome of vesicular stomatitis virus, *Cell,* 19, 415, 1980.
94. **Nair, C. N.,** Na and K changes in animal virus-infected HeLa cells, *J. Gen. Virol.,* 65, 1135, 1984.
95. **Garry, R. F. and Bostick, D. A.,** unpublished data, 1985.
96. **Francoeur, A. M. and Stanners, C. P.,** Evidence against the role of K^+ in the shut-off of protein synthesis by vesicular stomatitis virus, *J. Gen. Virol.,* 39, 551, 1978.
97. **Centrella, M. and Lucas-Lenard, J.,** Regulation of protein synthesis in vesicular stomatitis virus-infected mouse L-929 cells by decreased protein synthesis initiation factor 2 activity, *J. Virol.,* 41, 781, 1982.
98. **Jaye, M. C., Godchaux, W., and Lucas-Lenard, J.,** Further studies on the inhibition of cellular protein synthesis by vesicular stomatitis virus, *Virology,* 116, 148, 1982.
99. **Jaye, M. C., Wu, F.-S., and Lucas-Lenard, J. M.,** Inhibition of synthesis of ribosomal protein and ribosome assembly after infection of L cells with vesicular stomatitis virus, *Biochim. Biophys. Acta,* 606, 1, 1980.
100. **Thomas, J. R. and Wagner, R. R.,** Evidence that vesicular stomatitis virus produces double-stranded RNA that inhibits protein synthesis in a reticulocyte lysate, *J. Virol.,* 44, 189, 1982.
101. **Nishioka, Y. and Silverstein, S.,** Alteration in the protein synthetic apparatus of Friend erythroleukemia cells infected with vesicular stomatitis virus or herpes simplex virus, *J. Virol.,* 25, 422, 1978.
102. **Thomas, J. R. and Wagner, R. R.,** Inhibition of translation in lysates of mouse L cells infected with vesicular stomatitis virus: presence of a defective ribosome-associated factor, *Biochemistry,* 22, 1540, 1983.
103. **Kaempfer, R.,** Regulation of eukaryotic translation, in *Comprehensive Virology,* Vol. 19, Fraenkel-Conrat, H. and Wagner, R. R., Eds., Plenum Press, New York, 1984, 99.
104. **Baglioni, C., Lenz, J. R., Maroney, P. A., and Weber, L. A.,** Effect of double-stranded RNA associated with viral mRNA on *in vitro* protein synthesis, *Biochemistry,* 17, 3257, 1978.

105. **Marvaldi, J. and Lucas-Lenard, J.,** Differences in the ribosomal protein gel profile after infection of L cells with wild-type or temperature-sensitive mutants of vesicular stomatitis virus, *Biochemistry,* 16, 4320, 1977.
106. **Rutter, G. and Mannweiler, K.,** Alterations of actin containing structures in BHK-21 cells infected with Newcastle disease virus and vesicular stomatitis virus, *J. Gen. Virol.,* 37, 233, 1977.
107. **Wilcox, D. K., Whitaker-Dowling, P. A., Youngner, J. S., and Widnell, C. C.,** Rapid inhibition of pinocytosis in baby hamster kidney (BHK-21) cells following infection with vesicular stomatitis virus, *J. Cell Biol.,* 97, 1444, 1983.
108. **Genty, N.,** Analysis of uridine incorporation in chicken embryo cells infected with vesicular stomatitis virus and its temperature-sensitive mutants: uridine transport, *J. Virol.,* 15, 8, 1975.
109. **Garry, R. F. and Waite, M. R. F.,** Na^+ and K^+ concentrations and the regulation of the interferon system in chick cells, *Virology,* 96, 121, 1979.
110. **Lopez-Vancell, R., Beaty, G., Stefanie, E., Rodriguez-Boulin, E. E., and Cerelin, M.,** Changes in paracellular and cellular ionic permeabilities of monolayers of MDCK cells infected with influenza or vesicular stomatitis viruses, *J. Membr. Biol.,* 81, 171, 1984.
111. **Young, J. D.-E. and Lucas-Lenard, J.,** Interaction of enveloped viruses with planar bilayer membranes: observations on Sendai, influenza, vesicular stomatitis and Semliki Forest virus, *Virology,* 128, 186, 1983.
112. **McSharry, J. J. and Choppen, P. W.,** Biological properties of the vesicular stomatitis virus glycoprotein, *Virology,* 84, 172, 1978.
113. **Florkiewicz, R. A. and Rose, J. K.,** A cell line expressing vesicular stomatitis virus glycoprotein fuses at low pH, *Science,* 225, 721, 1984.
114. **Storey, D. G. and Kang, C. Y.,** Vesicular stomatitis virus-infected cells fuse when the intracellular pool of functional M protein is reduced in the presence of G protein, *J. Virol.,* 53, 374, 1985.
115. **Schlegel, R. and Wade, M.,** Biologically active peptides of the vesicular stomatitis glycoprotein, *J. Virol.,* 53, 319, 1985.
116. **Rose, J. K., Doolittle, R. F., Anilonis, A., Curtis, P. J., and Wunner, W. H.,** Homology between the glycoproteins of vesicular stomatitis and rabies virus, *J. Virol.,* 43, 361, 1982.
117. **Burrage, T. G., Tignor, G. H., and Smith, A. L.,** Rabies virus binding at neuromuscular junctions, *Virus Res.,* 2, 273, 1985.
118. **Stollar, V.,** Togaviruses in cultured arthropod cells, in *The Togaviruses,* Schlesinger, R. W., Ed., Academic Press, New York, 1980, 583.
119. **Stevens, T. M.,** Arbovirus replication in mosquito cell lines, (Singh) grown in monolayer or suspension culture, *Proc. Soc. Exp. Biol. Med.,* 134, 356, 1970.
120. **Riedel, B. and Brown, D. T.,** Novel antiviral activity found in the media of Sindbis virus persistently infected mosquito *(Aedes albopictus)* cell cultures, *J. Virol.,* 29, 51, 1979.
121. **Sarver, N. and Stollar, V.,** Sindbis virus-induced cytopathic effect in clones of *Aedes albopictus* (Singh) cells, *Virology,* 80, 390, 1977.
122. **Gillies, S. and Stollar, V.,** Conditions necessary for inhibition of protein synthesis and production of cytopathic effect in *Aedes albopictus* cells infected with vesicular stomatitis virus, *Mol. Cell. Biol.,* 2, 66, 1982.
123. **Tooker, P. and Kennedy, S. I. T.,** Semliki Forest virus multiplication in clones of *Aedes albopictus* cells, *J. Virol.,* 37, 589, 1981.
124. **Gillies, S. and Stollar, V.,** Protein synthesis in lysates of *Aedes albopictus* cells infected with vesicular stomatitis virus, *Mol. Cell. Biol.,* 2, 1174, 1982.
125. **Bratt, M. A. and Hightower, L. E.,** Genetics and paragenetic phenomena of paramyxoviruses, in *Comprehensive Virology,* Vol. 9, Fraenkel-Conrat, H. and Wagner, R. R., Eds., Plenum Press, New York, 1977, 457.
126. **Choppin, P. W. and Compans, R. W.,** Reproduction of paramyxoviruses, in *Comprehensive Virology,* Vol. 4, Fraenkel-Conrat, H. and Wagner, R. R., Eds., Plenum Press, New York, 1975, 95.
127. **Choppin, P. W. and Scheid, A.,** The role of viral glycoproteins in adsorption, penetration and pathogenicity of viruses, *Rev. Infect. Dis.,* 3, 40, 1980.
128. **Nagai, Y., Klenk, H-D., and Rott, R.,** Proteolytic cleavage of the viral glycoproteins and its significance for the virulence of Newcastle disease virus, *Virology,* 72, 494, 1976.
129. **Collins, P. L., Hightower, L. E., and Ball, L. A.,** Transcriptional map for Newcastle disease virus, *J. Virol.,* 35, 682, 1980.
130. **Collins, P. L., Wertz, G. W., Ball, L. A., and Hightower, L. E.,** Coding assignments of the five smaller messenger RNAs of Newcastle disease virus (NDV), *J. Virol.,* 43, 1024, 1982.
131. **Glazier, K., Raghow, R., and Kingsbury, D. W.,** Regulation of Sendai virus transcription: evidence for a single promoter *in vivo, J. Virol.,* 21, 863, 1977.
132. **Madansky, C. H. and Bratt, M. A.,** Relationships among virus spread, cytopathogenicity, and virulence as revealed by the noncytopathic mutants of Newcastle disease virus, *J. Virol.,* 40, 691, 1981.

133. **Madansky, C. H. and Bratt, M. A.,** Noncytopathic mutants of Newcastle Disease virus are defective in virus-specific RNA synthesis, *J. Virol.,* 37, 317, 1981.
134. **Hightower, L. E. and Bratt, M. A.,** Protein synthesis in Newcastle disease virus-infected chicken embryo cells, *J. Virol.,* 13, 788, 1974.
135. **Hightower, L. E. and Bratt, M. A.,** Protein metabolism during the steady state of Newcastle disease virus infection. I. Kinetics of amino acid and protein accumulation, *J. Virol.,* 15, 696, 1975.
136. **Clinkscales, C. W. and Bratt, M. A.,** Inhibition of protein synthesis by Newcastle disease virus, in *Negative Strand Viruses and the Host Cell,* Mahy, B. W. J. and Barry, R. D., Eds., Academic Press, New York, 1978.
137. **Harper, D. R. and Samson, A. C. R.,** Recovery of host cell protein synthesis during infection with a temperature-sensitive mutant of Newcastle disease virus, *J. Gen. Virol.,* 65, 2049, 1984.
138. **Portner, A., Scroggs, R. A., Marx, P. A., and Kingsbury, D. W.,** A temperature-sensitive mutant of Sendai virus with an altered hemaglutinin-neuraminidase polypeptide: consequences for virus assembly and cytopathology, *Virology,* 67, 179, 1975.
139. **Klenk, H-D., Rott, R., Nagai, Y., and Berk, W.,** The pathogenicity of Newcastle disease virus as depending on the structure of the viral glycoproteins, in *Negative Strand Viruses and the Host Cell,* Mahy, B. W. J. and Barry, R. D., Eds., Academic Press, New York, 1978.
140. **Smith, G. W. and Hightower, L. E.,** Revertant analysis of a temperature-sensitive mutant of Newcastle disease virus with defective glycoproteins: implication of the fusion glycoprotein in cell killing and isolation of a neuraminidase-deficient hemagglutinating virus, *J. Virol.,* 42, 659, 1982.
141. **Reeve, P. and Poste, G.,** Studies on the cytopathogenicity of Newcastle disease virus: relation between virulence, polykaryocytosis, and plaque size, *J. Gen. Virol.,* 11, 14, 1971.
142. **Bratt, M. A. and Gallaher, W. R.,** Preliminary analysis of the requirements for fusion from within and fusion from without by Newcastle disease virus, *Proc. Natl. Acad. Sci. U.S.A.,* 64, 539, 1969.
143. **Bashford, C. L., Adler, G. M., Gray, M. A., Micklem, K. J., Taylor, C. C., Turck, P. J., and Pasternak, C. A.,** Oxonal dyes as monitors of membrane potential: the effect of viruses and toxins on the plasma membrane potential of animal cells in monolayer culture and in suspension, *J. Cell. Physiol.,* 123, 326, 1985.
144. **Bashford, C. L., Micklem, K. J., and Pasternak, C. A.,** Sequential onset of permeability changes in mouse ascites cells induced by Sendai virus, *Biochim. Biophys. Acta,* 814, 247, 1985.
145. **Wyke, A. M., Impraim, C. C., Knutton, S., and Pasternak, C. A.,** Components involved in virally mediated membrane fusion and permeability changes, *Biochem. J.,* 190, 625, 1980.
146. **Bashford, C. L., Adler, G. M., Patel, K., and Pasternak, C. A.,** Common action of certain viruses, toxins, and activated complement: pore formation and its prevention by extracellular Ca^{2+}, *Biosci. Rep.,* 4, 797, 1984.
147. **Micklem, K. J., Nyaruwe, A., Adler, G. M., and Pasternak, C. A.,** The effect of Ca^{2+} or virus-cell fusion and permeability changes, *Cell Calcium,* 5, 537, 1984.
148. **Patel, K. and Pasternak, C. A.,** Permeability changes elicited by influenza and Sendai viruses: separation of fusion and leakage by pH-jump experiments, *J. Gen. Virol.,* 66, 767, 1985.
149. **Teich, N.,** Taxonomy of retroviruses, in *RNA Tumor Viruses,* Weiss, R., Teich, N., Varmus, H., and Coffin, J., Eds., Cold Spring Harbor Laboratory, Cold Spring Harbor, New York, 1982, 25.
150. **Coffin, J.,** Structure of the retroviral genome, in *RNA Tumor Viruses,* Weiss, R., Teich, N., Varmus, H., and Coffin, J., Eds., Cold Spring Harbor Laboratory, Cold Spring Harbor, New York, 1982, 261.
151. **Varmus, H. and Swanstrom, R.,** Replication of retroviruses, in *RNA Tumor Viruses,* Weiss, R., Teich, N., Varmus, H., and Coffin, J., Eds., Cold Spring Harbor Laboratory, Cold Spring Harbor, New York, 1982, 369.
152. **Bishop, J. M. and Varmus, H.,** Functions and origins of retroviral transforming genes, in *RNA Tumor Viruses,* Weiss, R., Teich, N., Varmus, H., and Coffin, J., Eds., Cold Spring Harbor Laboratory, Cold Spring Harbor, New York, 1982, 999.
153. **Ziegler, S. F., Whitlock, C. A., Goff, S. P., Gifford, A., and Witte, O. N.,** Lethal effect of the Abelson murine leukemia virus transforming gene product, *Cell,* 27, 477, 1981.
154. **Papkoff, J., Verma, I. M., and Hunter, T.,** Detection of a transforming gene product in Moloney murine sarcoma virus-transformed cells, *Cell,* 29, 109, 1982.
155. **Tarpley, W. G. and Temin, H. M.,** The location of v-src in a retrovirus vector determines whether the virus is toxic or transforming, *Mol. Cell. Biol.,* 4, 2653, 1984.
156. **Temin, H. M. and Kassner, V. K.,** Replication of reticuloendotheliosis viruses in cell culture: chronic infection, *J. Gen. Virol.,* 27, 267, 1975.
157. **Hoelzer, J. D., Franklin, R. B., and Bose, H. R.,** Transformation by reticuloendotheliosis viruses: development of a focus assay and isolation of a non-transforming virus, *Virology,* 93, 20, 1979.
158. **Temin, H. M. and Kassner, V. K.,** Replication of reticuloendotheliosis viruses in cell culture: acute infection, *J. Virol.,* 13, 291, 1974.

159. **Weller, S. K., Joy, A. E., and Temin, H. M.,** Correlation between cell killing and massive second-round superinfection by members of some subgroups of avian leukosis virus, *J. Virol.,* 33, 494, 1980.
160. **Herzog, N. K., Garry, R. F., and Bose, H. R.,** unpublished data, 1985.
161. **Moscovici, C., Chi, D., Gazzolo, L., and Moscovici, M. G.,** A study of plaque formation with avian RNA tumor viruses, *Virology,* 73, 181, 1976.
162. **Kawai, S. and Hanafusa, H.,** Plaque assay for some strains of avian leukosis virus, *Virology,* 48, 126, 1972.
163. **Battula, N. and Temin, H. M.,** Sites of integration of infectious DNA of avian reticuloendotheliosis viruses in different avian cellular DNAs, *Cell,* 13, 387, 1978.
164. **Keshet, E. and Temin, H. M.,** Cell killing by spleen necrosis virus is correlated with a transient accumulation of spleen necrosis virus DNA, *J. Virol.,* 31, 376, 1979.
165. **Battula, N. and Temin, H. M.,** Infectious DNA of spleen necrosis viruses is integrated at a single site in the DNA of chronically infected chicken fibroblasts., *Proc. Natl. Acad. Sci. U.S.A.,* 74, 281, 1977.
166. **Popovic, M., Sarngagharan, M. G., Read, E., and Gallo, R. C.,** Detection, isolation and continuous production of cytopathic retroviruses (HTLV-III) from patients with AIDS and preAIDS, *Science,* 224, 497, 1984.
167. **Rup, B. J., Hoelzer, J. D., and Bose, H. R., Jr.,** Helper viruses associated with avian acute leukemia viruses inhibit the cellular immune response, *Virology,* 116, 61, 1982.
168. **Rup, B. J., Spence, J. L., Hoelzer, J. D., Lewis, R. B., Carpenter, C. R., Rubin, A. S., and Bose, H. R., Jr.,** Immunosuppression induced by avian reticuloendotheliosis virus: mechanism of induction of the suppressor cell, *J. Immunol.,* 13, 1362, 1979.

Chapter 6

THE REGULATION OF TRANSLATION IN PICORNAVIRUS-INFECTED CELLS

L. Carrasco and J. L. Castrillo

TABLE OF CONTENTS

I. INTRODUCTION

The regulation of translation in eukaryotic cells plays an important part in gene expression. Two types of regulation of translation occur: the unspecific regulation that occurs when translation of all cellular mRNAs increases or decreases after a given stimulus, and the specific regulation when protein synthesis directed by some mRNAs declines, whereas the translation of other mRNAs remains unaltered or is even enhanced after a given cellular event. There are many instances of both types of regulation of translation in eukaryotic cells. Well-known examples of the former type are the increase in protein synthesis observed after stimulation of oocytes, or the inhibition of translation in cultured cells after serum depletion. As examples of specific regulation, the heat-shock response in mammalian cells and the shut-off of protein synthesis by viruses are among the best documented. Perhaps the most thoroughly studied example of regulation of translation in eukaryotic cells has been the inhibition of host protein synthesis by picornaviruses.

Picornaviruses are among the best studied animal viruses in molecular terms,[1-3] but many fascinating aspects of their biology still remain unclear. One topic that has attracted the interest of many laboratories for more than 20 years is the interference of picornaviruses with host cell metabolism. This interference involves a number of biochemical alterations in the cell that occur after infection. One of them is the blockade of host protein synthesis in such a way that cellular mRNA translation is inhibited, whereas at the same time viral proteins are synthesized. Soon after host translation has been blocked, viral protein synthesis also declines and stops a few hours after the peak of viral translation has been reached. (Figure 1). Therefore, in picornavirus-infected cells two inhibitory effects on translation can be clearly distinguished: the first one on cellular mRNAs and the second on viral protein synthesis. Most work on the regulation of translation in picornavirus-infected cells has focused on the first event, which is commonly referred to as the shut-off of host protein synthesis. However, in order to understand the different aspects that regulate translation of host and viral mRNAs in virus-infected cells, both events should be considered. In this review we have not attempted to discuss comprehensively all of the work done on the regulation of protein synthesis in picornavirus-infected cells. Rather, we prefer to show a critical view of the present status of the field. Several excellent reviews on the shut-off of host translation by animal viruses have been written throughout the years.[3-7]

II. ESTABLISHED FINDINGS ON THE SHUT-OFF OF TRANSLATION BY PICORNAVIRUSES

Early studies on the replication of phages in bacteria gave an experimental framework to undertake the analysis of animal virus development in mammalian cells. Once techniques to cultivate animal cells and to grow viruses became available, studies on animal virus replication were potentiated. Initial experiments in this field were directed to demonstrate that viruses reproduce intracellularly and use the cell synthetic machinery to make their own components. The idea was that viruses could incorporate the synthesis of their macromolecules into the host's own system. Thus, cell and viral macromolecular synthesis would run in parallel. In fact, Ackermann et al.[8] reported that in poliovirus-infected HeLa cells, the synthesis of proteins proceeded at a constant linear rate until the 7th hr p.i. (post infection). When isotopically labeled precursors became routinely used in different laboratories, it became clear that infection with a number of animal viruses led to a profound inhibition of cellular macromolecular synthesis soon after infection.[9] Picornaviruses were among the first viruses studied in this regard,[10] and ever since they have been the most studied with respect to viral interference with host metabolism.

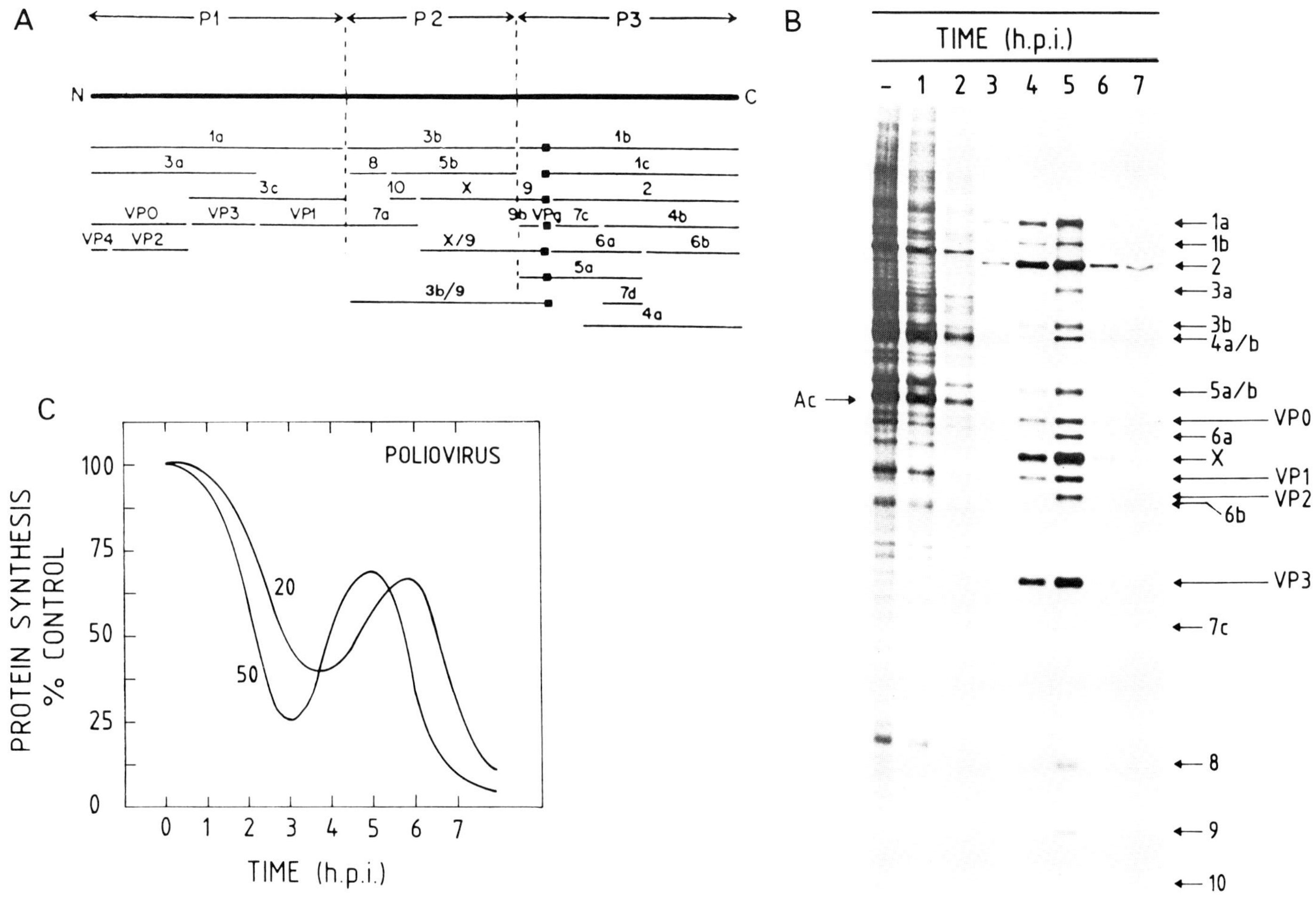

FIGURE 1. Kinetics of proteins synthesis in poliovirus-infected HeLa cells. (A) Processing map of poliovirus proteins. (B) Analysis by SDS-PAGE of the proteins synthesized in poliovirus-infected HeLa cells at various times after infection. The MOI was 50 PFU/cell. (C) Time course of protein synthesis at 20 and 50 PFU/cell. Protein synthesis was measured as (^{35}S)-methionine incorporation into TCA-precipitable material.

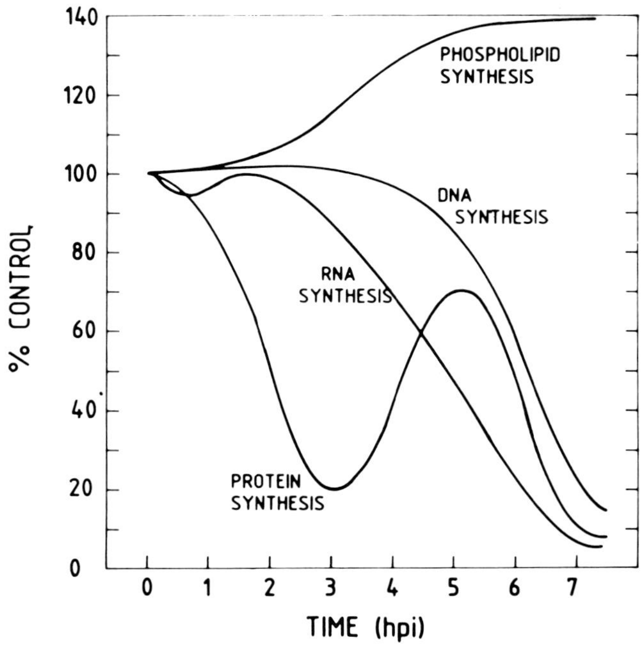

FIGURE 2. Idealized kinetics of protein, RNA, DNA, and phospholipid synthesis in poliovirus-infected HeLa-cells at a multiplicity of infection of about 50 PFU/cell.

A. A Number of Cell Functions Is Altered After Infection

The blockade of host translation by picornaviruses is only one of several cellular functions that is altered after infection (Figure 2). Cellular DNA synthesis, transcription, and energy production are also depressed, whereas the synthesis of lipids increases very much and the permeability barrier of the cell membrane is destroyed during infection. Some of these functions are primarily affected by viral replication, whereas the inhibition of cellular DNA synthesis is probably a consequence of the inhibition of cellular protein synthesis. This inhibition of translation is achieved earlier than inhibition of cellular RNA synthesis in poliovirus-infected HeLa cells,[11-13] whereas the contrary is true for mengovirus-infected L cells.[14,15] The inhibition of protein synthesis is not a consequence of transcription blockade because protein synthesis continues even in the presence of actinomycin D[16]

A large variation in the kinetics of inhibition of RNA synthesis by piconaviruses is observed, depending upon the virus/cell system analyzed and the conditions of infection used. For instance, mengovirus produces a drastic blockade of L929 cell RNA synthesis, whereas HeLa cells are affected to a much lower extent.[15,17,18] An inhibitor of the RNA polymerase appears in the cytoplasm of mengovirus-infected L cells.[19] Attempts to purify such an inhibitor from poliovirus-infected HeLa cells led to the identification of a lipid-containing compound that reduced cellular RNA synthesis in cell-free systems.[20] However, to our knowledge, these studies have not been further pursued.

Isolated nuclei from mengovirus-infected cells show a decrease in RNA polymerase activity.[14] This defect is associated with a preferential inhibition of polymerase II as compared to polymerases I and III.[21] Consequently, less RNA polymerase II molecules are engaged in chain elongation as infection proceeds.[22] However, when the RNA polymerases are purified from mengovirus-infected cells, they are as active as the enzymes obtained from control uninfected cells.[23,24] Similar results were found with encephalomyocarditis virus-infected plasmacytoma cells.[25] Cell-free extracts from poliovirus-infected cells inactive as regards to RNA polymerase II activity overcome the inhibition when supplemented with a crude preparation of transcription factors.[26] Thus the possibility exists that a factor involved in the initiation of transcription (specific for polymerase II) is destroyed after infection. In fact, poliovirus-coded proteins are present in the nuclei of poliovirus-infected cells.[27] However, infected cell extracts do not inhibit RNA synthesis in cell-free systems from uninfected cells.[26]

The cytoskeletal framework is also reorganized in poliovirus-infected HeLa cells.[27] This reorganization underlies, at least partially, the morphological changes induced by infection and occurs even when guanidine inhibits viral replication. The cell nucleus becomes deformed and a modified region of the cytoplasm occupies a central position where many vesicles are formed. This region contains intermediate filaments arranged in a unique pattern. When viral replication is allowed, the central region filaments become coated with dark staining material, which may be viral RNA.[28] Host cell mRNA, normally attached to the cytoskeletal framework, is released and replaced by viral polysomes. However, the connection between the dissociation of cellular mRNA from the cytoskeleton and the shut-off remains unclear, because attachment of mRNA to the cytoskeleton may not be the only requisite for translation.[29]

Lysosomes are also disrupted during infection,[30-32] but redistribution of lysosomal enzymes plays no part in cytopathic changes production.[33,34]

An increase in the synthesis of phospholipids is another alteration first studied in poliovirus-infected HeLa cells.[35] The synthesis of all phospholipids increases, lecithin and phosphatidyl ethanolamine synthesis being more than doubled 4 hr p.i.[36] The uptake of choline into trichloracetic acid insoluble material has been the most widely used test to analyze phospholipid synthesis in picornavirus-infected cells. Using this assay, an increase in phospholipid synthesis is observed from 2 to 3 hr p.i.[37] Inhibitors of protein synthesis block this stimulatory effect after infection. Guanidine, a known inhibitor of viral RNA synthesis, decreases, but does not suppress the enhanced uptake of choline after infection.[38,39] An increase in phospholipid synthesis is also observed with other precurosrs such as glycerol.[39]

The newly formed phospholipids are localized in the smooth endoplasmic reticulum, in newly formed cisternae, which proliferate in infected cells[40] and are in close association with viral RNA synthesis.[39,41] Viral transcription occurs exclusively in the so-called membrane-enclosed small bodies.[42,43] Of interest is the finding that the new cisternae also contain phospholipids of preformed membranes, suggesting a redistribution of these components between different membranes p.i. by a still unknown mechanism.[41] Within half an hour after poliovirus infection, the CTP level increases, and by 5 hr p.i., it reaches 300% of control levels. The total pools of choline, phosphocholine, CDP-choline, and diacylglycerol remain unchanged in infected cells, but the rate of the reaction catalyzed by CTP:phosphocholine cytidylyl-transferase is enhanced.[44] This is the rate limiting step in HeLa cells for the synthesis of phosphatidylcholine, and the rate of this reaction is determined by the CTP concentration.[44] Therefore, it was suggested that phospholipid synthesis increases after viral infection through an effect on CTP levels.

Changes in the function of membranes in picornavirus-infected cells also occur.[45] The possibility that modifications in membrane permeability to ions and metabolites may be directly involved in the virus-induced cytopathic effects and could play a part in the regulation of translation will be discussed later.

B. The Initiation of Cellular Protein Synthesis Is the Step Blocked After Viral Infection

Cellular polysomes disaggregate soon after infection and are replaced by polio-specific polyribosomes.[46] Therefore, it was inferred that ribosomes and other components of the protein synthesizing machinery remained active during infection since they must participate in the synthesis of viral proteins, but that cellular mRNAs were unable to associate with ribosomes.[47] This behavior of polysome run-off was in accordance with the idea that initiation

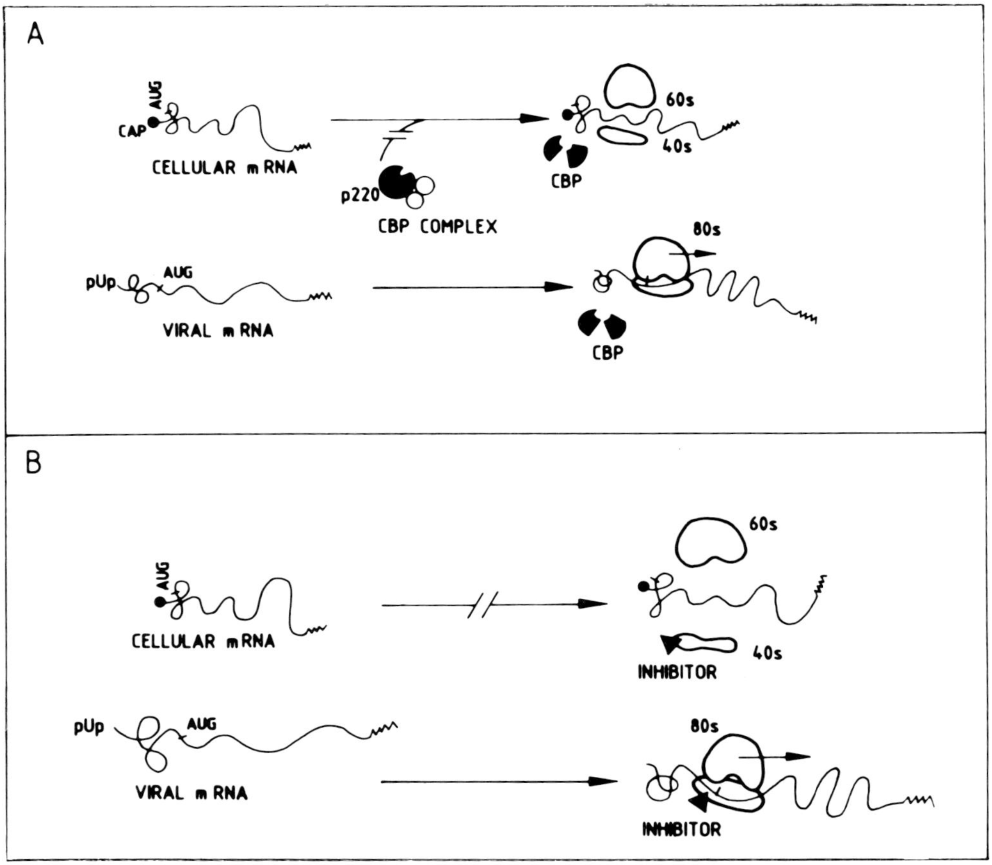

FIGURE 3. Two different models that have been proposed to account for the shut-off of protein synthesis in poliovirus-infected cells. (A) The proteolytic cleavage of p220 polypeptide that forms part of a CBP complex. (B) The generation of a selective inhibitor of cellular translation.

is the step blocked by poliovirus.[48] Treatment of cells with anysomycin, an inhibitor of peptide bond formation,[49] induces the accumulation of 80S initiation complexes resistant to dissociation by high salt. These complexes do not form after poliovirus infection, suggesting that the step blocked in the initiation of cellular protein synthesis is prior to the formation of 80S initiation complexes,[50] perhaps before the joining of the 60S subunit to the 40S initiation complex.[51] This would occur as a result of the inability of mRNA to participate in the initiation reaction. These results contrast with recent findings in mengovirus-infected L cells, where 80S ribosomes accumulate as a result of virus inhibition, but the 80S fraction contains methionyl-tRNA and mRNA, suggesting an accumulation of initiation complex formation on 80S ribosomes.[52] The mechanism of this inhibition is discussed in Section IV. Therefore, all the models directed to explain the shut-off phenomenon must take into account the findings that inhibition of host protein synthesis occurs at the initiation level and that polysome run-off takes place even if viral replication is blocked by means of guanidine.[48]

Figure 3 depicts the two different possibilities that exist at present to account for the shut-off phenomenon. Some researchers have proposed that some components of the protein synthesizing machinery are modified. For instance, polypeptide p220 can be cleaved after poliovirus infection (see Section III.A). Panel B explains the specific inhibition of the initiation of cellular mRNAs by invoking the generation of a selective inhibitor that preferentially blocks host protein synthesis. The modification of monovalent ion concentration after infection would fit in this second model (see Sections III.D. and IV).

C. Cellular mRNA Remains Unchanged After Shut-Off

Initial studies on the shut-off indicated that cellular mRNA was inactivated during infection, as it was unable to attach to ribosomes, but its sedimentation profile was similar to normal mRNA.[47] Subsequent experiments showed that cellular mRNA remains intact and unmodified, as regards to poly(A) content and the structure of the 5′-terminus cap.[53,54] Moreover, cellular mRNA extracted from a variety of picornavirus-infected cells is fully capable of directing protein synthesis in cell-free systems.[55-60] The mRNAs extracted from poliovirus-infected cells translate proteins that migrate as cellular polypeptides, as determined by polyacrylamide gel electrophoresis.[60] Even in cells treated with interferon and infected with poliovirus, where a drastic inhibition of translation ensues, the cellular mRNAs remain translatable.[60]

In cells infected with VSV and superinfected with high multiplicites of poliovirus, VSV protein synthesis is blocked, whereas VSV mRNA synthesis continues unabated. This leads to accumulation of high quantities of intact VSV mRNAs that direct the synthesis of VSV proteins in cell-free systems.[61] These results further support the idea that mRNA is not inactivated after picornavirus infection.

D. Viral Gene Expression Is Required for Shut-Off

It seems clear that for the virus-induced shut-off and cytopathic changes, no new expression of cellular genes is required since these effects are induced by poliovirus in the presence of actinomycin D and in enucleated cells.[33] Therefore, the action of a virion component or a virus-coded protein must be responsible for this interference.

Although it was initially suggested that a protein component of the virus input was able to shut-off cellular protein synthesis,[62] it was soon realized that viral protein synthesis was necessary. Thus, inhibition of virus decapsidation, or inhibitors that block the cleavage or synthesis of the picornavirus polyprotein, suppresses the blockade of host macromolecular synthesis. Rhodanine,[63,64] arildone,[65] and S-7,[66,67] compounds that inhibit virus decapsidation block the virus-induced shut-off. Inactivation of the input virion RNA with proflavine or UV-light[38,68,69] renders the genomic RNA inactive for translation and therefore cellular protein synthesis continues. Inhibitors of polypeptide processing (i.e., amino acid analogues[70,71] or cystatin, a protein inhibitor of cysteine proteases,[72]) also prevent shut-off. Direct inhibitors of translation (i.e., puromycin or cycloheximide) are also efficient blockers of the shut-off effect.[38,73-75] After removal of the translation inhibitor, cellular protein synthesis recovers, and the shut-off of protein synthesis by poliovirus starts immediately afterwards.[75]

The availability of compounds that selectively interfered with picornavirus RNA synthesis (i.e., guanidine,[76] gliotoxin,[77] and 2-(α-hydro xybenzyl)-benzimidazole (HBB)[78,79]) prompted investigations to elucidate the role of virus replication in the inhibition of cell functions. The inhibition of cellular protein or RNA synthesis was not prevented even if guanidine or HBB were present from the beginning of infection.[12,13,80] However, it is not yet certain that viral RNA synthesis is totally prevented by these drugs, and the possiblity that some viral dsRNA is synthesized, even in the presence of guanidine, has been documented.[81]

In recent years, more inhibitors of piconavirus functions have been found,[65,82-90] most of which are particularly active against rhinoviruses. We have recently worked out the mode of action of 3-methylquercetin (3-MQ), a selective inhibitor of poliovirus growth.[91] 3-MQ abolishes viral RNA synthesis and has no effect on any other viral function. The shut-off of host translation occurs even though no new viral mRNAs are made (Figure 4). It is likely that translation of the input viral RNA is sufficient to block host protein synthesis. Another conclusion that can be established from the experiments done with several of these picor-

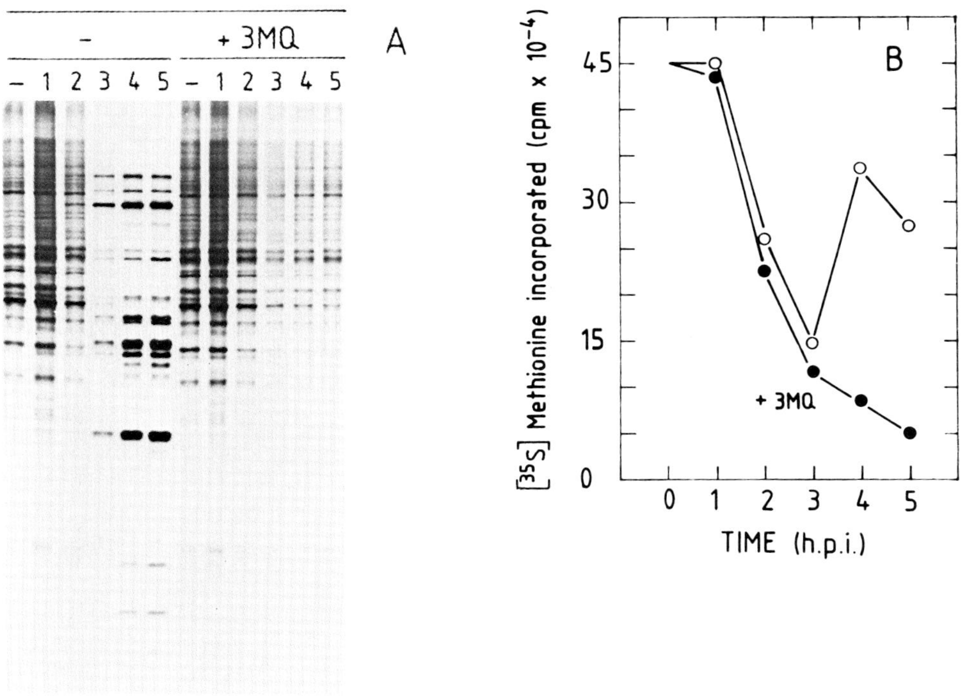

FIGURE 4. (A) Analysis by SDS-PAGE of the proteins synthesized in poliovirus-infected HeLa cells untreated (−), or treated with 20 μg/mℓ 3-methylquercetin (+3MQ). (B) Kinetics of protein synthesis in poliovirus-infected cells untreated (O), or treated with 3MQ (●).

navirus inhibitors is that shut-off occurs even in the absence of significant amounts of viral mRNAs. This assertion is reinforced by the existence of poliovirus ts mutants (temperature sensitive) able to cause shut-off in the absence of significant RNA synthesis.[92] This fact strongly argues against the possibility that competition between cellular and viral mRNAs is responsible for shut-off.

E. The Precise Viral Protein Involved in Shut-Off Still Remains Unknown

Very few poliovirus mutants are available to analyze the shut-off phenomenon.[92-94] This situation, together with the fact that picornavirus proteins are made as a huge precursor that is cleaved to render all the viral native proteins,[70] has hampered the exact location in the picornavirus genome of the function involved in the shut-off. Poliovirus ts mutants unable to block protein synthesis have been claimed to map in the region that codes for viral capsid proteins.[95] However, this conclusion has not been adequately documented, and the exact biological significance of these mutants remains uncertain. More interesting was the finding that defective-interference (DI) particles are generated during poliovirus replication.[96] Those particles lacked a portion of the genome located in the 5′ region that comprises most of the capsid protein coding capacity. Such DI particles efficiently block host protein synthesis.[97,98] These results suggest, although do not yet prove, that the shut-off function may be located in the P2 or P3 region of the poliovirus genome. It is conceivable that site-directed mutagenesis and molecular cloning of selected portions of the picornavirus genome will clarify this problem in the near future, indicating unequivocally which viral protein is responsible for the shut-off.

III. REGULATION OF TRANSLATION IN POLIOVIRUS-INFECTED CELLS

When HeLa cells are infected by poliovirus, there is a drastic decline in host translation, which is dependent on the multiplicity of infection used. This is followed by a peak where viral proteins are synthesized. Soon after, viral translation also declines and finally stops (see Figure 1). The speed of these events varies, depending upon several factors (i.e., it is more rapid in suspension cells as compared to HeLa cells grown in monolayers, and the presence of actinomycin D also favors the shut-off of cellular protein synthesis). As discussed in the previous section, the translation of viral mRNA is necessary for the shut-off to occur, but compounds that interfere with viral RNA synthesis (i.e., guanidine and 3-methylquercetin) do not block the shut-off. Therefore, they provide a good model system to study the mechanisms involved in the shut-off of host translation in the absence of viral replication.

Although cellular protein synthesis declines p.i., not all cellular proteins are affected to the same extent. For instance, the heat-shock proteins from HeLa cells are more resistant to inhibition, both after poliovirus infection and by hypertonic media.[99]

Since more efforts have concentrated on the elucidation of the poliovirus-induced shut-off by means of cell-free systems derived from poliovirus-infected cells, we will commence the following section by reviewing such work.

A. Cell-Free Systems From Poliovirus-Infected Cells Do Not Translate Capped mRNAs

The mechanism of translation in eukaryotic cells has been reviewed and will not be described.[100,101] Details of the mechanism of the initiation of translation and the function of all the factors involved have also been recently described in detail.[102]

The availability of active cell-free systems that faithfully translated endogenous and exogenous mRNAs encouraged several groups to analyze the translation capacity of cell-free systems from poliovirus-infected cells. A powerful inhibitor of translation that was found in such lysates was identified as dsRNA.[103] The involvement of this macromolecule in the inhibition of host translation by poliovirus was discarded on the basis that it had no selectivity in the inhibition of protein synthesis directed by viral or cellular mRNAs. However, dsRNA has a selective effect on viral and cellular protein synthesis in virus-infected cells.[104] The reason for this specificity has not been discovered. More recently, it was reported that translation of picornavirus RNA was resistant to dsRNA in a reticulocyte lysate under conditions in which translation of cellular mRNA was inhibited.[105] The resistance of picornavirus RNA translation to dsRNA is due to the prevention of the formation of a dsRNA-activated inhibitor that phosphorylates eIF2, while globin mRNA has no effect.[105] The use of F-Met-tRNAsMet as an indicator of new rounds of initiation taking place in the cell-free system allowed Celma and Ehrenfeld[106] to show that S10 lysates from control and poliovirus-infected (3 hr p.i.) HeLa cells translate endogenous cellular or viral mRNA, respectively.

When lysates from poliovirus-infected cells are fractionated into several components and the translation of exogenous cellular mRNA is tested, the ribosomal wash fraction where most initiation factors are located is inactive. This led Kaufmann et al.[57] to suggest that a translation inhibitor generated during infection associates with polysomes and is removed after high-salt washing. Following this line of research, two laboratories presented evidence that cell-free systems from poliovirus-infected cells were deficient in the initiation of cellular, but not viral, mRNAs. Helentjaris and Ehrenfeld[107] showed that the ribosomal wash fraction from poliovirus-infected cells was even inhibitory towards the initiation of translation on uninfected polysomes. More puzzling was the finding that S10 lysates obtained from infected cells harvested at the peak of viral protein synthesis were inactive even to translate viral mRNA, whereas if the S10 was prepared from cells harvested later when viral translation has stopped, the lysate was highly active. This was in agreement with initial results obtained with mengovirus-infected Ehrlich cells.[108] This inconsistency of cell lysates to reflect the translation capacity observed in vivo was interpreted to be a consequence of the method used to prepare the S10 extract. This procedure discarded the majority of active membrane-bound polysomes, whereas at later times they will be released from the membranes. If this were so, it would be tempting to suggest that no active ribosomal wash fraction could be generated if most groups use such S10 lysates which were almost devoid of polysomes. Following a different approach, Rose et al.[109] assayed the translation of VSV mRNAs in lysates from poliovirus-infected HeLa cells 3 hr after the absorption period, i.e., after shut-off had occurred and the peak of viral translation had taken place. It should be pointed out that those groups worked with HeLa cells grown in suspension, where the shut-off and poliovirus replication occur much faster than in monolayer cultures. They observed that capped VSV mRNAs behaved as cellular mRNAs and, therefore, were not translated in these lysates. The addition of either one of the reticulocyte initiation factors eIF-1, eIF-2, eIF-3, eIF-4A, or eIF4C + 5 did not restore translation, whereas in the presence of eIF-4B, some VSV proteins were clearly synthesized.

Curiously enough, the translation of these VSV mRNAs in the uninfected lysate was also very poor and increased by almost 20-fold after the addition of eIF-2. The method used to purify the initiation factors was that described by Schreier et al.,[110] and it is now accepted by workers in this field that many of these initiation factors were contaminated with other polypeptides that participate in the initiation of translation, but were unknown at the time. Ehrenfeld's group performed further fractionation of the inactive ribosomal wash from poliovirus-infected cells by ammonium sulfate precipitation into two fractions named A and B.[111] The A fraction obtained in the 0 to 40% cut was the one responsible for the observed specificity in the in vitro translation. In contradiction with Rose et al.,[109] they made the following observations: the inactivity of the infected cell fraction was not reversed by further addition of active wash fractions from uninfected cells or purified reticulocyte initiation factors (eIF-3 or eIF-4B). Moreover, eIF-4B was present in an active form, both in preparations from infected or uninfected cells. Of interest was the finding that no eIF-3 activity was present in fractions from infected cells. Also, VSV and polio mRNAs were translated simultaneously in reticulocyte lysates supplemented with ribosomal salt wash from uninfected cells, but addition of ribosomal salt wash from infected cells stimulated only the translation of polio RNA, suggesting that it contains a component that restricts VSV translation in this system.[51]

A breakthrough in this research was the finding of a small polypeptide of M_r 24,000 involved in the initiation of protein synthesis[112] responsible for restoring the ability of poliovirus-lysates to translate VSV, Sindbis, and reovirus mRNAs.[113,114] This 24 kdalton protein contaminated previous eIF-4B and eIF-3 preparations and cross-linked to the 5′-terminal cap on mRNA. This factor would be necessary only to translate capped mRNAs. Therefore, these findings were clearly in agreement with the suggestion advanced by Fernández-Muñoz and Darnell[54] that the lack of a cap structure in poliovirus RNA could constitute the molecular basis for the selective inhibition of protein synthesis p.i. However, further research showed that things were not so simple. Cross-linking of the 5′ cap group of mRNA to the 24 kdalton protein was found in ribosomal salt washes from both uninfected and infected cells.[115]

By sucrose gradient centrifugation, two forms of cap binding proteins were found in rabbit reticulocytes.[116] One slowly sedimenting fraction of 6S containing the 24 kdalton cap binding protein known as CBP I did not restore translation of capped mRNAs in poliovirus lysates. Another fraction of 8-10S containing several proteins (CBP II or eIF-4F) as well as the 24 kdalton protein was able to restore translation in the above systems. For a recent review on the different cap binding proteins, the reader is referred to the Shatkin review.[117] In the presence of ATP and Mg^{++}, two proteins of 28 and 50 kdaltons cross-linked to oxidized reovirus mRNA.[118]

These proteins were different from the previously identified 24 kdalton polypeptide that cross-links in the absence of ATP/Mg^{++}. Lee et al.[119] suggested a model in which those proteins bind to the 5′ end in the presence of ATP and unwind capped eukaryotic mRNAs and suggested that this step should be necessary for translation.[119,120] eIF-4A also causes a structural change in mRNA, as evidenced by means of a nuclease sensitivity test. Addition of high quantities of eIF-4A increases the nuclease sensitivity of the mRNA, indicating that this factor unwinds mRNA secondary structure. This activity is ATP-dependent and enhanced by eIF-4A. It is also more efficient when the eIF-4A is part of the eIF-4F complex.[121,122] These results were in agreement with previous findings suggesting that the efficiency of mRNA binding to ribosomes decreases as the stability of the secondary structure increases.[123] The sucrose gradient sedimentation profiles of cap binding proteins from ribosomal washes of uninfected and poliovirus-infected HeLa cells were different. Thus, most CBPs sedimented faster and were associated with eIF-3, whereas no CBP from infected cells sedimented with this factor and remained at the top of the gradient. These findings supported the idea that the association of the CBP with eIF-3 is impaired in poliovirus-infected cell lysates.[124]

Not only are there less CBPs able to cross-link to reovirus mRNAs in poliovirus-lysates, but also these lysates inhibit the restoring activity of initiation factors obtained from mock-infected cells.[125] The use of an antibody against reticulocyte eIF-3 served to identify an antigen of 220 kdaltons (p220) present in uninfected HeLa cell lysates, but absent from infected ones.[126] Smaller polypeptides of M_r 130 and 100 kdalton antigenically related to the p220 were present in poliovirus-infected lysates. The p220 associates with eIF-3 and may be a component of the so-called cap binding complex (CBPII). Its addition restores translation of capped mRNAs in poliovirus lysates. Analysis of the integrity of p220 in the infected cells at various times p.i. roughly correlates with the inhibition of protein synthesis.[126] However, it was noted that at low multiplicity of infection, even though most p220 disappeared from infected cells at 2 to 2.5 hr p.i., there was almost no inhibition of protein synthesis. To explain this lack of correlation between shut-off and p220 cleavage, the authors pointed out that it "may be due to the ability of cells to synthesize a limited amount of protein in the absence of p220."

The use of high-resolution, two-dimensional gel electrophoresis allowed Duncan and Hershey[127] to identify virtually all known polypeptides that form part of the initiation factors. Using this technique, no proteolytic degradation was observed of any polypeptide involved in initiation, including a 3p220 polypeptide that forms part of eIF-3.[128] The relationship between this 3p220 and the p220 that belongs to the cap binding complex remains obscure. Apparently, they are antigenically different, although the cap binding protein p220 was first identified by means of an antibody raised against eIF-3.[129] More detailed analyses of initiation factors from lysates of poliovirus-infected HeLa cells are now available. It seems that, contrary to previous suggestions, both eIF-3 and eIF-4B are active, whereas the CBP complex is inactive. In agreement with previous experiments, eIF-3 activity is undetectable unless active CBP is present.[130] The activity that restores the translation of capped mRNAs and the cap-specific mRNA-protein interaction in extracts of poliovirus-infected cells copurify with CBP II.[131] The only initiation factor found to be cleaved is p220. Efforts to show which protease is responsible for this cleavage indicate that the virus-coded protease 7c (polypeptide 3c) is not involved.[132,133] At least two possibilities remain open, one is that another putative (still unidentified) virus protease is involved in this cleavage, or that the activity of a very selective cellular protease is induced after poliovirus infection.

Several important controls are still lacking in these studies. For instance, degradation of the p220 in the presence of inhibitors of poliovirus replication that allows shut-off to occur should be tested. Also, analysis of the p220 protein should be done under conditions of inhibition of the initiation of protein synthesis in uninfected HeLa cells by inhibitors such as hypertonic medium, NaF, etc. It could well be that dissociation of the CBP complex is the consequence, and not the cause, of the inhibition of translation by poliovirus.

A poliovirus mutant has been recently obtained by insertion of an additional codon in an infectious cDNA clone. This extra triplet is located in the region of the genome that codes for the amino terminal of protein 8. The mutant has a defect in replication such that the synthesis of the bulk of viral proteins does not take place. It produces a full yield of RNA in CV 1 cells, but in HeLa cells there is an inhibition of viral RNA synthesis.[134] Infection of HeLa cells with this mutant does not cleave p220 under conditions where 80% inhibition of cellular translation has occurred. Some replication of mutant viral RNA takes place, since the shut-off is prevented by guanidine. These results suggest that protein 8 (or protein 2A) may play a part in viral RNA synthesis and provide genetic evidence against a role for p220 cleavage in the shut-off of host translation.

Active research in this field continues, and more experiments will be needed to clarify the involvement, if any, of p220 cleavage in poliovirus-induced shut-off and also to ascertain many inconsistencies found with cell-free systems in different laboratories.

Recent evidence supports the idea that cleavage of p220 also occurs in rhinovirus 14-infected HeLa cells.[135,136] Therefore, this proteolytic degradation could also be applicable to members of the *Rhinovirus* genus. Inactivation of eIF-4F has also been implicated in the inhibition of protein synthesis after heat-shock (HS) of Ehrlich cells.[137] However, it must be remembered that the selectivity in translation observed after HS treatment is reversible even in the presence of actinomycin D, i.e., a normal pattern of protein synthesis reappears when cells are returned to their physiological temperature.[99,138] Also, the synthesis of HS proteins is inhibited in poliovirus-infected cells, even if viral RNA synthesis is blocked.[99] These results clearly suggested that inactivation of CBP-II is the result, but not the cause, of translation inhibition in mammalian cells.

B. Intact Poliovirus-Infected Cells Translate Capped Viral mRNAs

A more direct confrontation of the suggestion that poliovirus infection destroys an initiation factor needed to translate capped mRNAs takes advantage of cells doubly-infected with poliovirus and another virus that has capped mRNAs. Previous experiments in these systems indicated that poliovirus was able to shut-down the translation of viral mRNAs efficiently. Perhaps the only exception was found with SV5 infected cells.[139] However, in those studies, a clear advantage was given to poliovirus replication. For example, experiments were reported in which cells infected with VSV at a multiplicity of ten were superinfected with poliovirus at 100 PFU/cell.[61] Obviously, poliovirus blocked VSV mRNA translation very efficiently. However, in other experiments where the competition was in cells infected with poliovirus and Semliki Forest virus, the translation of the capped SFV 26S mRNA continued even after poliovirus translation had stopped. A possible explanation of this result is that SFV uses a strategy similar to that of poliovirus to inhibit cellular translation, i.e., SFV blocks an initiation factor necessary to translate capped mRNAs. This possibility seems very unlikely, first, because the SFV 26S mRNA itself has a cap structure in its 5′ end and, second, because cells doubly infected with VSV and SFV translate simultaneously the mRNAs from both viruses for several hours.[140]

The use of conditions that allow the shut-off of cellular translation and inhibit poliovirus replication has strengthened the idea that even VSV mRNAs, which have been used as the archtype of capped mRNAs, are translated at control levels in poliovirus-infected HeLa cells. In the experiment shown in Figure 5, poliovirus replication has been blocked with guanidine, but the shut-off of cellular proteins still takes place. If the advantage is given to VSV replication, it is able to translate its own mRNAs in poliovirus-infected cells when the translation of host mRNAs has been depressed.

Adenovirus is another example of how viral capped mRNAs are efficiently translated after poliovirus infection. If HeLa cells are infected with adenovirus and superinfected with poliovirus at the time when late proteins are synthesized, translation of several adenovirus

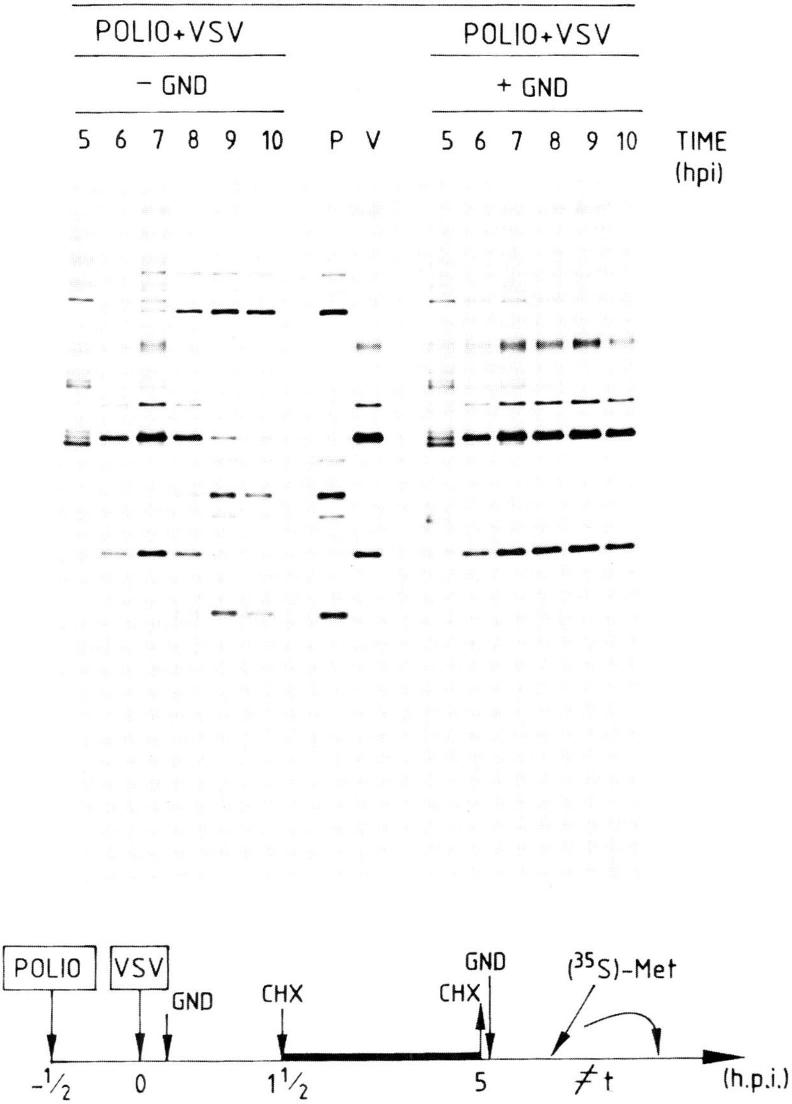

FIGURE 5. Analysis by SDS-PAGE of the proteins synthesized in HeLa cells infected with poliovirus (10 PFU/cell) and super-infected with vesicular stomatitis virus (VSV) at an MOI of 100 PFU/cell. The cells were treated with guanidine 3 m*M* (GND) since zero time and cycloheximide $5 \times 10^{-5}M$ (CHX) from $1^1/_2$ to 5 hr p.i. (h.p.i.) (see scheme). The controls of protein synthesis in poliovirus- (P) and VSV- (V) infected HeLa cells are shown.

proteins from capped mRNAs continues (Figure 6). These experiments on cells doubly infected with poliovirus and another virus cannot easily be explained by the suggestion that an initiation factor required to translate capped mRNAs is inactivated. On the contrary, they are more in agreement with a model where an inhibitor of initiation is generated (membrane derived or not) that has a selective effect on cellular mRNAs. However, much more work is needed on these systems to ascertain if, under mixed infection conditions, any initiation factor activity is absent from cell-free systems. Also, we need to analyze the state of the cell membrane regarding its components and activities, in order to clarify the regulatory role that membrane integrity could play on viral and cellular protein synthesis.

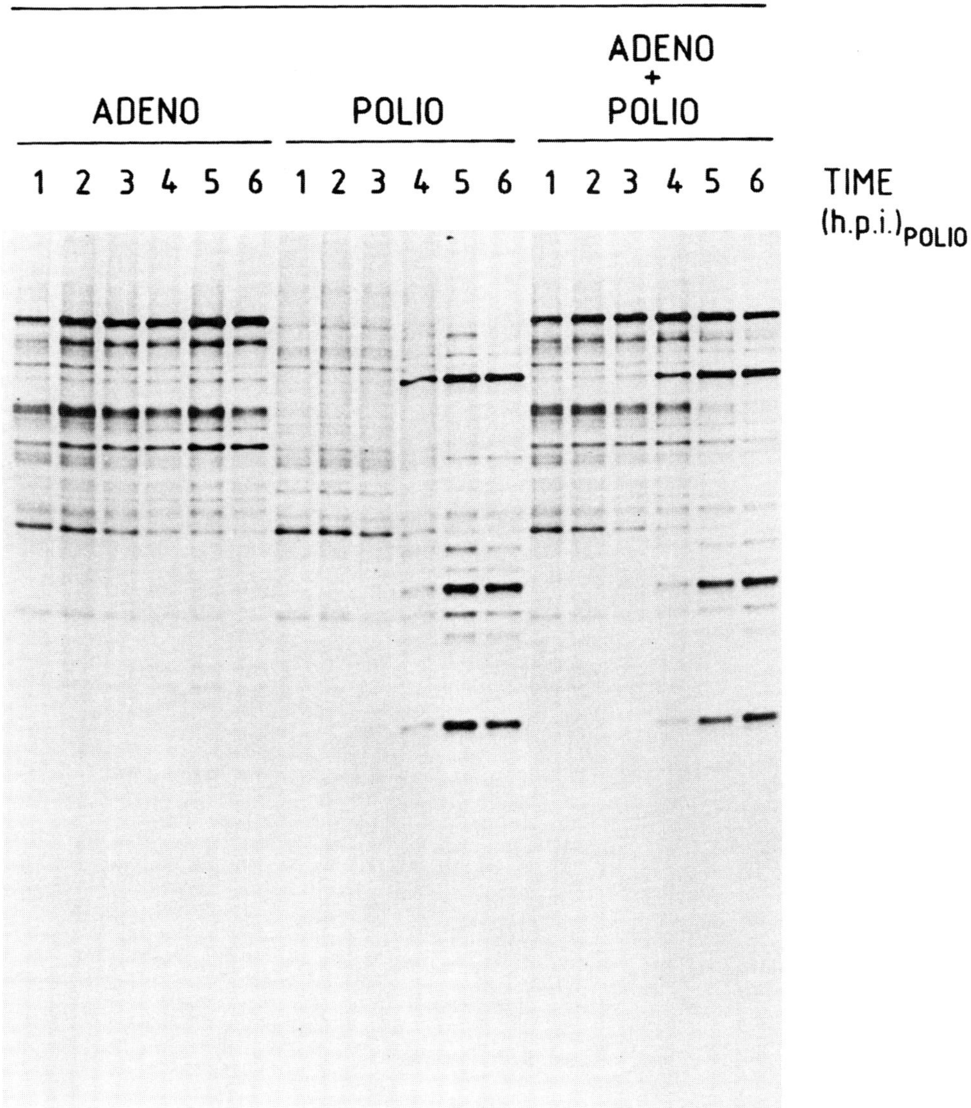

FIGURE 6. Analysis by SDS-PAGE of the proteins synthesized in HeLa cells infected with adenovirus type 3 (10 PFU/cell) and super-infected with poliovirus (10 PFU/cell) after 16 hr p.i. (h.p.i.) with adenovirus. The times refer to poliovirus infection.

C. Cell-Free Systems from Several Virus-Infected Cells Are Translation-Deficient

Soon after the suggestion that poliovirus shut-off was mediated by modifications of initiation factors, other laboratories investigated this possibility with other picornaviruses. The main approach used consisted of the analysis of cell-free systems obtained from virus-infected cells to translate viral or cellular mRNAs. As will be discussed in Section IV, attempts were made to translate capped or uncapped mRNAs in mengovirus and EMC virus infected-cell lysates, with the conclusion that the discrimination observed in the infected cells was lost; that is, both cellular and viral mRNAs are translated in lysates from mengovirus-infected cells. The conclusion that EMC virus and mengovirus follow a different mechanism to shut-off host protein synthesis is not validated by these observations, because the EMC-infected cell-lysates were made before viral protein synthesis was apparent, as analyzed by polyacrylamide gels. It must be remembered that the inability of poliovirus-infected cell lysates to translate capped mRNAs is observed when active replication of poliovirus and its protein synthesis is underway. It was most surprising to find that lysates from VSV-infected cells (a virus with typical capped mRNAs) were also deficient in the translation of cellular mRNAs.

The general rule that applies to cell-free systems for translation is that they are less active when obtained from cells that are not actively synthesizing proteins, either because they have been manipulated with translation inhibitors or after viral infection.[57,109,141,142] Translation of low affinity mRNAs in these lysates requires the addition of external initiation factors.

Cells infected with vesicular stomatitis virus are a good example. Lysates prepared from VSV-infected cells are impaired in their ability to translate cellular or viral mRNAs.[143-145] Stimulation of translation was obtained by addition of eIF-2 and the guanine exchange factor (GEF).[143] In one case, maximal stimulation was attained by addition of eIF-4B and eIF-4F,[144] while Thomas and Wagner[144] assigned the defect of VSV-infected cell lysates to eIF-3 and eIF-4B. These results indicate that the lysates are certainly defective in those factors, but not necessarily that VSV-infection inactivates them in the intact cell.

Competition has also been implicated as the main cause of VSV-induced shut-off of host translation.[146,147] However, this suggestion is not supported by recent findings indicating that shut-off by VSV does not parallel the amount of VSV mRNA.[148] Also, in interferon-treated cells, shut-off by high multiplicities of VSV occurs in the absence of mRNA synthesis and, thus, without competition.[149]

Detailed analyses have also been performed with Semliki Forest virus and reovirus-infected cell-free systems. Semliki Forest virus belongs to the Togaviridae family and also causes a profound and rapid inhibition of host translation after infection.[2] Translation of early 42S mRNA and late 26S mRNA from SFV has been accomplished in a fractionated pH 5 system from neuroblastoma cells.[150] Polysomal 26S mRNA isolated from infected cells[151] was capped. The crude ribosomal wash obtained at 8 hr p.i. exhibited a reduced capacity to support translation of early 42S mRNA or host mRNA, although it was only slightly less active with late 26S mRNA or EMC RNA. Addition of purified initiation factors eIF-4B and CBP 24 kdalton from rabbit reticulocytes reversed the inhibition. The possibility that those initiation factors were inactivated after SFV infection was discarded, since both eIF-4B and CBP 24 kdalton were partially purified from infected cells with no loss in their activity.[151] Therefore, it was suggested that a virus-derived component that interfered with the translation of host mRNA was present in the ribosomal washes. Efforts to purify this inhibitor led to the identification of the virus capsid protein of M_r 33 kdalton, which is known as the C (core) protein.[152] Addition of this protein to cell-free systems selectively inhibits the translation of host and early viral 42S mRNA, but has no effect on protein synthesis directed by late 26S mRNA or EMC RNA.[152] Therefore, a model is proposed in which the C protein is responsible for shut-off, probably by inhibiting the binding of host mRNAs to ribosomes. Surprisingly, when initiation complexes using labeled host mRNA were analyzed, the C protein did not inhibit its interaction with the 40S ribosomal subunit and only a partial inhibition of the formation of 80S initiation complexes was found.[152] These findings make it more difficult to find an explanation for the specificity of translation inhibition observed in vivo. Recent advances in the cloning and expression of SFV proteins have resulted in the efficient expression of the C protein in CV-1 cells transformed with a eukaryotic vector.[153] These cells were able to faithfully synthesize the SFV C protein with no apparent signs of cytotoxicity, although a direct measurement of host protein synthesis was not made. Undoubtedly, future studies with cells bearing a vector capable of expressing the SFV C protein will be helpful in determining whether the C protein plays a part in the shut-off phenomenon in SFV-infected cells.

The possibility that SFV destroyed a factor necessary for translation of capped mRNAs

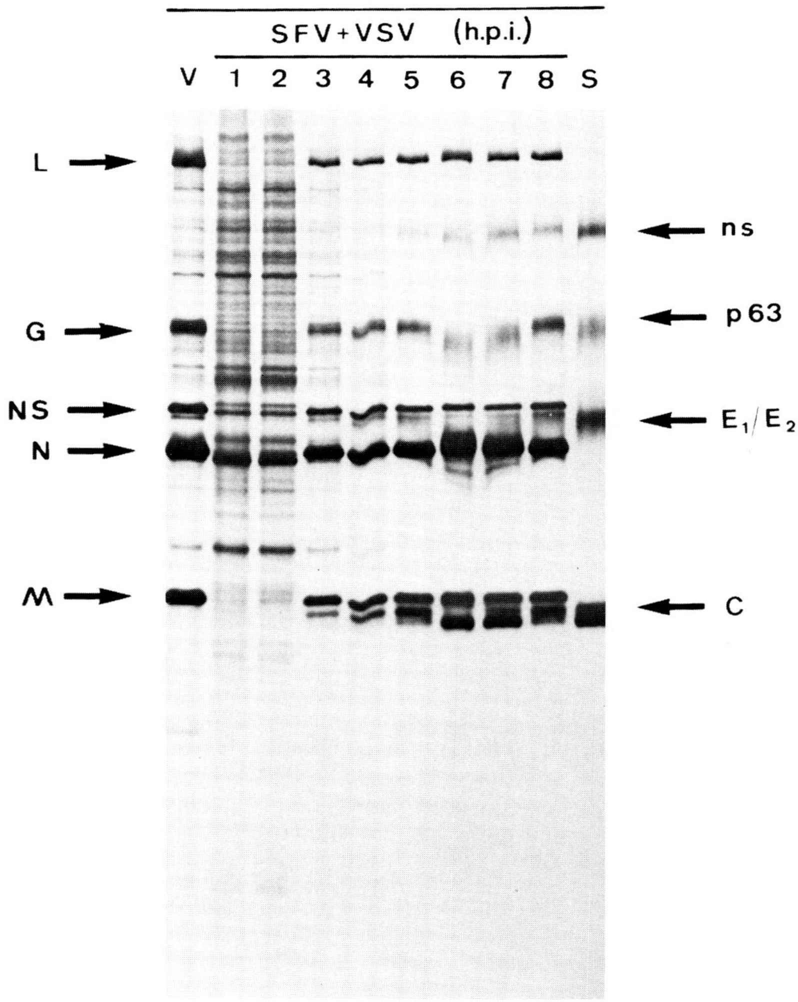

FIGURE 7. Analysis by SDS-PAGE of proteins synthesized in HeLa cells infected with SFV (10 PFU/cell) and super-infected with VSV (20 PFU/cell) after 30 min. The controls of protein synthesis in SFV- (S) and VSV- (V) infected HeLa cells are shown.

was discarded because cells doubly-infected with SFV and VSV efficiently synthesized proteins of both viruses for long periods of time[140] (Figure 7). Also, the SFV was the prototype of capped mRNA used in studies to show that it was not translated in poliovirus-infected cell lysates.[109]

There is now well-documented evidence that togaviruses (SFV and Sindbis virus) cause a profound alteration of cell membrane permeability to monovalent cations at the time when shut-off occurs. Also, evidence from cell-free systems indicates that monovalent ions discriminate between late viral mRNA, as compared to host and early viral mRNA. Therefore, both in vivo and in vitro evidence suggest that the shut-off in this system is mediated by membrane modification.[45,154]

Reoviruses were also suspected as modifying the protein synthesis machinery p.i. Reoviruses cause the shut-off of host protein synthesis depending on the virus and cell line analyzed. Thus, reovirus infection inhibits translation in mouse L cells more efficiently than that in HeLa cells.[155] Three serotypes of human reoviruses are known;[156,157] serotype 1 (Lang strain) does not significantly affect the synthesis of L cell proteins, whereas under similar conditions, serotype 3 (Dearing strain) completely inhibits host translation at late times.[158] Reovirus serotype 2 is the most efficient inhibitor of L cell protein synthesis.[159] It is interesting to note that the rate of virus multiplication, but not the final yield, correlates with the extent of virus-induced shut-off of cellular translation.[159] Several lines of evidence point to protein 3 coded by gene S4 as being responsible for shut-off.[159,160]

Extracts from uninfected L cells efficiently translate capped reovirus mRNAs, but uncapped reovirus mRNAs are poorly translated. In contrast, extracts from reovirus-infected L cells translate these uncapped mRNAs very efficiently in a manner that is insensitive to cap analogues.[161] A model was proposed for reovirus-induced shut-off in which reovirus late mRNAs were not capped, and the protein synthesizing machinery underwent a transition throughout infection from cap dependence to cap independence.[162] Evidence was presented that reovirus progeny subviral particles possessed an active RNA polymerase and masked guanylyl-transferase and methylase enzymes and, therefore, produced in vitro uncapped mRNAs.[162,163] Moreover, direct analysis of the 5′ terminal structure of reovirus mRNA associated with polysomes supports this model, since these mRNAs are uncapped at early times p.i., whereas at late times essentially all viral mRNA is uncapped.[164] The fact that monoclonal antibodies against the 24 kdalton CBP do not inhibit translation of uncapped reovirus mRNAs lends support to the notion that translation of these mRNAs is cap-independent as occurs with picornavirus mRNAs.[129] However, this model has been questioned based on two main findings. One is that extracts from reovirus-infected mouse SC-1 cells translate capped or uncapped globin mRNAs equally well, and translation of both cellular or viral mRNAs was equally sensitive to cap analogues.[165] The second finding was that HeLa cells doubly infected with reovirus translated capped or uncapped viral mRNAs together with reovirus mRNAs at both early and late stages of infection.[155] In a recent report, results from Lemieux et al.[166] contradict their previous claim in that uncapped late reovirus mRNA was now found to be sensitive to cap analogues in lysates from infected cells. They conclude that reovirus infection does not induce a transition from cap-dependent to cap-independent translation. In fact, sensitivity of mRNA translation to cap analogues is not related to the presence of a cap structure, but rather to the affinity of a given mRNA to participate in the initiation process.[167] Speculations about possible similarities between the reovirus and the poliovirus model have been advanced;[7] however, once again the conclusion that we can reach at present time on the reovirus system is that results obtained with cell-lysates are sometimes misleading, and great care must be taken when extrapolating the findings obtained in cell-free systems to the in vivo situation.

D. Is Membrane Damage Involved in Poliovirus Shut-Off?

Saborio et al.[168] and Nuss et al.[169] found that hypertonic medium preferentially inhibited cellular protein synthesis in poliovirus-infected HeLa cells. However, no physiological significance for this finding was proposed. Instead, they suggested that hypertonic treatment could represent a new method to map proteins in the poliovirus genome. Furthermore, with hypertonic medium, the background of cellular protein synthesis in many virus-infected cell systems was suppressed. Subsequently, Carrasco and Smith[170] found that sodium ions could affect differentially viral and cellular translation in cell-free systems. Physiological concentrations of 20 to 30 m*M* NaCl in vitro inhibited cellular protein synthesis by 80%, whereas picornavirus RNA translation was greatly stimulated. This was the first indication that ions were able to inhibit one type of mRNA and at the same time stimulate translation of another mRNA in cell-free systems.[170,171] Analysis of the initial di- and tri-peptide formation on viral and cellular mRNAs by the sparsomycin technique showed that the effect of monovalent ions was at the level of initiation.[2] Subsequently, the binding of eIF-2 to mengovirus mRNA was found to be more resistant to increasing salt concentrations than its binding to globin mRNA.[172]

We proposed that a plausible physiological significance of these findings might be that poliovirus mRNA is optimally translated under high ionic conditions because it is translated in the infected cell under different ionic concentrations. Moreover, it is possible that if the distribution of monovalent ions varies during infection, it could cause the shut-off of host protein synthesis.

Very early during picornavirus infection there are alterations in the cell membrane, i.e., an increase in lipid fluidity[173] and increased membrane permeability.[174,177] However, these changes must not be directly related with shut-off because they occur even in the presence of inhibitors of gene expression.

Support for this idea also comes from the effects of external cations on translation in intact poliovirus-infected cells. In addition to the previous findings reported by Saborio et al. and Nuss et al., we have found that hypotonic medium added after poliovirus entry blocks the shut-off, and that HeLa cells exclusively synthesize cellular proteins as determined by polyacrylamide gel electrophoresis. Moreover, when hypotonic medium is added just at the time when poliovirus proteins appear on gels, it reverses the shut-off in such a way that viral protein synthesis stops and cellular translation continues.

Attempts to measure the concentration of sodium and potassium ions in poliovirus-infected cells suggested that the shut-off of host protein synthesis occurred prior to the redistribution of monovalent ions, but it was coincident with the bulk of viral protein synthesis. These results were confirmed by using radioactive $^{86}Rb^{+}$ as a potassium analogue. The use of guanidine indicated that poliovirus shut-off occurred in the absence of any substantial decrease in the concentration of intracellular potassium. Therefore, we may conclude that the redistribution of monovalent ions occurs at a time when the bulk of viral mRNA translation is taking place. These viral mRNAs have evolved a particular structure that makes them translatable under such unusual ionic conditions. However, if the concentration of sodium continues to increase in the infected cell as a consequence of continuous membrane damage, then viral translation will also be inhibited.

The rate of release of $^{86}Rb^{+}$ from preloaded cells increases as infection proceeds, and it is not sensitive to ouabain or quinidine, which are inhibitors of the Na^{+}/K^{+} ATPase and Ca^{++}-induced K^{+} systems, respectively. The release of $^{86}Rb^{+}$ is only slightly sensitive to furosemide, an inhibitor of the $Na^{+}/K^{+}/Cl^{-}$ cotransport system, suggesting that the mechanism of potassium release involves an increased passive permeability. The possibility that divalent cations or changes in cytoplasmic pH could be involved in the early shut-off has also been examined in our laboratory.[178] We have not found significant alterations in these parameters, at least during the first 3 hr p.i.

All these results suggest that redistribution of ions is not involved in the early shut-off of host translation. Nevertheless, it is still possible that membrane damage might be responsible for this early shut-off in poliovirus-infected cells. Bearing in mind that the picornavirus genome has a single initiation site for the initiation of translation, the same viral proteins are made throughout the course of infection. If one of these viral components is responsible for the modification of cell membranes observed late in infection, it must be synthesized from the moment that viral mRNA translation starts and acts on the membrane from the very beginning of viral infection, perhaps through a mechanism akin to that of the S protein of λ phage.[179] At low concentrations of this putative poliovirus protein, the membrane is still able to maintain the gradients of ions, but other membrane functions might be altered.

Recent results from our laboratory show that there is a correlation between shut-off by poliovirus and the alteration of some membrane enzyme activities, such as the increase in phospholipase A_2 induced by the inophore A23187. Increases in choline released into the medium also correlate with shut-off.[201] However, it is still not known whether there is any connection between alterations in phospholipid synthesis and the inhibition of translation. Phospholipids alone are inhibitors of protein synthesis in cell-free systems, and they block the initiation of translation.[180] The possibility that a membrane constituent may be involved in shut-off, i.e., a phospholipid or a product derived from it, remains unexplored.

E. Why Does Poliovirus Translation Stop?

As indicated above, poliovirus translation does not last for long periods of time in the cell. Rather, it declines abruptly around 2 hr after the shut-off of cellular translation has occurred and ceases soon after. If viral translation is inhibited because the membrane is unable to maintain the gradients of monovalent ions, the blockade of viral protein synthesis should be prevented by modifying the external ionic conditions. This seems to be the case when poliovirus-infected HeLa cells are maintained in a medium containing a concentration of potassium ions similar to that found in the cytoplasm of uninfected cells, in such a way that potassium is kept in the cell throughout infection. Under these conditions, viral translation is affected in two ways: (1) there is an increase in total viral translation, suggesting that even though viral mRNAs are translated in a cell with a leaky membrane, they are not translated at their optimum, and (2) viral translation can be substantially prolonged (Figure 8). Perhaps, as a consequence of continued membrane damage and energy depletion, viral translation will also be abated after prolonged incubation. These findings suggest that monovalent ions do play a significant role in the regulation of viral translation in the infected cell.

It is noteworthy that even though many more viral proteins are synthesized under conditions of high external potassium, the cytopathic effect induced by poliovirus replication is largely prevented (Figure 9). This supports the idea that alterations in the concentration of ions in the cytoplasm are a consequence of virus-induced membrane-damage and may play a fundamental part in the modifications of cellular morphology and cell toxicity observed during viral infection.

IV. INHIBITION OF HOST TRANSLATION BY ENCEPHALOMYOCARDITIS VIRUS AND MENGOVIRUS

Encephalomyocarditis (EMC) virus and mengovirus belong to the *Cardiovirus* genus and are often taken as models to study the shut-off of host translation by picornaviruses. Comparative studies of the morphological modifications that they cause in L cells have been carried out.[181] Both viruses cause a profound inhibition of host macromolecular synthesis, although the speed of inhibition translation is lower than that for poliovirus.[182,183] Cell-free systems have been prepared from EMC virus or mengovirus infected cells[184-186] and although lysates from infected cells are generally less active in the translation of exogenous mRNAs, neither control for virus-infected cell lysates discriminates between the translation of viral or cellular mRNAs.

Contrary to the studies done with poliovirus, lysates from cardiovirus-infected cells have been routinely performed after the early shut-off has taken place, but before viral protein synthesis is apparent by polyacrylamide gel analysis. Therefore, a direct comparison of these results with those obtained with poliovirus-infected cells is difficult, and the only valid conclusion is that no modification in the translation specificity of cell lysates is observed after the early shut-off by cardiovirus has occurred.

Other in vitro studies have indicated that EMC RNA has a high affinity for translation components and can out compete cellular mRNAs under given conditions.[56,187] Thus, studies of translation of cellular mRNA in the presence of EMC RNA showed that addition of a ribosomal wash fraction had a differential effect on globin mRNA or EMC RNA translation,[186] and that EMC RNA competed for a message discriminatory initiation factor that was

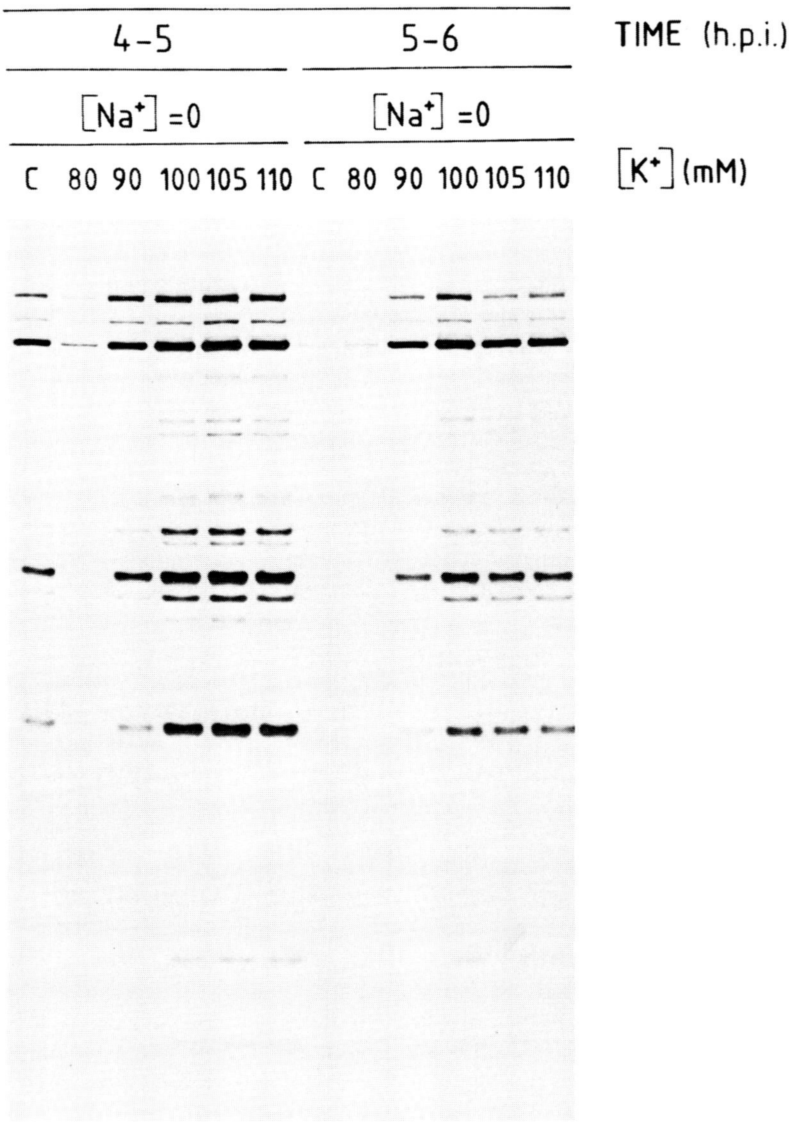

FIGURE 8. Analysis by SDS-PAGE of the proteins synthesized in poliovirus-infected HeLa cells (25 PFU/cell) in different media. E4D2-NaCl-$NaCO_3H$ medium (without sodium ion) plus 20 m*M* HEPES pH 7.2 was supplemented with 80, 90, 100, 105, and 110 m*M* KCL. These media were added at 90 min p.i. and proteins were labeled with (^{35}S)-methionine at 4 to 5 and 5 to 6 h.p.i.

first identified as initiation factor eIF-4B (known then as IF-M3).[187] Using a direct assay to analyze mRNA binding to eIF-4B, it was shown that EMC RNA had a higher affinity than globin mRNA.[188] This result was puzzling because eIF-4B was involved in cap recognition, and EMC RNA does not possess a cap structure.[189] It was later realized that the preparations of initiation factors were rather impure and contained other components involved in mRNA recognition.[121] The biological significance of the higher requirement of EMC RNA translation for eIF-4B still remains unclear.[188,190] Binding of eIF-2 to mengovirus RNA has also been demonstrated in cell-free systems.[102] This factor binds with relative specificity to mengovirus

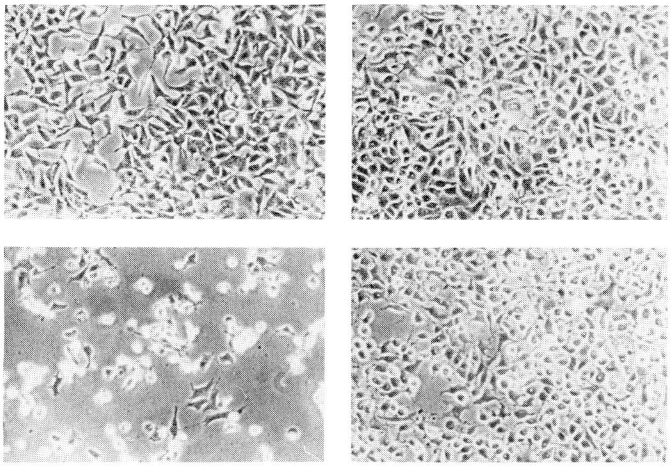

FIGURE 9. HeLa cell monolayers infected with poliovirus in control medium or medium without sodium ion plus 90 m*M* KCl (0/90). The MOI was 15 PFU/cell. 0/90-Medium was added at 90 min p.i. and the cytopathic effect recorded 6 $^1/_2$ hr later.

RNA in two widely separated domains located at the 3′ side of the poly(C) tract.[100] Competition between globin mRNA and mengovirus RNA in nuclease-treated reticulocyte lysates is influenced by the addition of purified eIF-2.[172] Under conditions where globin synthesis is very much inhibited by mengovirus RNA, addition of eIF-2 restores globin synthesis with no overall stimulation of translation.[172] Of interest is the finding that eIF-2 binding to mengovirus RNA is significantly more salt-resistant than that to globin mRNA.[172] This result is in good agreement with our finding that picornavirus RNA requires a higher concentration of monovalent ions for optimal in vitro translation than do cellular mRNAs.[170] Translation of poly(A) containing mRNAs from EMC virus-infected Krebs-II cells in uninfected Krebs-II cell lysates shows a preferential translation of EMC RNA as compared to cellular mRNAs, reproducing the in vivo situation.[191] Unfortunately, no studies were conducted in this work to examine the regulatory role that monovalent ions might have in this system. Such analyses would be of interest to mimic the situation found in EMC-infected cells where hypotonic medium favors cellular translation and eliminates viral protein synthesis.[192]

The main conclusion reached with these in vitro experiments is that cardiovirus-mRNAs are good competitors against cellular mRNAs for translation. Although the exact message discriminatory component in the initiation machinery for which picornavirus RNA competes still remains in dispute, the physiological significance of these findings seems clear; once the picornavirus genome is delivered into the cell cytoplasm at the beginning of infection, it will compete efficiently with cellular mRNAs for translation. Later in the replication cycle when the bulk of viral mRNAs has been synthesized, competition will be facilitated by the modified salt concentrations that occur at these stages in the infected cell. The idea that competition plays a part in the shut-off phenomenon itself can be ruled out for two important reasons: first, shut-off in mengovirus or EMC virus-infected cells still occurs when viral mRNA synthesis is blocked (see Figure 10) and, second, because a decrease in total protein synthesis occurs during shut-off, a finding that is not compatible with the replacement of poor mRNAs by more efficient viral mRNAs.

Some in vitro and in vivo evidence available at present points to membrane modifications being responsible for the inhibition of host translation. As described above, monovalent ions discriminate between viral and cellular mRNA in such a way that under optimal conditions for picornavirus RNA translation, the synthesis of cellular mRNAs is severely depressed.[170]

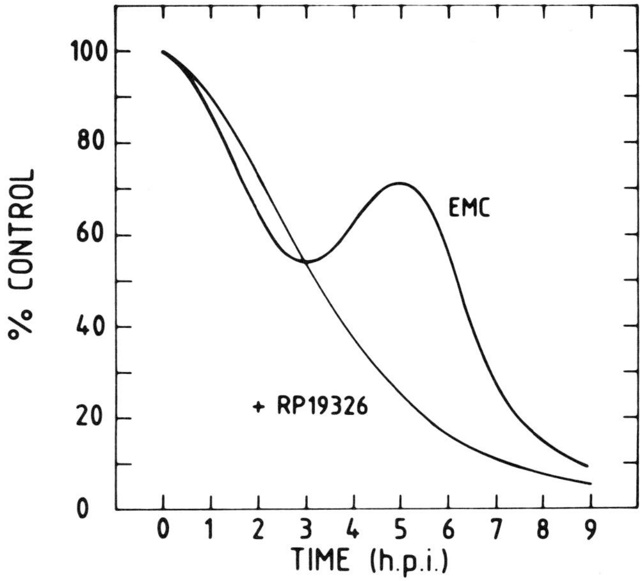

FIGURE 10. Time course of protein synthesis in EMC virus-infected L929 cells in the presence or absence of 100 μg/mℓ RP-19326 from zero time. The MOI was 100 PFU/cell. Protein synthesis was measured as (^{35}S)-methionine incorporated into TCA-precipitable material.

This discriminatory effect of monovalent ions is exerted at the level of the initiation of translation.[2,45] A simple change in monovalent cations directs the protein synthesizing system to translate mainly cellular or exclusively viral proteins when exogenous poly(A) containing mRNAs from polyoma-infected cells are exogenously added to the cell-free system.[171] More studies with mixtures of viral and cellular mRNAs from picornavirus-infected cells are required to evaluate the role that monovalent ions play in translation. Also, studies similar to those reported by Rosen et al.[172] on the effect of salt on eIF-2 binding to mengovirus RNA using other initiation factors known to be involved in the binding of mRNAs are necessary.

Compelling evidence obtained in EMC-infected cells indicates that monovalent ions play a central role in the regulation of translation. Thus, the kinetics of shut-off of host translation by EMC virus correlate with a decrease in the $^{86}Rb^{+}$ content of L cells.[180] The kinetics of shut-off of protein synthesis by EMC virus vary with different cell lines (L 929, HeLa, Vero, BHK-21), a decrease in ^{86} Rb^{+} content being observed p.i. in all cases[180] (Figure 11). The concentration of monovalent ions in the culture medium also has drastic effects on the proportion of viral or cellular proteins synthesized in EMC-infected HeLa cells.[192] As initially observed in poliovirus-infected cell, hypertonic media favor the shut-off of host translation, and have less effect on viral protein synthesis.[168] On the other hand, hypotonic media suppress EMC protein synthesis and are even able to reverse the shut-off in HeLa cells.[192] Another piece of evidence obtained from intact cells that implicates membrane modification in the shut-off phenomenon is that amphotericin B accelerates the inhibition of host protein synthesis, but has no effect on viral translation.[193] These results clearly implicate modifications in both the membrane and concentration of intracellular monovalent ions in the phenomenon of EMC virus-induced shut-off. However, the exact mechanism by which this occurs remains unknown, and its clarification must await the identification of the viral protein involved in shut-off, its location in the infected cell, and its precise mode of action.

Direct analysis of translation of capped mRNAs in EMC virus-infected cells conclusively

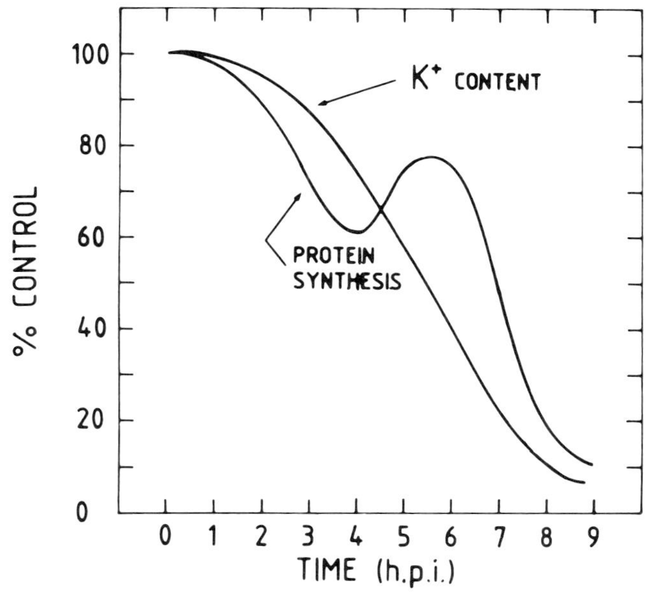

FIGURE 11. Kinetics of protein synthesis and K^+ content in EMC-infected HeLa cells. The MOI was 20 PFU/cell. The potassium content was measured after the infection of prelabeled cells maintained overnight in the presence of 1 uCi/mℓ $^{86}Rb^+$ in E4D2 medium.

indicated that the capped 26S mRNA from SFV was translated simultaneously with EMC RNA in HeLa cells doubly infected with both viruses.[194] This system of double infection is very neat in studying the factors that regulate the translation of capped and uncapped mRNAs, since infected cells exclusively translate two kinds of messengers, namely the picornavirus mRNA which is uncapped and the SFV late mRNA which is capped. Analysis of the influence of external ionic conditions in this system gave interesting results. Hypertonic medium shifted the translation towards picornavirus mRNAs, whereas hypotonic medium favored the synthesis of SFV late proteins.[194] Alternative changes to hypo- or hypertonic media shifted the protein translation machinery from capped to uncapped mRNA translation, respectively, and vice versa. As discussed in Section III.B, similar results were obtained with polio infected cells.[139]

In mengovirus-infected Krebs-II cells the shut-off of host translation precedes the loss of potassium ions, whereas the decrease in cellular ATP occurs after shut-off, as observed in other picornavirus-infected cells.[195] Recent studies in mengovirus-infected L cells describe an accumulation of 80S initiation complexes.[52] Cell-free translation systems have a decreased synthetic capacity, which is not relieved by the addition of reticulocyte initiation factors. The main inactive components found were the ribosomes that recovered their activity after high-potassium washing. This salt wash contained a ts and nuclease-resistant inhibitor. Based on these results, Pensiero and Lucas-Lenard[52] proposed a model whereby mengovirus infection induces an inhibitor that binds to ribosomes and blocks protein synthesis after the formation of 80S initiation complexes. The relationship between their findings, and those reported by Wright and Cooper[196] describing the association between structural poliovirus proteins VPO, VP1, and VP3 with ribosomes, has not yet been established.

V. ARE THERE DISTINCT MECHANISMS INVOLVED IN SHUT-OFF BY DIFFERENT PICORNAVIRUSES?

There are at least two different possibilities to account for the inhibition of translation in virus-infected cells and, more particularly, in picornavirus-infected cells. One is that the basic and primary mechanism common to almost all cytolytic viruses involves cell membrane damage.[193] An alternative view is that multiple mechanisms for inhibition of host translation exist, depending not only upon the animal virus studied, but also on the host cell used. However, it is still premature to decide between these two extreme possibilities, although the present evidence favors the multiple mechanism hypothesis. For example, if we accept, as established, that poliovirus infection inactivates an initiation factor and that this phenomenon is the primary event responsible for shut-off,[7] then it will clearly differ from the inhibition of cellular translation in EMC virus-infected cells which is mediated by membrane modification.[45] However, we believe that this suggestion needs to be substantiated with more convincing evidence from intact cell studies. The possibility that EMC, mengo-, and poliovirus block host translation by an indirect mechanism on the cell membrane also lacks evidence with respect to the molecular basis by which membrane modification occurs. Clearly, it is of fundamental importance to characterize the viral component responsible for shut-off in the different systems and its mode of action, in order to elucidate the mechanisms used by different picornaviruses.

It is known that the kinetics of inhibition of host translation by various picornaviruses, i.e., EMC virus and poliovirus, varies.[183,198] However, the kinetics depend not only on the virus used, but, more importantly, on the multiplicity of infection, the host cell, and several other undetermined events such as the state of cell growth, etc. In general, EMC virus shuts-down host translation less effectively than poliovirus. Thus, ten times more EMC virus gives a kinetics of inhibition similar to that of poliovirus.[149] However, the kinetics of inhibition doesn't say anything about the mechanism involved in shut-off. In this respect, the finding that HeLa cells doubly-infected with EMC virus and poliovirus support the simultaneous translation of both kinds of viral mRNAs is of particular interest.[199,200] The pattern of polypeptides synthesized was dependent on the time of addition of each virus and the relative multiplities of EMC and poliovirus used.[195] These results indicate that although the mechanisms employed by each picornavirus to shut-off translation may be different, they are compatible with each other. However, if advantage is given to the replication of one virus, it will interfere with translation of the other. This effect is not simply due to competition of mRNAs from each virus, because it is observed even if poliovirus RNA synthesis is blocked with guanidine.[199]

VI. CONCLUDING REMARKS

The mechanisms used by picornaviruses to interfere with host translation have been analyzed in great detail since the shut-off phenomenon was described more than 20 years ago. Cell-free systems obtained from poliovirus-infected cells are unable to translate capped cellular or viral mRNAs, and the defect has been assigned to the proteolytic cleavage of a 220 kdalton polypeptide that forms part of a cap binding complex. However, poliovirus-infected cells are able to translate several species of viral capped mRNAs. Therefore, the physiological significance of this cleavage remains obscure. The involvement of p220 cleavage in shut-off is questioned by the finding that this polypeptide remains intact when shut-off is induced by a poliovirus mutant that is unable to replicate.

Cardioviruses, such as EMC virus or mengovirus, may block cellular translation by causing alterations in the cell membrane that result in a redistribution of the gradients of ions that it maintains. It is still controversial whether the shut-off caused by different picornaviruses follows a similar mechanism or not, because the viral component responsible for depression of host translation remains unidentified and, consequently, its exact mode of action is unknown.

The availability of a number of inhibitors that block picornavirus replication without effect on the shut-off provides suitable model systems to study the shut-off in the absence of other secondary effects originated by virus replication.

Future work in this area will benefit from progress in cloning the region of the genome responsible for shut-off. Once this viral protein is identified, its obtainment in great quantities will help the elucidation of its mode of action in intact cells without the interference of the other viral components that are synthesized as a result of virus replication.

ACKNOWLEDGMENTS

FIS and CAICYT are acknowledged for financial support. We thank Ms. M. Teresa Aguado for kindly typing this manuscript. J. L. Castrillo is the holder of a Ministerio de Educación y Ciencia Fellowship.

REFERENCES

1. **Sangar, D. V.,** The replication of picornaviruses, *J. Gen. Virol.,* 45, 1, 1979.
2. **Carrasco, L. and Smith, A. E.,** Molecular biology of animal virus infection, *Pharm. Ther.,* 9, 331, 1980.
3. **Koch, G. and Koch, F.,** The Molecular Biology of Poliovirus, Springer-Verlag, Wien, 1985.
4. **Martin, E. M. and Kerr, I. M.,** Virus-induced changes in host cell macromolecular synthesis, *Symp. Soc. Gen. Microbiol.,* 187, 15, 1968.
5. **Roizman, B. and Spear, P. G.,** Macromolecular biosynthesis in animal cells infected with cytolytic viruses, *Curr. Top. Biol.,* 4, 79, 1969.
6. **Lucas-Lenard, J. M.,** Inhibition of cellular protein synthesis after virus infection, in *The Molecular Biology of Picornaviruses,* Pérez-Bercoff, R., Ed., Plenum Press, New York, 1979, 73.
7. **Ehrenfeld, E.,** Picornavirus inhibition of host cell protein synthesis, in *Viral Citopathology,* Fraenkel-Conrat, H. and Wagner, R. R., Eds., Plenum Press, New York, 1984; as cited in *Comprehensive Virol.,* 19, 177, 1984.
8. **Ackermann, W. W., Loh, P. C., and Payne, F. E.,** Studies of the biosynthesis of protein and ribonucleic acid in HeLa cells infected with poliovirus, *Virology,* 7, 170, 1959.
9. **Salzman, N. P., Lockart, R. Z., Jr., and Sebring, E. D.,** Alterations in HeLa cell metabolism resulting from poliovirus infection, *Virology,* 9, 244, 1959.
10. **Kerr, I. M., Martin, E. M., Hamilton, M. G., and Work, T. S.,** The initiation of virus protein synthesis in Krebs ascites tumor cells infected with EMC virus, *Cold Spring Harbor Symp. Quant. Biol.,* 27, 259, 1962.
11. **Zimmerman, E. F., Heeter, M., and Darnell, J. E.,** RNA synthesis in poliovirus-infected cells, *Virology,* 19, 400, 1963.
12. **Bablanian, R., Egger, H. J., and Tamm, I.,** I. The relation between poliovirus-induced metabolic and morphological alterations in cultured cells, *Virology,* 26, 100, 1965.
13. **Bablanian, R., Eggers, H. J., and Tamm, I.,** II. The relation between poliovirus growth and virus-induced morphological changes in cells, *Virology,* 26, 114, 1965.
14. **Franklin, R. M. and Baltimore, D.,** Patterns of macromolecular synthesis in normal and virus-infected mammalian cells, *Cold Spring Harbor Symp. Quant. Biol.,* 27, 175, 1962.
15. **Baltimore, D. and Franklin, R. M.,** The effect of mengovirus infection on the activity of the DNA-dependent RNA polymerase of L cells, *Proc. Natl. Acad. Sci. U.S.A.,* 48, 1383, 1962.
16. **Holland, J. J. and Peterson, J. A.,** Nucleic acid and protein synthesis during poliovirus infection of human cells, *J. Mol. Biol.,* 8, 556, 1964.
17. **McCormick, W. and Penman, S.,** Inhibition of RNA synthesis in HeLa and L cells by mengovirus, *Virology,* 31, 135, 1967.
18. **Bienz, K., Egger, D., Rasser, Y., and Loeffler, H.,** Differential inhibition of host cell RNA synthesis in several picornavirus-infected cell lines, *Intervirology,* 10, 209, 1978.
19. **Balandin, I. G. and Franklin, R. M.,** The effect of mengovirus infection on the activity of the DNA-dependent RNA polymerase of L-cells. II. Preliminary data on the inhibitory factor, *Biochem. Biophys. Res. Commun.,* 15, 27, 1964.
20. **Ho, P. K. and Washington, A. L.,** Evidence for a cellular ribonucleic acid synthesis inhibitor from poliovirus-infected HeLa cells, *Biochemistry,* 10, 3646, 1971.
21. **Miller, H. I. and Penhoet, E. E.,** Differential inhibition of nuclear RNA polymerases in L cells infected with mengovirus, *Proc. Soc. Exp. Biol. Med.,* 140, 435, 1972.

22. **Flores-Otero, G., Fernández-Tomás, C., and Gariglio-Vidal, P.,** DNA-bound RNA polymerases during poliovirus infection: reduction in the number of form II enzyme molecules, *Virology,* 116, 619, 1982.
23. **Apriletti, J. W. and Penhoet, E. E.,** Recovery of DNA-dependent RNA polymerase activities from L cells after mengovirus infection, *Virology,* 61, 597, 1974.
24. **Apriletti, J. W. and Penhoet, E. E.,** Cellular RNA synthesis in normal and mengovirus-infected L-929 cells, *J. Biol. Chem.,* 253, 603, 1978.
25. **Schwartz, L. B., Lawrence, C., Thach, R. E., and Roeder, R. G.,** Encephalomyocarditis virus infection of mouse plasmacytoma cells. II. Effect on host RNA synthesis and RNA polymerases, *J. Virol.,* 14, 611, 1974.
26. **Crawford, N. M., Fire, A., Samuels, M., Sharp, P. A., and Baltimore, D.,** Inhibition of transcription factor activity by poliovirus, *Cell,* 27, 555, 1981.
27. **Fernández-Tomás, C.,** The presence of viral induced proteins in nuclei from poliovirus infected HeLa cells, *Virology,* 116, 629, 1982.
28. **Lenk, R. and Penman, S.,** The cytoskeletal framework and poliovirus metabolism, *Cell,* 16, 289, 1979.
29. **Bonneau, A. M., Darveau, A., and Sonenberg, N.,** Effect of viral infection on host protein synthesis and mRNA association with the cytoplasmic cytoskeletal structure, *J. Cell Biol.,* 100, 1209, 1985.
30. **Wolff, D. A. and Bubel, H. C.,** The disposition of lysosomal enzymes as related to specific viral cytopathic effects, *Virology,* 24, 502, 1964.
31. **Guskey, L. E., Smith, P. C., and Wolff, D. A.,** Patterns of cytopathology and lysosomal enzyme release in poliovirus-infected HEp-2 cells treated with either 2-(hydroxybenzyl)-benzimidazole or guanidine HCl, *J. Gen. Virol.,* 5, 151, 1970.
32. **Bienz, K., Egger, D., and Wolff, D. A.,** Virus replication, cytopathology, and lysosomal enzyme response of mitotic and interphase Hep-2 cells infected with poliovirus, *J. Virol.,* 11, 565, 1973.
33. **Bossart, W. and Bienz, K.,** Virus replication, cytopathology, and lysosomal enzyme response in enucleated Hep-2 cells infected with poliovirus, *Virology,* 92, 331, 1979.
34. **Hotham-Iglewski, B. and Ludwig, E. H.,** Effect of cortisone on activation of lysosomal enzymes resulting from mengovirus infection of L929 cells, *Biochem. Biophys. Res. Commun.,* 22, 181, 1966.
35. **Miroff, G., Cornatzer, W. E., and Fischer, R. G.,** The effect of poliomyelitis virus type I (Mahoney strain) on the phosphorus metabolism of the HeLa cell, *J. Biol. Chem.,* 228, 255, 1957.
36. **Cornatzer, W. E., Sandstrom, W., and Fischer, R. G.,** The effect of poliomyelitis virus type I (Mahoney strain) on the phospholipid metabolism of the HeLa cell, *Biochim. Biophys. Acta,* 49, 414, 1961.
37. **Penman, S.,** Stimulation of the incorporation of choline in poliovirus infected cells, *Virology,* 25, 148, 1965.
38. **Penman, S. and Summers, D.,** Effects on host cell metabolism following synchronous infection with poliovirus, *Virology,* 27, 614, 1965.
39. **Mosser, A. G., Caliguiri, L. A., and Tamm, I.,** Incorporation of lipid precursors into cytoplasmic membranes of poliovirus infected HeLa cells, *Virology,* 47, 39, 1972.
40. **Skinner, M. S., Halperen, S., and Harkin, J. C.,** Cytoplasmic membrane-bound vesicles in echovirus 12-infected cells, *Virology,* 36, 241, 1968.
41. **Amako, K. and Dales, S.,** Cytopathology of mengovirus infection. II. Proliferation of membranous cisternae, *Virology,* 32, 201, 1967.
42. **Dales, S., Eggers, H. J., Tamm, L., and Palade, G. E.,** Electron microscopic study of the formation of poliovirus, *Virology,* 26, 379, 1965.
43. **Bienz, K., Egger, D., Rasser, Y., and Bossart, W.,** Kinetics and location of poliovirus macromolecular synthesis in correlation to virus-induced cytopathology, *Virology,* 100, 390, 1980.
44. **Vance, D. E., Trip, E. M., and Paddon, H. B.,** Poliovirus increases phosphatidylcholine biosynthesis in HeLa cells by stimulation of the rate-limiting reaction catalyzed by CTP: phosphocoline cytidylyltransferase, *J. Biol. Chem.,* 255, 1064, 1980.
45. **Carrasco, L. and Lacal, J. C.,** Permeabilization of cells during animal virus infection, *Pharm. Ther.,* 23, 109, 1983.
46. **Penman, S., Scherrer, K., Becker, Y., and Darnell, J. E.,** Polyribosomes in normal and poliovirus infected HeLa cells and the relationship to messenger RNA, *Proc. Natl. Acad. Sci. U.S.A.,* 49, 654, 1963.
47. **Willems, M. and Penman, S.,** The mechanism of host cell protein synthesis inhibition by poliovirus, *Virology,* 30, 355, 1966.
48. **Leibowitz, R. and Penman, S.,** Regulation of protein synthesis in HeLa cells. III. Inhibition during poliovirus infection, *J. Virol.,* 8, 661, 1971.
49. **Vázquez, D.,** Inhibitors of protein biosynthesis, in *Molecular Biology, Biochemistry and Biophysics,* Kleinzeller, A., Springer, G., and Wittmann, H., Eds., Springer-Verlag, Berlin, 1979, 30.
50. **Ehrenfeld, E. and Manis, S.,** Inhibition of 80S initiation complex formation by infection with poliovirus, *J. Gen. Virol.,* 43, 441, 1979.
51. **Brown, B. A. and Ehrenfeld, E.,** Initiation factor preparations from poliovirus-infected cells restrict translation in reticulocyte lysates, *Virology,* 103, 327, 1980.

52. **Pensiero, M. N. and Lucas-Lenard, J. M.,** Evidence for the presence of an inhibitor on ribosomes in mouse L cells infected with mengovirus, *J. Virol.*, 56, 161, 1985.
53. **Koschel, K.,** Poliovirus infection and poly-(A) sequences of cytoplasmic cellular RNA, *J. Virol.*, 13, 1061, 1974.
54. **Fernández-Muñoz, R. and Darnell, J. E.,** Structural difference between the 5′ termini of viral and cellular mRNA in poliovirus-infected cells: possible basis for the inhibition of host protein synthesis, *J. Virol.*, 18, 719, 1976.
55. **Colbi, D. S., Finnerty, V., and Lucas-Lenard, J.,** Fate of mRNA of L-cells infected with mengovirus, *J. Virol.*, 13, 858, 1974.
56. **Lawrence, C. and Thach, R. E.,** Encephalomyocarditis virus infection of mouse plasmacytoma cells. I. Inhibition of cellular protein synthesis, *J. Virol.*, 14, 598, 1974.
57. **Kaufmann, Y., Golstein, E., and Penman, S.,** Poliovirus-induced inhibition of polypeptide initiation in vitro on native polyribosomes, *Proc. Natl. Acad. Sci. U.S.A.*, 73, 1834, 1976.
58. **Hackett, P. B., Egberts, E., Buchwald, I., and Traub, P.,** Coding capacity of poly(A)$^+$ mRNA isolated from mengovirus-infected Ehrlich ascites tumor cells, *Mol. Biol. Rep.*, 3, 387, 1977.
59. **Gallwitz, D., Traub, U., and Traub, P.,** Fate of histone messenger RNA in mengovirus infected Ehrlich ascites tumor cells, *Eur. J. Biochem.*, 81, 387, 1977.
60. **Muñoz, A., Harvey, R., and Carrasco, L.,** Cellular RNA is not degraded in interferon treated HeLa cells after poliovirus infection, *FEBS Lett.*, 160, 87, 1983.
61. **Ehrenfeld, E. and Lund, H.,** Untranslated vesicular stomatitis virus messenger RNA after poliovirus infection, *Virology*, 80, 297, 1977.
62. **Holland, J. J.,** Inhibition of host cell macromolecular synthesis by high multiplicities of poliovirus under conditions preventing virus synthesis, *J. Mol. Biol.*, 8, 574, 1964.
63. **Eggers, H. J., Koch, M. A., Furst, A., Daves, G. D., Jr., Wilczynsk, J. J., and Folkers, K.,** Rhodanine: a selective inhibitor of the multiplication of echovirus 12, *Science*, 167, 294, 1970.
64. **Rosenwirth, B. and Eggers, H. J.,** Early process of echovirus 12 infection: elution, penetration, and uncoating under the influence of rhodanine, *Virology*, 97, 241, 1979.
65. **McSharry, J. J., Caliguiri, L. A., and Eggers, H. J.,** Inhibition of uncoating of poliovirus by arildone, a new antiviral drug, *Virology*, 97, 307, 1979.
66. **La Colla, P., Marcialis, M. A., Mereu, G. P., and Loddo, B.,** Specific inhibition of poliovirus induced blockade of cell protein synthesis by a thiopyrimidine derivate, *J. Gen. Virol.*, 17, 13, 1972.
67. **Lonber-Holm, K., Gosser, L. B., and Kauer, J. C.,** Early alteration of poliovirus in infected cells and its specific inhibition, *J. Gen. Virol.*, 27, 328, 1975.
68. **Bablanian, R.,** Depression of macromolecular synthesis in cells infected with guanidine-dependent poliovirus under restrictive conditions, *Virology*, 47, 255, 1972.
69. **Helentjaris, T. and Ehrenfeld, E.,** Inhibition of host cell protein synthesis by UV-inactivated poliovirus, *J. Virol.*, 21, 259, 1977.
70. **Butterworth, B. E.,** Proteolytic processing of animal virus proteins, *Curr. Top. Microb. Immunol.*, p. 1, 1977.
71. **Korant, B. D., Lonberg-Holm, K., and La Colla, P.,** Picornaviruses and Togaviruses, in *Targets for the Design of Antiviral Agents*, De Clerg, E. and Walker, R. T., Eds., Plenum Press, New York, 1984, 61.
72. **Korant, B. D., Brzin, J., and Turk, V.,** Cystatin, a protein inhibitor of cysteine proteases alters viral protein cleavages in infected human cells, *Biochem. Biophys. Res. Commun.*, 127, 1072, 1985.
73. **Borgert, K., Koschel, K., Tauber, H., and Wecker, E.,** Effect of hydroxylamine on early functions of poliovirus, *J. Virol.*, 8, 1, 1971.
74. **Collins, F. D. and Roberts, W. K.,** Mechanism of mengo virus-induced cell injury in L cells: use of inhibitors of protein synthesis to dissociate virus-specific events, *J. Virol.*, 10, 969, 1972.
75. **Muñoz, A. and Carrasco, L.,** Effect of interferon treatment on blockade of protein synthesis induced by poliovirus infection, *Eur. J. Biochem.*, 137, 623, 1983.
76. **Rightel, W. A., Dice, J. R., McAlpine, R. J., Timm, E. A., McLean, I. W., Jr., Dixon, G. J., and Schabel, F. M.,** Antiviral effect of guanidine, *Science*, 134, 558, 1961.
77. **Rightsel, W. A., Schneider, H. G., Sloan, B. J., Graf, P. R., Miller, F. A., Bartz, Q. R., Ehrlich, J., and Dixon, G. J.,** Antiviral activity of gliotoxin and gliotoxin acetate, *Nature (London)*, 204, 1333, 1964.
78. **Hollinshead, A. C. and Smith, P. K.,** Effects of certain purines and related compounds on virus propagation, *J. Pharmacol. Exp. Ther.*, 123, 54, 1958.
79. **Tamm, I. and Eggers, H. J.,** Differences in the selective virus inhibitory action of 2-(alpha-hydroxibenzyl)-benzimidazole and guanidine-HCl, *Virology*, 18, 439, 1962.
80. **Baltimore, D., Eggers, H. J., Franklin, R. M., and Tamm, I.,** Poliovirus induced RNA polymerase and the effects of virus specific inhibitors on its production, *Biochemistry*, 49, 843, 1963.

81. **Noble, J. and Levintow, L.,** Dynamics of poliovirus-specific RNA synthesis and the effects of inhibitors of virus replication, *Virology,* 40, 634, 1970.
82. **Bonina, L., Orzalesi, G., Merendino, R., Arena, A., and Mastroeni, P.,** Structure activity relationships of new antiviral compounds, *Antimicrob. Agents Chemother.,* 22, 1067, 1982.
83. **Powers, R. D., Gwaltney, J. M., Jr., and Hayden, F. G.,** Activity of 2-(3,4-dicholorophenoxy)-5-nitrobenzonitrile (MDL-860) against picornaviruses in vitro, *Antimicrob. Agents Chemother.,* 22, 639, 1982.
84. **Torney, H. L., Dulworth, J. K., and Steward, D. L.,** Antiviral activity and mechanism of action of 2-(3,4-dicholorophenoxy)-5-nitrobenzonitrile (MDL-860), *Antimicrob. Agents Chemother.,* 22, 635, 1982.
85. **Eggers, H. J.,** Antiviral agents against picornaviruses, *Antiviral Res.,* Supp. 1, 57, 1985.
86. **Van Hoof, L., Vanden Berghe, D. A., Hatfield, G. M., and Vlietinck, A. J.,** Plant antiviral agents. V. 3-Methoxiflavones as potent inhibitors of viral induced block of cell synthesis, *Planta Med.,* 6, 513, 1984.
87. **Ishitsuka, H., Ohsawa, C., Ohiwa, T., Umeda, I., and Suhara, Y.,** Antipicornavirus flavone R0 09-0179, *Antimicrob. Agents Chemother.,* 22, 611, 1982.
88. **Ishitsuka, H., Ninomiya, Y., Ohsawa, C., Fujiu, M., and Suhara, Y.,** Direct and specific inactivation of rhinovirus by chalcone Ro 09-0410, *Antimicrob. Agents Chemother.,* 22, 617, 1982.
89. **Bauer, D. J., Selway, J. W. T., Batchelor, J. F., Tisdale, M., Caldwell, I., and Young, D.,** 4′,6-Dichloroflavan (BW683C), a new anti-rhinovirus compound, *Nature (London),* 292, 369, 1981.
90. **Merryman, P., Jaffe, I. A., and Ehrenfeld, E.,** Effect of D-penicillamine on poliovirus replication in HeLa cells, *J. Virol.,* 13, 881, 1974.
91. **Castrillo, J. L., Vanden Bergue, D., and Carrasco, L.,** 3-Methylquercetin is a potent and selective inhibitor of poliovirus RNA-synthesis, *Virology,* 152, 219, 1986.
92. **Hewlett, M. J., Axelrod, J. H., Antinoro, N., and Feld, R.,** Isolation and preliminary characterization of temperature sensitive mutants of poliovirus type 1, *J. Virol.,* 41, 1089, 1982.
93. **Young, S. A. and Radloff, R. J.,** Functional characterization of cleavage-defective mutants of EMC virus, *J. Virol.,* 42, 814, 1982.
94. **Radloff, R. J.,** Isolation and preliminary characterization of temperature-sensitive mutants of EMC virus, *J. Virol.,* 27, 182, 1978.
95. **Steiner-Pryor, A. and Cooper, P. D.,** Temperature-sensitive poliovirus mutants defective in repression of host protein synthesis are also defective in structural protein, *J. Gen. Virol.,* 21, 215, 1973.
96. **Cole, Ch. N. and Baltimore, D.,** Defective interfering particles of poliovirus. II. Nature of the defect, *J. Mol. Biol.,* 76, 325, 1973.
97. **Cole, Ch. N. and Baltimore, D.,** Defective interfering particles of poliovirus. III. Interference and enrichment, *J. Mol. Biol.,* 76, 345, 1973.
98. **Anderson-Sillman, K., Bartal, S., and Tershak, D. R.,** Guanidine-resistant poliovirus mutants produce modified 37-kilodalton proteins, *J. Virol.,* 50, 922, 1984.
99. **Muñoz, A., Alonso, M. A., and Carrasco, L.,** Synthesis of heat-shock proteins in HeLa cells: inhibition by virus infection, *Virology,* 137, 150, 1984.
100. **Perez-Bercoff, R. and Kaempfer, R.,** Genomic RNA of mengovirus. V. Recognition of common features by ribosomes and eucaryotic initiation factor 2, *J. Virol.,* 41, 30, 1982.
101. **Ochoa, S. and de Haro, C.,** Regulation of protein synthesis in eukaryotes, *Annu. Rev. Biochem.,* 45, 549, 1979.
102. **Kaempfer, R.,** Regulation of eukaryotic translation, in *Viral Citopathology,* Fraenkel-Conrat, H. and Wagner, R. R., Eds., Plenum Press, New York, 1984.
103. **Hunt, T. and Ehrenfeld, E.,** Cytoplasmic from poliovirus infected HeLa cells inhibits cell-free haemoglobin synthesis, *Nature London New Biol.,* 230, 91, 1971.
104. **Cordell-Stewart, B. and Taylor, M. W.,** Effect of double-stranded viral RNA on mammalian cells in culture, *Proc. Natl. Acad. Sci. U.S.A.,* 68, 1326, 1971.
105. **Rosen, H., Knoller, S., and Kaempfer, R.,** Messenger ribonucleic acid specificity in the inhibition of eukaryotic translation by double-stranded ribonucleic acid, *Biochemistry,* 20, 3011, 1981.
106. **Celma, M. L. and Ehrenfeld, E.,** Effect of poliovirus double-stranded RNA on viral and host-cell protein synthesis, *Proc. Natl. Acad. Sci. U.S.A.,* 71, 2440, 1974.
107. **Helentjaris, T. and Ehrenfeld, E.,** Control of protein synthesis in extracts from poliovirus-infected cells. I. mRNA Discrimination by crude initiation factors, *J. Virol.,* 26, 510, 1978.
108. **Egberts, E., Hackett, P. B., and Traub, P.,** Protein synthesis in postnuclear supernatants from mengovirus-infected Ehrlich ascites tumor cells, *Physiol. Chem.,* 357, 1779, 1976.
109. **Rose, J. K., Traschsel, H., Leong, K., and Baltimore, D.,** Inhibition of translation by poliovirus: inactivation of a specific initiation factor, *Proc. Natl. Acad. Sci. U.S.A.,* 75, 2732, 1978.
110. **Schreier, M. H., Erni, B., and Staehelin, T.,** Initiation of mammalian protein synthesis. I. Purification and characterization of seven initiation factors, *J. Mol. Biol.,* 116, 727, 1977.

111. **Helentjaris, T., Ehrenfeld, E., Brown-Luedi, M. L., and Hershey, J. W. B.,** Alterations in initiation factor activity from poliovirus infected HeLa cells, *J. Biol. Chem.,* 254, 10973, 1979.
112. **Parets-Soler, A., Reibel, L., and Schapira, G.,** Differential stimulation of alpha and beta globin mRNA translation by M_r 50000 and 28000 polypeptide containing fractions isolated from reticulocyte polysomes, *FEBS Lett.,* 136, 259, 1981.
113. **Trachsel, H., Sonenberg, N., Shatkin, A. J., Rose, J. K., Leong, K., Bergmann, J. E., Gordon, J., and Baltimore, D.,** Purification of a factor that restores translation of vesicular stomatitis virus mRNA in extracts from poliovirus-infected HeLa cells, *Proc. Natl. Acad. Sci. U.S.A.,* 77, 770, 1980.
114. **Sonenberg, N., Trachsel, H., Hecht, S., and Shatkin, A. J.,** Differential stimulation of capped mRNA translation in vitro by cap binding proteins, *Nature (London),* 285, 331, 1980.
115. **Hansen, J. and Ehrenfeld, E.,** Presence of the cap-binding protein in initiation factor preparations from poliovirus-infected HeLa cells, *J. Virol.,* 38, 438, 1981.
116. **Tahara, S. M., Morgan, M. A., and Shatkin, A. J.,** Two forms of purified m^7G-cap binding protein with different effects on capped mRNA translation in extracts of uninfected and poliovirus-infected HeLa cells, *J. Biol. Chem.,* 256, 7691, 1981.
117. **Shatkin, A. J.,** mRNA Cap binding proteins: essential factors for initiating translation, *Cell,* 40, 223, 1985.
118. **Sonenberg, N.,** ATP/Mg^{++}-dependent cross-linking of cap binding proteins to the 5′ end of eukaryotic mRNA, *Nucl. Acid Res.,* 9, 1643, 1981.
119. **Lee, K. A. W., Guertin, D., and Sonenberg, N.,** mRNA Secondary structure as determinant in cap recognition and initiation complex formation, *J. Biol. Chem.,* 258, 707, 1983.
120. **Edery, I., Lee, K. A. W., and Sonenberg, N.,** Functional characterization of eukaryotic mRNA cap binding protein complex: effects on translation of capped and naturally uncapped RNAs, *Biochemistry,* 23, 2456, 1984.
121. **Ray, B. K., Brendler, T. G., Adya, S., Daniels-McQueen, S., Miller, J. K., Hersmey, J. W. B., Grifo, J. A., Merrick, W. C., and Thach, R. E.,** Role of mRNA competition in regulating translation: further characterization of mRNA discriminatory initiation factors, *Proc. Natl. Acad. Sci. U.S.A.,* 80, 663, 1983.
122. **Ray, B. K., Lawson, T. G., Kramer, J. C., Cladaras, M. H., Grifo, J. A., Aramson, R. D., Merrick, W. C., and Thach, R. E.,** ATP-dependent unwinding of messenger RNA structure by eukaryotic initiation factors, *J. Biol. Chem.,* 260, 7651, 1985.
123. **Kozak, M.,** Influence of mRNA secondary structure on binding and migration of 40S ribosomal subunits, *Cell,* 19, 79, 1980.
124. **Hansen, J., Etchison, D., Hershey, J. W. B., and Ehrenfeld, E.,** Association of cap-binding protein with eucaryotic initiation factor 3 in initiation factor preparations from uninfected and poliovirus-infected HeLa cells, *J. Virol.,* 42, 200, 1982.
125. **Lee, K. A. W. and Sonenberg, N.,** Inactivation of cap-binding proteins accompanies the shut-off of host protein synthesis by poliovirus, *Proc. Natl. Acad. Sci. U.S.A.,* 79, 3447, 1982.
126. **Etchison, D., Milburn, S. C., Edery, I., Sonenberg, N., and Hershey, J. W. B.,** Inhibition of HeLa cell protein synthesis following poliovirus infection correlates with the proteolysis of a 220K polypeptide associated with eIF-3 and CBP, *J. Biol. Chem.,* 257, 14806, 1982.
127. **Duncan, R. and Hershey, J. W. B.,** Identification and quantitation of levels of protein synthesis initiation factors in crude HeLa cell lysates by 2-D polyacrilamide electrophoresis, *J. Biol. Chem.,* 258, 7228, 1983.
128. **Duncan, R., Etchison, D., and Hershey, J. W. B.,** Protein synthesis eukaryotic initiation factors 4A and 4B are not altered by poliovirus infection of HeLa cells, *J. Biol. Chem.,* 258, 7236, 1983.
129. **Sonenberg, N., Guertin, D., Cleveland, D., and Trachsel, H.,** Probing the function of the eucaryotic 5′ cap structure by using a monoclonal antibody directed against cap-binding proteins, *Cell,* 27, 563, 1981.
130. **Etchison, D., Hansen, J., Ehrenfeld, E., Edery, I., Sonenberg, N., Milburn, S., and Hershey, J. W. B.,** Demonstration in vitro that eucaryotic initiation factor 3 is active but that a cap-binding protein complex is inactive in poliovirus-infected cells, *J. Virol.,* 51, 832, 1984.
131. **Lee, K. A. W., Edery, I., and Sonenberg, N.,** Isolation and structural characterization of cap-binding proteins from poliovirus-infected HeLa cells, *J. Virol.,* 54, 515, 1985.
132. **Lee, K. A. W., Edery, I., Hanecak, R., Wimmer, E., and Sonenberg, N.,** Poliovirus protease 3C (P3-7c) does not cleave p220 of the eucaryotic mRNA cap-binding protein complex, *J. Virol.,* 55, 489, 1985.
133. **Lloyd, R. E., Etchison, D., and Ehrenfeld, E.,** Poliovirus protease does not mediate cleavage of the 220,000-Da component of the cap binding protein complex, *Proc. Natl. Acad. Sci. U.S.A.,* 82, 2723, 1985.
134. **Bernstein, H. D., Sonenberg, N., and Baltimore, D.,** A mutant of poliovirus that does not selectively inhibit host cell protein synthesis, *Mol. Cell. Biol.,* 5, 2913, 1985.
135. **Mosenkis, J., Daniels-McQueen, S., Janovec, S., Duncan, R., Hershey, J. W. B., Grifo, J. A., Merrick, W. C., and Thach, R. E.,** Shut-off of host translation by encephalomyocarditis virus infection does not involve cleavage of the eucaryotic initiation factor 4F polypeptide, *J. Virol.,* 54, 643, 1985.

136. **Etchison, D. and Fout, S.,** Human rhinovirus 14 infection of HeLa cells results in the proteolytic cleavage of the p220 cap-binding complex subunit and inactive globin mRNA translation, *J. Virol.*, 54, 634, 1985.
137. **Panniers, R., Stewart, E. B., Merrick, W. C., and Henshaw, E. C.,** Mechanism of inhibition of polypeptide chain initiation in heat-shocked Ehrlich cells involves reduction of eukaryotic initiation factor 4F activity, *J. Biol. Chem.*, 260, 9648, 1985.
138. **Schlesinger, M. J., Ashburner, M., and Tissieres, A.,** *Heat-Shock. From Bacteria to Man*, Cold Spring Harbor Laboratory, Cold Spring Harbor, N.Y., 1982.
139. **Choppin, P. W. and Holmes, K. V.,** Replication of SV5 RNA and the effects of superinfection with poliovirus, *Virology*, 33, 442, 1967.
140. **Alonso, M. A. and Carrasco, L.,** Translation of capped viral mRNAs in poliovirus-infected HeLa cells, *EMBO J.*, 1, 913, 1982.
141. **Bossart, W. and Bienz, K.,** Virus replication, cytopathology, and lysosomal enzyme response in enucleated Hep-2 cells infected with poliovirus, *Virology*, 92, 331, 1979.
142. **Hackett, P. B., Egberts, E., and Traub, P.,** Translation of ascites and mengovirus RNA in fractionated cell-free systems from uninfected and mengovirus-infected Ehrlich-ascites-tumor cells, *Eur. J. Biochem.*, 83, 341, 1978.
143. **Centrella, M. and Lucas-Lenard, J.,** Regulation of protein synthesis in vesicular stomatitis virus-infected mouse L-929 cells by decreased protein synthesis initiation factor 2 activity, *J. Virol.*, 41, 781, 1982.
144. **Thomas, J. R. and Wagner, R. R.,** Inhibition of translation in lysates of mouse L cells infected with vesicular stomatitis virus: presence of a defective ribosome-associated factor, *Biochemistry*, 22, 1540, 1983.
145. **Dratewka-Kos, E., Kiss, I., Lucas-Lenard, J., Mehta, H. B., Woodley, C. L., and Wahba, A. J.,** Catalytic utilization of eIF-2 and mRNA binding proteins are limiting in lysates from vesicular stomatitis virus infected L cells, *Biochemistry*, 23, 6184, 1984.
146. **Lodish, H. F. and Porter, M.,** Translation control of protein synthesis after infection by vesicular stomatitis virus, *J. Virol.*, 36, 719, 1980.
147. **Lodish, H. F. and Porter, M.,** Vesicular stomatitis virus mRNA and inhibition of translation of cellular mRNA. Is there a P function in vesicular stomatitis virus?, *J. Virol.*, 38, 504, 1981.
148. **Schnitzlein, W. M., O'Banion, M. K., Poirot, M. K., and Reichmann, M. E.,** Effect of intracellular vesicular stomatitis virus mRNA concentration on the inhibition of host cell protein synthesis, *J. Virol.*, 45, 206, 1983.
149. **Muñoz, A. and Carrasco, L.,** Action of human lymphoblastoid interferon on HeLa cells infected with RNA-containing animal viruses, *J. Gen. Virol.*, 65, 377, 1984.
150. **Van Steeg, H., Thomas, A., Verbeek, S., Kasperaitis, M., Voorma, H. O., and Benne, R.,** Shutoff of neuroblastoma cell protein synthesis by Semliki forest virus: loss of ability of crude initiation factors to recognize early SFV and host mRNA's, *J. Virol.*, 38, 728, 1981.
151. **Van Steeg, H., Grinsven, M. V., Mansfeld, F. V., Voorma, H. O., and Benne, R.,** Initiation of protein synthesis in neuroblastoma cells infected by SFV. A decreased requirement of late viral mRNA for eIF-4B and cap binding protein, *FEBS Lett.*, 129, 62, 1981.
152. **Van Steeg, H. V., Kasperaitis, M., Voorma, H. O., and Benne, R.,** Infection of neuroblastoma cells by SFV. The interference of viral capsid protein with the binding of host RNAs into initiation complexes is the cause, *Eur. J. Biochem.*, 138, 473, 1984.
153. **Jalanko, A.,** Expression of Semliki forest virus capsid protein from SV40 recombinant virus, *FEBS Lett.*, 186, 59, 1985.
154. **Simizu, B.,** Inhibition of host cell macromolecular synthesis by togavirus, in *Viral Citopathology*, Fraenkel-Conrat, H. and Wagner, R. R., Eds., Plenum Press, New York, 1984, 465.
155. **Munoz, A., Alonso, M., and Carrasco, L.,** The regulation of translation in reovirus-infected cells, *J. Gen. Virol.*, 66, 2161, 1985.
156. **Joklik, W. K.,** Structure and function of the reovirus genome, *Microbiol. Rev.*, 45, 483, 1981.
157. **Joklik, W. K.,** The reovirus particle, in *The Viruses, the Reoviridae*, Fraenkel-Conrat, H. and Wagner, R. R., Eds., Plenum Press, New York, 1983, 9.
158. **Munemitsu, S. M. and Samuel, C. E.,** Biosynthesis of reovirus specified polypeptides, *Virology*, 136, 133, 1984.
159. **Sharpe, A. E. and Fields, B. N.,** Reovirus inhibition of cellular RNA and protein synthesis: role of the S4 gene, *Virology*, 122, 381, 1982.
160. **Fields, B. N. and Greene, M. I.,** Genetic and molecular mechanisms of viral pathogenesis: implications for prevention and treatment, *Nature (London)*, 300, 19, 1982.
161. **Skup, D. and Millward, S.,** mRNA Capping enzymes are masked in reovirus progeny subviral particles, *J. Virol.*, 34, 490, 1980.
162. **Skup, D. and Millward, S.,** Reovirus induced modification of cap dependent translation in infected cells, *Proc. Natl. Acad. Sci. U.S.A.*, 77, 152, 1980.
163. **Zarbl, H., Skup, D., and Millward, S.,** Reovirus progeny subviral particles synthesize uncapped mRNA, *J. Virol.*, 34, 497, 1980.

164. **Skup, D., Zarbl, H., and Millward, S.,** Regulation of translation in L-cells infected with reovirus, *J. Mol. Biol.*, 151, 35, 1981.
165. **Morgan-Detjen, B., Walden, W. E., and Thach, R. E.,** Translational specificity in reovirus-infected mouse fibroblasts, *J. Biol. Chem.*, 257, 9855, 1982.
166. **Lemieux, R., Zarbl, H., and Millward, S.,** mRNA Discrimination in extracts from uninfected and reovirus infected L cells, *J. Virol.*, 51, 215, 1984.
167. **Fresno, M. and Vázquez, D.,** Inhibitory effects of "cap" analogues on globin mRNA and encephalomyocarditis RNA translation in a reticulocyte cell-free system, *Eur. J. Biochem.*, 103, 125, 1980.
168. **Saborio, J. L., Pong, S. S., and Koch, G.,** Selective and reversible inhibition of initiation of protein synthesis in mammalian cells, *J. Mol. Biol.*, 85, 195, 1974.
169. **Nuss, D. L., Oppermann, H., and Koch, G.,** Selective blockage of initiation of host protein synthesis in RNA-virus-infected cells, *Proc. Natl. Acad. Sci. U.S.A.*, 72, 1258, 1975.
170. **Carrasco, L. and Smith, A. E.,** Sodium ions and the shut-off of host cell protein synthesis by picornaviruses, *Nature (London)*, 264, 807, 1976.
171. **Carrasco, L., Harvey, R., Blanchard, C., and Smith, A. E.,** Regulation of translation of eukaryotic virus mRNAs, in *Modern Trends in Human Leukemia*, Neth, R., Hofscheneider, P., and Mannweller, K., Eds., Springer-Verlag, Basel, 1979, 1889.
172. **Rosen, H., Di Segni, G., and Kaempfer, R.,** Translation control by messenger RNA competition for eukaryotic initiation factor 2, *J. Biol. Chem.*, 257, 946, 1982.
173. **Levanon, A., Kohn, A., and Inbar, M. F.,** Increase in lipid fluidity of cellular membranes induced by adsorption of RNA and DNA virions, *J. Virol.*, 22, 353, 1977.
174. **Fernández-Puentes, C. and Carrasco, L.,** Viral infection permeabilizes mammalian cells to protein toxins, *Cell*, 20, 769, 1980.
175. **Carrasco, L.,** Modification of membrane permeability induced by animal viruses early in infection, *Virology*, 113, 623, 1981.
176. **Muñoz, A., Castrillo, J. L., and Carrasco, L.,** Modification of membrane permeability during Semliki Forest virus infection, *Virology*, 146, 203, 1985.
177. **Castrillo, J. L. and Carrasco, L.,** Increased inhibition of cellular RNA synthesis by alpha-amanitin during entry of viruses into animal cells, *FEMS Microb. Lett.*, 26, 221, 1985.
178. **Lopez-Rivas, A., Castrillo, J. L., and Carrasco, L.,** Cation content in poliovirus-infected HeLa cells, *J. Gen. Virol.*, in press.
179. **Wilson, D. B.,** Effect of the lambda S gene product on properties of the *Escherichia coli* inner membrane, *J. Bacteriol.*, 151, 1403, 1982.
180. **De-Herreros, A. G., De-Haro, C., and Ochoa, S.,** Mechanism of activation of the heme-stabilized translational inhibitor of reticulocyte lysates by calcium ions and phospholipid, *Proc. Natl. Acad. Sci. U.S.A.*, 82, 3119, 1985.
181. **Dales, S. and Franklin, R. M.,** A comparison of the changes in fine structure of L cells during single cycles of viral multiplication, following their infection with Mengo and EMC viruses, *J. Cell. Biol.*, 14, 281, 1962.
182. **Martin, E. M., Malec, J., Sved, S., and Work, T. S.,** Studies on protein and nucleic acid metabolism in virus infected mammalian cells. I. Encephalomyocarditis virus in Krebs II mouse ascites tumour cells, *Biochem. J.*, 80, 585, 1961.
183. **Lacal, J. C. and Carrasco, L.,** Relationship between membrane integrity and the inhibition of host translation in virus-infected mammalian cells. Comparative studies between EMC and poliovirus, *Eur. J. Biochem.*, 127, 359, 1982.
184. **Svitkin, Yu. V., Ugarova, T. Yu., Ginevskaya, V. A., Kalinina, N. O., Scarlat, I. V., and Agol, V. I.,** Efficiency of translation of viral and cellular mRNAs in extracts from cells infected with EMC virus, *Intervirology*, 4, 214, 1974.
185. **Abreu, S. L. and Lucas-Lenard, J.,** Cellular protein synthesis shut-off by mengovirus: translation of nonviral and viral mRNA's in extracts from uninfected and infected Ehrlich ascites tumor cells, *J. Virol.*, 18, 182, 1976.
186. **Lebleu, B., Nudel, U., Falcoff, E., Prives, C., and Revel, M.,** A comparison of the translation of mengo virus RNA and globin mRNA in Krebs ascites cell-free extracts, *FEBS Lett.*, 25, 97, 1972.
187. **Golini, F., Thach, S. S., Birge, C. H., Safer, B., Merrick, W. C., and Thach, R. E.,** Competition between cellular and viral mRNAs in vitro is regulated by a messenger discriminatory initiation factor, *Proc. Natl. Acad. Sci. U.S.A.*, 73, 3040, 1976.
188. **Baglioni, C., Simili, M., and Shafritz, D. A.,** Initiation activity of EMC virus RNA, binding to initiation factor eIF-4B and shut-off of host cell protein synthesis, *Nature (London)*, 275, 240, 1978.
189. **Shatkin, A. J.,** Capping of eukaryotic mRNAs, *Cell*, 9, 645, 1976.
190. **Wigle, D. T. and Smith, A. E.,** Specificity in initiation of protein synthesis in a fractionated mammalian cell-free system, *Nature (London)*, 242, 136, 1973.

191. **Svitkin, Yu. V., Ginevskaya, V. A., Ugarova, T. Yu., and Agol, V. I.,** A cell-free model of the encephalomyocarditis virus induced inhibition of host cell protein synthesis, *Virology,* 87, 199, 1978.
192. **Alonso, M. A. and Carrasco, L.,** Reversion by hypotonic medium of the shut-off of protein synthesis induced by encephalomyocarditis virus, *J. Virol.,* 37, 535, 1981.
193. **Alonso, M. A. and Carrasco, L.,** Selective inhibition of cellular protein synthesis by amphotericin B in EMC virus-infected cells, *Virology,* 114, 247, 1981.
194. **Alonso, M. A. and Carrasco, L.,** Translation of capped virus mRNA in encephalomyiocarditis virus-infected cells, *J. Gen. Virol.,* 60, 315, 1982.
195. **Egberts, E., Hackett, P. B., and Traub, P.,** Alteration of the intracellular energetic and ionic conditions by mengovirus infection of Ehrlich ascites protein synthesis in the midphase of infection, *J. Virol.,* 22, 591, 1977.
196. **Wright, P. J. and Cooper, P. D.,** Poliovirus proteins associated with ribosomal structures in infected cells, *Virology,* 59, 1, 1974.
197. **Carrasco, L.,** The inhibition of cell functions after viral infection, *FEBS Lett.,* 76, 11, 1977.
198. **Jen, G., Morgan, B., and Thach, R. E.,** Shut-off on HeLa cell protein synthesis by encephalomyocarditis virus and poliovirus: a comparative study, *J. Virol.,* 35, 150, 1980.
199. **Alonso, M. A. and Carrasco, L.,** Protein synthesis in HeLa cells double infected with encephalomyocarditis virus and poliovirus, *J. Gen. Virol.,* 61, 15, 1982.
200. **Detjen, B. M., Jen, G., and Thach, R. E.,** Encephalomyocarditis viral RNA can be translated under conditions of polio induced translation shut-off in vivo, *J. Virol.,* 38, 777, 1981.
201. **Lopez-Rivas, A. and Carrasco, L.,** unpublished observations.

Chapter 7

EFFECTS OF INTERFERON ON VIRAL REPLICATION AND ON THE CYTOPATHIC EFFECTS INDUCED BY ANIMAL VIRUSES

A. Muñoz and L. Carrasco

TABLE OF CONTENTS

I. INTRODUCTION

Interferons are a group of proteins produced by vertebrate cells in response to viruses or other stimuli that exert a wide range of biological activities, among which their ability to interfere with the replication of many viruses has been one of the most studied.

Different molecules of interferon (IFN) exist: over eight subtypes are synthesized by human leukocytes (IFN-αs) stimulated by viruses;[1] at least one other is produced by fibroblasts (IFN-β) when they are induced by double-stranded RNAs or viruses; and finally another type composed of two glycosylated forms is released by lymphocytes (IFN-γ) stimulated by lectins or mitogens.[2] All of these molecules display the same general biological activities, although some minor differences can be observed in some particular systems when their mechanism of action is analyzed in detail. It should be pointed out the IFN types and subtypes differ in their cross-species activity, their capacity to protect a given cell type against different viruses, and their spectrum of cell specificities.[3,4] In addition to their capacity to block viral replication, interferons reduce the rate of cell growth[5-7] and have profound effects on the activity of a number of cells of the immune systems.[8-10] Therefore, a great deal of interest has been concentrated during recent years on the potential of interferons as antiviral and antitumor agents.[11-13] However, the mechanisms by which IFN exerts these effects are not yet clear. In this chapter, we will review the data concerning the effect of interferon on the infection of mammalian cells by animal viruses, with special emphasis on its effect on the interference of the latter with the host cell and the molecular basis of this activity.

In addition to the blockade of the replication of many viruses in infected cells, interferons have an indirect antiviral action in the whole organism by acting on several cells of the immune system responsible for the defense against virus infections, i.e., interferons enhance NK, macrophage, or killer T cell activities, and they also augment expression of Fc receptors, histocompatibility antigens, etc.[14]

Although some of the effects of different IFNs observed in cell-extracts with cultured cells cannot be directly extrapolated to the organism level, the results obtained during recent years with different virus/cell systems and different IFN preparations might be relevant to the understanding of the mechanism of the antiviral action of IFN in the organism, as we will see below.

II. MECHANISM OF ANTIVIRAL ACTION OF INTERFERON: IN VITRO MODELS

A few years after Isaacs and Lindenmann[15] initially described IFN as a substance released by virus-infected cells capable of conferring resistance against viruses in other cells, it became clear that IFN acted as a complex regulator of cell function. IFN was able to trigger cellular mechanisms responsible for the establishment of the so-called antiviral state that appeared to be dependent on the presence of the cell nucleus[16,17] and RNA and protein synthesis.[18-20]

Treatment of cells with IFNs induces the synthesis of several mRNAs that code for at least 6 to 12 proteins,[21] some of which are specific for IFN.[22-24] The availability of cloned cDNAs corresponding to several of the mRNAs induced by IFN will allow us to characterize these proteins in more detail and to further our understanding of the molecular basis of the antiviral state.[21]

The first step in the action of interferon is the binding to high affinity receptors on the cell surface. While IFN-α and IFN-β share a common receptor, IFN-γ binds to a different cell-surface receptor.[25,26] The signal that induces IFN action after this initial binding to the receptor is still unknown. One possibility is that IFN, like many polypeptides, exerts its action from the cell surface generating a second messenger that migrates from the membrane to the nucleus and activates specific cellular genes. This hypothetical second messenger has not been identified yet, since cyclic GMP or cAMP do not appear to be involved in this process.[27] Alternatively, the interferon molecule internalized as a complex with its receptor, or a cleavage product of one of them, may trigger the cellular responses directly.[28,29] According to the first hypothesis, Vengris et al.[30] suggested that for IFN to induce the antiviral state, it must be secreted to the external medium. Moreover, Huez et al.[31] have recently observed that direct microinjection into the cell nucleus of IFN does not induce the antiviral state. However, it may be that interferon is only "activated" after entering the cell through physiological pathways.

Early attempts to explain the mechanism of IFN-induced antiviral activity were made by Marcus and Salb,[32] who proposed the existence of an antiviral protein induced by interferon that bound to ribosomes specifically inhibiting viral mRNA translation. In agreement with this hypothesis, Carter and Levy[33,34] reported the inability of mengovirus mRNA to bind to ribosomes obtained from IFN-treated mouse L cells. These data, however, have not been reproduced by other laboratories whose results argued strongly against an effect of IFN treatment on the binding of viral mRNA to ribosomes.[35,36]

A. The Endonuclease Pathway

The study of IFN mechanism of action was undoubtedly stimulated by the development of cell-free protein synthesizing systems.[37] Several groups[38-41] reported the inhibition of viral mRNA translation in extracts from IFN-treated cells. Friedman and co-workers[42] claimed that this inhibition occurred only in extracts from IFN-treated virus-infected cells. With only one exception,[40] in which a specific blockade of viral translation was observed, the translation of both cellular and viral mRNAs was found to be inhibited in such extracts.[43-45]

Kerr et al.[46] observed that extracts from IFN-treated cells were more sensitive to the inhibition of protein synthesis by exogenous double-stranded RNA (dsRNA) as compared to control cell extracts. This inhibitory effect also required the addition of ATP. This inhibition persisted, even after the removal of these agents (dsRNA, ATP), suggesting that some stable inhibitors were generated. Different groups[47-50] showed the activation of a protein kinase and a latent endonuclease in IFN-treated cell extracts incubated with dsRNA and ATP. At least two proteins were phosphorylated: one of about 67 to 73 kdalton (P1), which appeared to be the result of an autophosphorylation,[51] and a 35 to 38 kdalton polypeptide that was identified as the α subunit of the translation initiation factor eIF-2.[47] Phosphorylation of this factor rendered it inactive, and as a consequence, protein synthesis was blocked (Figure 1). Recently, Hovanessian et al.[52] found that this protein kinase system was composed of at least two proteins: a p68 kinase and an eIF-2 kinase, both of which were induced by IFN. The p68 kinase is a 48 kdalton M_r protein (p48), which phosphorylates in the presence of dsRNA a protein kinase (eIF-2 kinase) that in turn is capable of phosphorylating eIF-2. The kinase activity, which could also act on several histones,[47] was shown to be cAMP-independent.

On the other hand, the endonuclease degrades cellular and viral RNAs[53,54] and is activated by a series of oligonucleotides with a general structure of ppp (A2′p5′)nA where n varies from 2 to 4, which are referred to as 2′-5′ A[55,56] (see Figure 1). These oligonucleotides, unique for the 2′-5′-phosphodiester bonds, are the product of a 2-5A synthetase [ATP:(2′5′) oligoadenililtransferase, EC.2.7.7.7.19], which uses ATP as the substrate and is activated by dsRNA (Figure 2).

The discovery of dsRNA replicative-intermediates formed during the growth of RNA viruses[57,58] and even during DNA virus infections[59-69] gave support to the idea that these dsRNAs could activate both pathways that lead to the inhibition of translation (eIF-2 phosphorylation and mRNA degradation) in intact cells.

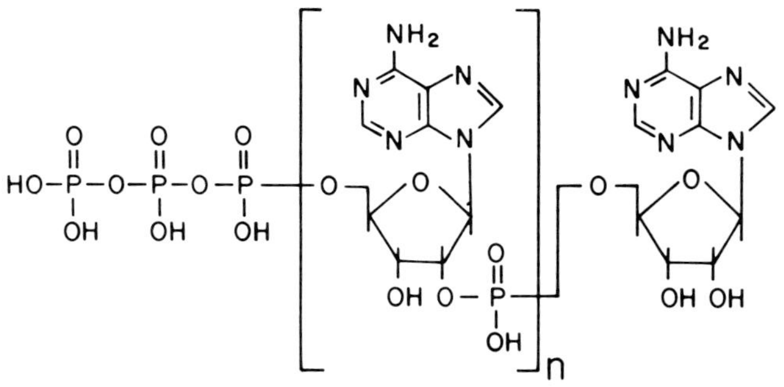

FIGURE 1. Structure of the 2-5 A: ppp(A2′p5′)nA.

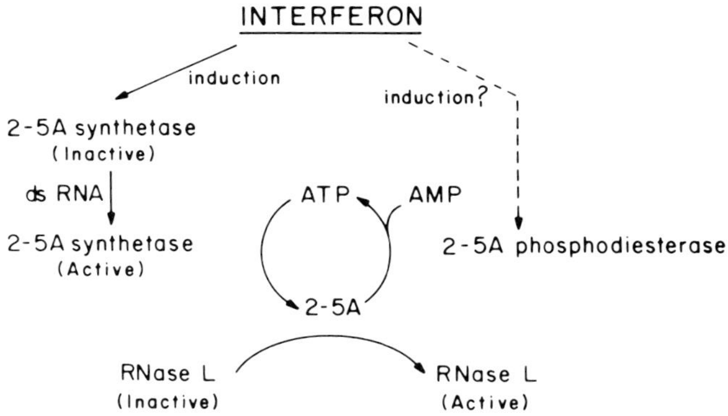

FIGURE 2. Pathway of RNase L activation.

The 2-5A synthetase uses ATP as the substrate giving rise to the dimer, trimer, and tetramer forms of 2-5A as the most abundant in vitro products. Moreover, the synthetase can also 2′adenylate other metabolites such as NAD^+, ADP ribose, ApppppA, and tRNA[61,62] in vitro. The activation of the synthetase requires the presence of dsRNA molecules higher than 30 bp (base pair), being optimal when this is 65 bp and its concentration as compared to the enzyme is approximately 1:2. Under these conditions, there is 1 molecule of the synthetase for every 80 bp fragments of dsRNA.[64,65] It appears, however, that regions of secondary structure in single stranded RNA molecules and replicative intermediates of viral RNA can also function as in vitro activators of the synthetase.[65]

The 2-5A synthetase is widely distributed in a variety of cells and tissues from higher organisms and induced by all human IFN types. The levels found in uninduced cells vary enormously, and consequently IFN treatment results in increases of 10- to 10,000-fold.[66,67] This induction requires gene expression[68] and is transient and sensitive to other agents apart from interferon.[66-69,70]

In cell extracts, the 2-5A oligomers bind to a latent endoribonuclease (RNase L) that thus becomes activated (see Figure 2). RNase L is constitutive of the cell, and its level is not affected by IFN treatment. Activation of RNase L leads to messenger RNA and ribosomal RNA cleavage that, in turn, block translation. The nuclease has little RNA sequence specificity,[71,72] which explains the indiscriminate degradation of viral and cellular RNAs in cell-free extracts.[73,74] As occurs with the synthetase, nuclease activation is transient because 2-5A is rapidly degraded by a 2′phosphodiesterase, which is also induced by IFN treatment in some cell lines[75] but not in others.[76]

Therefore, the function of these enzymes, the protein kinase and the 2-5A synthetase/ RNase L, may give account for the inhibition of protein synthesis by exogenous dsRNA observed in extracts from IFN-treated virus-infected cells. Thus, dsRNA molecules, which are one of the best IFN inducers,[77-79] are also the only known activators of these enzymatic pathways. The central role that dsRNA plays in these systems, together with its formation during the replication cycle of several animal viruses, has led to the idea that it may act as the signal that alerts the cell after viral infection. However, the existence of dsRNA of the required size in virus-infected cells and its modulatory role of the above pathways have not been definitively demonstrated in culture cells.[80]

On the other hand, the broad distribution of the 2-5A synthetase, its activation by agents other than IFN, its ability to adenylate different metabolites, and finally the existence of a great variability in enzyme levels depending on the tissue, differentiation, or growth state of the cell, have led to the hypothesis that the 2-5A system may act as the regulatory mechanism for the inhibition of cell growth by IFN.[5] The 2-5A system can even regulate other processes not related to IFN action such as the turnover of cellular RNAs.

Moreover, as the only well-established function of the 2-5A is the activation of RNase L (an enzyme which cleaves RNAs with no apparent specificity), and double-stranded stretches existing in some cellular RNAs[81,82] that can activate the system,[83] possible cytotoxic effects can be envisaged if the 2-5A system works in uninfected cells. Nilsen and Baglioni[84] proposed a model whereby the compartmentalization of the activated synthetase and the rapid degradation of 2-5A oligomers by the phosphodiesterase would create a gradient of 2-5A concentration that was maximal at the site of 2-5A synthetase-dsRNA complex formation. This could limit the potential toxicity of this system. These authors also tried to explain with this model the discrimination observed in the inhibition of cellular and viral mRNA translation in some virus-infected cells (see below). This suggestive hypothesis is, however, very difficult to test. A series of experiments was designed with cell-free extracts in which a preferential cleavage of RNA molecules covalently linked to dsRNA was detected.[84] The specificity of degradation of RNA molecules linked to dsRNAs is lower when their concentration is gradually increased. This result was explained by a high level of RNase L activity due to 2-5A synthetase multisite activation, and consequently, an elevated 2-5A concentration.[84] This mechanism, however, has not yet been definitively demonstrated to operate in intact virus-infected cells, as will be discussed below.

B. The Protein-Kinase Pathway

The IFN-induced increase in protein kinase activity observed in vitro has also been implicated in the antiviral action of IFN. Its exact role in the intact cell remains as controversial as that of the 2-5A system (see Section III). The inhibition of translation caused by dsRNAs in IFN-treated cell extracts as well as in rabbit reticulocyte lysates, was postulated to be a consequence of impaired eIF-2 activity. One hypothesis was that phosphorylation of the subunit affects its normal interaction with factor eIF-2B (also named GEF) resulting in the formation of an inactive complex.[86-88] The eIF-2B is responsible for the GDP/GTP-exchange reaction necessary for the recycling of active eIF-2 from eIF-2-GDP complexes[89] (Figure 3).

In favor of this hypothesis, Kaempfer et al.[90] detected a reversion of the initiation blockade of protein synthesis by addition of purified eIF-2 to IFN-treated cell extracts. If the eIF-2B pool size is smaller than that of eIF-2, as occurs in reticulocyte lysates,[87] a high level of phosphorylated eIF-2 will sequester the available free eIF-2B to act on nonphosphorylated eIF-2.[91] However, it is difficult to explain by that mechanism the specific inhibition of viral protein synthesis observed in several in vivo systems.[44,92,93]

Nilsen and Baglioni[84] again hypothesized that the protein kinase that binds to dsRNA may be bound to viral replicative complexes in infected cells.[85] Thus, its activation would result

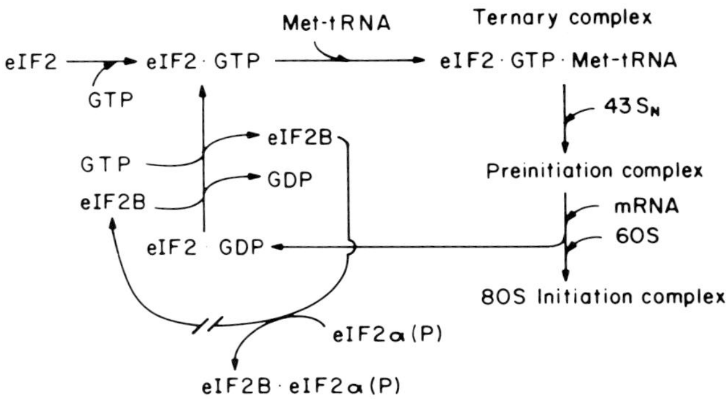

FIGURE 3. Inhibition of the initiation of translation by phosphorylation.

in a high level of phosphorylated eIF-2 just where viral mRNA is synthesized. It has, indeed, been reported that mRNA covalently linked to dsRNA is preferentially inhibited from binding to initiation complexes by a localized activation of the protein kinase.[95] As in the case of the 2-5A system, the localized activation of the kinase could originate a gradient of phosphorylated eIF-2 that in turn might be responsible for the sequestration of viral mRNAs into inactive complexes. The possibility that compartmentalization could explain the specific inhibition of viral translation seems unlikely, as there is no evidence that viral protein synthesis is compartmentalized.

C. Other In Vitro Activities

A number of mechanisms unrelated to the dsRNA activated systems discussed above also act in IFN-treated cell extracts. Several groups reported that addition of tRNA reversed the inhibition of translation observed in such extracts.[36,45,96-98] This apparent tRNA deficiency for translation leads to the production of incomplete polypeptide chains and could be corrected with a specific minor tRNA species that in some cases was different for cellular or viral mRNA translation.[45,99-101] Although these results are compatible with a discriminative blockade of viral mRNA translation since viral mRNAs might require rare isoacceptor tRNAs, the biological significance of this phenomenon is not clear. Thus, some observed specific tRNA deficiencies appear only if cell-free extracts are filtered or dialyzed prior to being used for translating exogenous mRNAs.[102] Furthermore, Colby et al.[103] and Sen et al.[53] found no differences in the content of isoacceptor tRNAs between control and IFN-treated cells that, besides, were equally able to restore translation in IFN-treated cell extracts.

The picture that is now arising is more complicated than that. For example, the inactivation of certain tRNAs in IFN-treated cell extracts could be caused by the phosphodiesterase that degrades 2-5A oligomers that has been found to be a strong inhibitor of tRNA aminoacylation.[104] This enzyme removes the CCA 3′ terminus of tRNA molecules,[105] and only tRNAs which are deacylated would be degraded.[105] Furthermore, the phosphodiesterase inhibits protein synthesis in extracts from untreated cells, and its action is reversed by adding tRNA.[104] The selectivity in the tRNA deficiency could be achieved by the action of tRNA-nucleotidyl transferases on the tRNA pool, but the minor species would be more readily inactivated. Since the phosphodiesterase does not cause an irreversible inactivation and the CCA terminus can be repaired, the observed capability of tRNA purified from IFN-treated cells to restore translation would also be expected. In spite of these considerations, this mechanism has never been shown to operate in intact cells, where even the induction of the phosphodiesterase by IFN is still controversial.[76] Thus, it might not play a role in the inhibition of viral replication by interferon.

The methylation at the 5′ terminus (cap structure) of viral mRNAs was also impaired in extracts from IFN-treated cells.[106-108] As the presence of the cap structure in mRNAs gives them stability and a higher translation efficiency, the undermethylation of the cap could explain the low level of protein synthesis detected in those systems. dsRNA does not affect the extent of mRNA methylation. Undermethylation seems to be caused by an inhibitor generated in IFN-treated cell lysates capable of reversible inactivation of the methylase, without affecting the integrity of mRNAs. However, the proposed inhibitor has never been identified or purified,[107] neither has the impairment in cap methylation in extracts from IFN-treated cells been reproduced in other laboratories.[109,110] The contradictory results obtained in intact cells (see below) make it difficult to propose a possible physiological significance, if any, of this phenomenon at present.[111-113]

Sharma and Goswani[114] observed that 2-5A inhibited vaccinia virus mRNA cap methylation in vitro, but at very high doses. Moreover, different patterns of cap methylation are observed in cell lysates and in intact cells.[115] Therefore, it is not clear whether the inhibition of translation occurs as a consequence of impairment of RNA cap methylation or inhibition by some other mechanism.[115]

III. INTERFERON ACTION ON VIRUS-INFECTED CELLS

IFN was discovered as an antiviral factor that afforded resistance against viral infection. Since then, it has been traditionally considered to protect cells from the death and cytopathic effect (CPE) caused by cytocidal animal viruses. This protection should be based on a selective blockade of viral functions while host activity is unaffected.

In contrast to this initial idea, such discrimination does not occur in several systems where cells do not remain viable p.i. (postinfection), although virus multiplication is restricted. Thus, the concept of the antiviral protection by interferon in some instances only occurs at the level of cell population, tissue, or whole organism.

Based on the data obtained from the in vitro experiments discussed in Section II, much effort was concentrated on the study of translation in virus-infected cells as the probable target of IFN action.

The discovery of the dsRNA-activated enzymes (2-5A synthetase and protein kinase) and their activity in cell-free extracts induced many scientists to establish their potential role in the antiviral activity of IFN in intact cells. Results obtained during the past few years support the view that certainly they can be involved in virus inhibition in some systems, while in others they do not seem functional or at least responsible for the IFN action. The involvement of other mechanisms (impairment of cap methylation, tRNA deficiency) detected in cell extracts has also been examined. In addition, several groups reported inhibitions in different IFN-treated cells of virus entry, viral transcription, glycosylation of viral proteins, virus assembly, budding or release, etc.

As expected from its broad antiviral effects on many viruses, IFN seems to trigger a number of different mechanisms. Depending on the cell line or virus studied, a blockade at different stages of viral replication has been detected. Because IFN blocks the replication of animal viruses, it will also affect the cytotoxic effects developed after their replication. To give an idea of the present status of research on this issue and on the different mechanisms induced by IFN in intact cells, we will review some of the results obtained during the last two decades in cell cultures infected with a variety of animal viruses.

A. Effect of Interferon on DNA Viruses

1. Papovaviruses

Simian vacuolating virus 40 (SV40) is the best known and probably most representative member of this group. Its genome consists of a small dsDNA molecule. SV40 is capable of a lytic cycle of replication in permissive cell lines (monkey cells), but infection of other nonpermissive cells, i.e., human or rodent cells, leads to stable integration of the viral DNA into host genome and cellular transformation.[117] The replication cycle has two characteristics not found in most viral systems in which IFN action has been studied: first that SV40 infection does not lead to the inhibition of cellular macromolecular synthesis (shut-off phenomenon), and second, its cycle is long enough (72 hr) to allow the analysis of IFN treatment both before infection and p.i.[105]

Half of the SV40 genome is expressed at an early phase of the lytic cycle in monkey cells. At least two mRNAs that code for two different proteins, known as large T (90 kdalton M_r) and small t (20 kdalton M_r) antigens, are present in the cytoplasm of transformed or lytic infected cells during this early phase.[118] These proteins are responsible for full transformation.[119,120] In contrast, the expression of small t in productive infection is dispensible, but large T is essential.[120,121] Its function is probably involved in the initiation of viral and cellular DNA synthesis after binding to the origin of replication.[122] Besides, T antigen is involved in the regulation of viral DNA transcription and the switch from the early to the late phase.[122] SV40 DNA replication begins 20 to 24 hr p.i., and is followed by transcription of the sequences that code for late proteins. Three cytoplasmic mRNAs derived from splicing events on the primary RNA transcripts code for all three coat proteins VP1, VP2, and VP3. Synthesis of these late proteins starts 24 hr p.i. and is maximal at 40 hr.[105]

Interferon protects cells from virus-induced CPE only if they are treated previously to infection.[123] The level of protection decreases at higher virus multiplicities.[125] Cells maintained in IFN-containing medium for 12 days do not develop CPE even after the IFN is removed.[126]

Treatment with IFN prior to SV40 infection inhibits the synthesis of early antigens in monkey cells.[124,127-129] probably as a consequence of a reduction in the accumulation of early mRNA.[127,130] However, Graessman et al.[131] observed an inhibition of T antigen synthesis in IFN-treated monkey cells microinjected with SV40 mRNA, indicating a direct block of viral translation in such cells. An effect on early viral gene expression is also derived from the experiments of Kingsman and Samuel[132] using a ts (temperature sensitive) mutant of SV40. Under conditions where only early viral functions are expressed, the synthesis of both large T and small t antigens was inhibited. Yakobson et al.[133] also found an inhibition of early viral RNA synthesis when IFN was given prior to infection, while it was not reduced when IFN treatment was carried out p.i. This effect should be primary and on the parental SV40 genome for two reasons: (1) inhibitors of protein synthesis (cycloheximide) do not cause a similar blockade, and (2) it occurs even when SV40 DNA replication is blocked.[105] In contrast, Yamamoto et al.[125] proposed that IFN inhibits SV40 uncoating. They based this claim on the finding that viral RNA production is not affected by IFN when infection is carried out with SV40 DNA (using a large number of molecules) instead of virions. However, Mozes and Defendi,[124] using the same approach, reported just the opposite result. Neither of these groups took into account the influence of the multiplicity of infection of SV40 in IFN-treated cells.

Both the requirement for IFN pretreatment and the decreased inhibition at high multiplicity of infection can be explained by the IFN-induced synthesis of a cellular repressor of SV40 transcription, or by the blockade of another viral process previous to viral gene expression such as uptake, transport to the nucleus, or uncoating. Brennan and Stark[126] reported that treatment with human leukocyte IFN prior to infection of CV-1 monkey cells causes a marked inhibition of viral mRNA accumulation. When T antigen is present in IFN-treated cells before infection (i.e., by a previous infection with a ts mutant at nonpermissive temperature, or in COS cells which are transformed by a SV40 mutant and constitutively express this antigen) viral DNA replication is only slightly inhibited, suggesting that most virions are transported to the cell nucleus in a form that can be replicated but not transcribed.

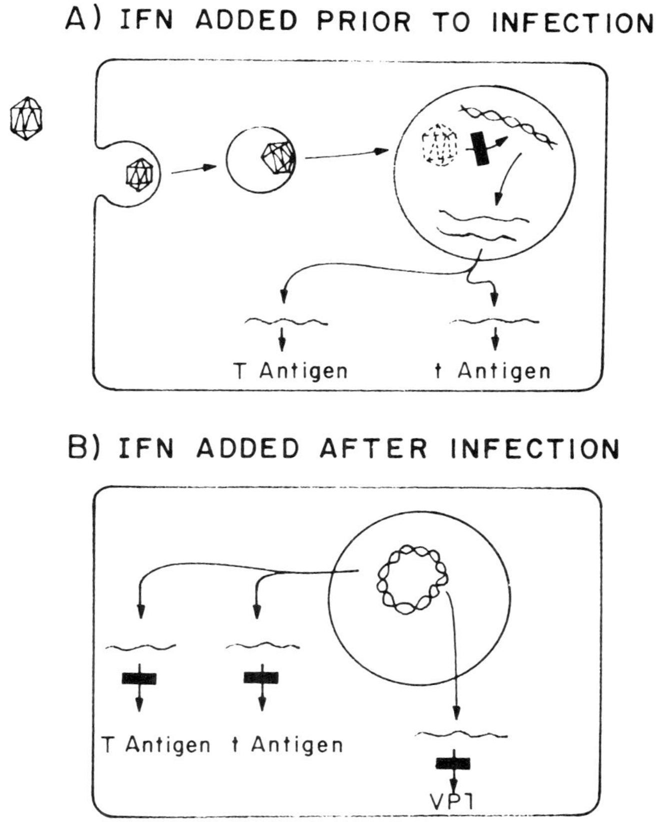

FIGURE 4. Effect of interferon on SV40 replication in permissive cells.

Transcription of infecting SV40 virions is blocked in IFN-treated cells, but this inhibition is quite low when transcription has already started before treatment with IFN. Infection with viral DNA is not inhibited by IFN pretreatment of cells.[126] These authors considered that a direct block of virus uncoating instead of the action of any cellular repressor blocking transcription was more probable. They proposed that deficiently uncoated virions formed in IFN-pretreated cells are not able to initiate transcription, but retain their capability to replicate. This would explain why IFN pretreatment is required to get inhibition, whereas if it is added after uncoating, the ongoing transcription is less affected. The resistance to the antiviral action at high multiplicities of infection may occur because the uncoating of some infecting virions is not affected, giving rise to virus replication.

Yakobson et al.[128,133] analyzed the effect of IFN treatment after SV40 infection. They observed over 90% inhibition in the production of virus progeny in BSC-1 monkey cells treated with IFN even 24 hr p.i. However, viral DNA synthesis was inhibited only slightly. Neither transcription nor the processing of viral RNA (polyadenylation, splicing) is diminished when IFN is added 24 hr p.i. At this stage SV40 DNA replication and late protein synthesis in these cells have already started. No massive degradation of newly synthesized viral mRNA can be detected. However, synthesis of both early (T and t antigens) and late proteins (VP1) are blocked, suggesting a direct inhibition of viral translation in such cells. At 20 hr after IFN addition, while host mRNAs remain bound to polysomes, viral mRNAs are found in small ribonucleoprotein particles (see Figure 4). However, this selective inhibition of viral mRNA translation observed in intact cells is not detected in cell lysates.

Thus, a block of total protein synthesis takes place in extracts from IFN-treated cells. The question remains as to why only viral mRNA translation is blocked in whole cells. From these observations, a role cannot be assigned to dsRNA-activated enzymatic systems. Viral mRNAs do not seem to be cleaved, although Kahana et al.[113] described an overmethylation (in internal positions and in the cap) of SV40 mRNAs in IFN-treated cells. However, these mRNAs are fully active and are translated in vitro, making the biological significance of this phenomenon unclear.

As hypothesized, the absence of dsRNA during the early phase of SV40 replication could be responsible for the lack of inhibition on T antigen synthesis in transformed cells (COS). It has also been suggested that the resistance of late transcription to IFN added p.i. can be explained by differences in the mechanisms involved in the early transcription.[105]

Daher and Samuel[134,135] did not detect inhibition of T antigen synthesis in cells treated with IFN after SV40 infection. This treatment also failed to reduce the production of infectious viral progeny. One possible explanation for these contradictory results could be the use of different viral strains. They also tested the effect of IFN treatment on cells doubly infected with SV40 and reovirus. When IFN was added p.i., the synthesis of reovirus polypeptides was significantly inhibited, while that of SV40 T antigen was unaffected. This specific blockade of reovirus translation is difficult to explain by the action of the RNase L. Based on their own results and on those describing an impairment in viral RNA and protein synthesis observed by prior IFN treatment, these authors suggested that IFN also affected a very early stage of the viral cycle, probably after virus uptake, but previous to early transcription.

On the other hand, IFN pretreatment has been shown to inhibit SV40-transformation of nonpermissive cells (mouse, hamster), presumably as a result of the inhibition of T antigen synthesis. In SV40-transformed cells, only early viral functions are expressed, but T antigen synthesis is now resistant to IFN treatment p.i.[136-139] This different behavior may be due to the integration of SV40 DNA into the host genome. Similarly, SV40 T antigen production is resistant to IFN in cells infected with adenovirus — SV40 hybrids. As adenovirus T antigen expression is hardly affected by IFN,[138] synthesis of SV40 T antigen must have acquired the IFN sensitivity of the adenovirus genome, as T antigen synthesis remains insensitive to IFN in transformed cells, even when VSV replication is effectively blocked.[129] A common molecular mechanism may be responsible for the resistance of T antigen synthesis to IFN in transformed cells and during the early phase of the lytic cycle, if cells treated with IFN late in infection have the SV40 DNA integrated. However, this analogy has not yet been demonstrated.[140] In contrast to the action of IFN on the lytic SV40 cycle, inhibiting the cytopathic effect, when IFN is added prior to infection, transformed cells develop CPE independently of the stage of treatment.[126]

The effect of IFN on another papovavirus, BK virus, has been reported by Cheeseman et al.[141] Pretreatment of human fibroblast cultures with 620 U/mℓ leukocyte IFN reduced BK virus titers by 1 log. This reduction was higher (1.5 to 2.9 log) when cells were maintained in medium with IFN p.i.

2. Poxviruses

This family includes complex viruses that contain a large dsDNA molecule of about 160 $\times$ 10^6 M_r as genome, able to code for about 200 proteins of medium size. The virion particle consists of an internal core where the genome is located, two lateral bodies, and an external envelope.[142] Vaccinia and smallpox viruses are representative of this group. They develop in the cytoplasm of infected cells, producing a drastic CPE and inhibition of host macromolecular synthesis within 2 to 4 hr p.i.[143,144]

After virus entry, partial uncoating produces a DNase-resistant virion core containing several enzymes, among them an RNA polymerase that transcribes the parental DNA giving rise to the early mRNA. The synthesis of early viral mRNA has been implicated in the inhibition of host protein synthesis,[144] although the vaccinia virion itself has been claimed to be responsible for inhibition.[145] Around 30 viral proteins are synthesized before the initiation of DNA replication, including various enzymes involved in the metabolism of DNA and some structural proteins.[146]

Synthesis of late viral mRNAs, coding for over 50 proteins (mostly virion structural components and some virion enzymes), takes place after viral DNA replication. Many of the vaccinia-induced polypeptides undergo different posttranslational modifications such as cleavage, glycosylation, and phosphorylation.[118] Mature virions result from the assembly of *de novo* formed virus envelopes and nucleoprotein cores located beneath the plasma membrane.[142]

The study of the mode of action of IFN on vaccinia virus infection started soon after its discovery. Several initial reports showed an inhibition of the induction of early thymidine kinase by vaccinia virus in chicken cells[147-150] but not mouse cells.[151] Joklik and Merigan[152] observed a drastic inhibition of both viral and cellular protein synthesis in IFN-treated mouse L cells grown in suspension p.i. with vaccinia virus. This inhibition led to polysomal run-off even though the synthesis of viral mRNAs was enhanced.[152,153] IFN-treated cells developed a profound CPE and were destroyed even more rapidly than untreated cells.[152,154,155] Joklik and Merigan proposed that this phenomenon could be due to profound and irreversible alterations to the cell membrane induced by virus infection. However, it should be pointed out that they used low doses of IFN and very high multiplicities of infection (50 to 750 PFU/cell).

Subsequently, Metz et al.[156] described the inability of viral mRNAs obtained from IFN-treated cells to bind to the small ribosomal subunit, and suggested that this phenomenon might be the cause of viral translation inhibition without discarding the possibility of an additional blockade of polypeptide chain elongation. A differential effect on viral proteins has been reported in chicken cells.[157] In this system, some viral proteins are strongly inhibited by only 10 U/mℓ IFN, whereas the synthesis of other polypeptides is still detected after treatment with a 100-fold IFN dose. Different results were obtained by Bialy and Colby,[158] who observed an inhibition of vaccinia virus early mRNA synthesis in IFN-treated chicken fibroblasts.

The fate of vaccinia virus mRNA in IFN-treated mouse L cells where it is transcribed but not translated has been analyzed by several groups. Klessel et al.[159] reported a normal polyadenylation in the 3′ end, and Kroath et al.[112] found a slightly reduced methylation at the 5′ terminal cap. As small amounts of dsRNA can be formed during the vaccinia virus replication cycle,[59] the inhibition of viral mRNA translation that takes place in vivo could be the consequence of the activation of the enzymatic pathways thought to be involved in the inhibition of protein synthesis in extracts from IFN-treated cells (2-5A synthetase/RNase L and protein kinase). Following this line of thinking, Esteban et al.[160] analyzed the integrity of cellular and viral RNAs in IFN-treated mouse L cells grown in suspension. In agreement with previous observations, they found an enhanced primary transcription of vaccinia virus. Besides, viral mRNAs extracted from IFN-treated cells were translated in lysates, suggesting that they were biologically active. Nevertheless, it should be emphasized that while comparable levels of protein synthesis take place in such systems at saturating concentrations of RNA extracted from control and IFN-treated cells, when lower concentrations of mRNA are used, a decline is observed in the amount of polypeptides, especially those of large size synthesized by RNA from IFN-treated cells. Esteban and co-workers[160] detected an extensive cleavage of the 28S and 18S ribosomal RNA at early stages in IFN-treated virus-infected cells that was dependent on viral transcription and can account for the total blockade of translation in those cells. The pattern of this degradation is characteristic of the activity of the 2-5A-activated endoribonuclease L. This specific cleavage of cellular rRNAs, as compared to vaccinia messenger RNAs, has tentatively been explained by enhanced viral transcription, in agreement with several reports on the stimulation of vaccinia virus RNA synthesis by translation inhibitors.[161,162]

In a subsequent study using Northern blot hybridization, these authors detected a reduction in the amount of all species of vaccinia mRNAs in IFN-treated mouse L cells growing in suspension.[163] This reduction does not appear to be due to an inhibition of virus entry into cells, and is more pronounced in the case of large viral mRNAs according to in vitro observations. In conclusion, in suspension cultures of mouse L cells treated with IFN, vaccinia virus transcription is not inhibited, but rather is prolonged. An inhibition of both viral and cellular translation takes place that is accomplished by a severe breakdown of ribosomal RNA and a reduction in the amount of viral RNAs, suggesting the activation of the 2-5A-dependent endoribonuclease L. Thus, in this system, IFN may act by inducing an indiscriminate degradation of RNAs that inhibits protein synthesis and blocks viral replication. This effect is toxic to cells that might also be affected in their plasma membrane and die rapidly.

Surprisingly, a very different picture arises when the effect of IFN on vaccinia virus infection of monolayer cell cultures is analyzed. In mouse L cells,[164] mouse L929 cells, monkey CV-1 cells,[165] and human HeLa cells[166] grown in monolayers (see Figure 5) vaccinia virus transcription and translation are quite resistant to IFN. Very high doses (thousands of units per milliliter) are necessary to achieve significant inhibition of virus proteins.

This resistance to IFN is correlated with a severe impairment of IFN-mediated enzyme activities (2-5A synthetase and protein kinase) in the extracts from a variety of cells.[167] High levels of 2-5A were detected in different IFN-treated cell lines in which vaccinia virus translation and multiplication was only slightly reduced.[165] Therefore, it seems unlikely that the 2-5A system is responsible for the inhibition of virus observed at very high IFN doses.[165] A role of the protein kinase in this reduction of virus yield has also been discarded.[167] When intact cells are doubly-infected with vaccinia virus and other IFN-sensitive viruses (VSV, polio, or EMC), a reversal of IFN-mediated inhibition of protein synthesis and even virus yields is obtained. The basis for this reversion could be a virus-mediated induction or alteration of some cellular product(s) and/or a virus-specified product(s) that affects IFN action. Indeed, the in vivo resistance of vaccinia virus RNA and protein synthesis to the IFN-mediated 2-5A synthetase/RNase L system can be correlated in vitro with the inhibition and inactivation of 2-5A by at least two viral-induced functions: an ATPase that degrades the ATP available to the synthetase, and a phosphatase that dephosphorylates 2-5A.[167] Although a definitive role of these viral enzymes in the in vivo inhibition of 2-5A synthetase remains to be determined, such data represent the first indication of IFN-antagonistic virus genes.

Youngner and co-workers[170,171] have detected a specific kinase inhibitory factor (SKIF) in extracts from vaccinia virus-infected cells capable of inhibiting the IFN-induced protein kinase. Moreover, vaccinia gene expression also causes an alteration different from phosphorylation of eIF-2.

Esteban[172,173] has further analyzed the effect of IFN-treatment of mouse L cell monolayers after vaccinia virus infection. Interferon treatment inhibited both the initiation and elongation of viral DNA synthesis. A reduction in the level of synthesis of some specific viral proteins was also detected, suggesting that some of them might be required for vaccinia virus DNA synthesis. In fact, reduced levels of proteins that bind to single-stranded DNA have been found in IFN-treated chicken embryo fibroblasts infected with vaccinia virus.[174] Moreover, as the amount of mature virions is reduced to a greater extent than the synthesis of viral DNA, IFN could also induce a defect in virion assembly. A detailed analysis of the virus particles formed in IFN-treated cells has revealed alterations in the posttranslational modifications of proteins (phosphorylation and glycosylation). The mature vaccinia virus particles

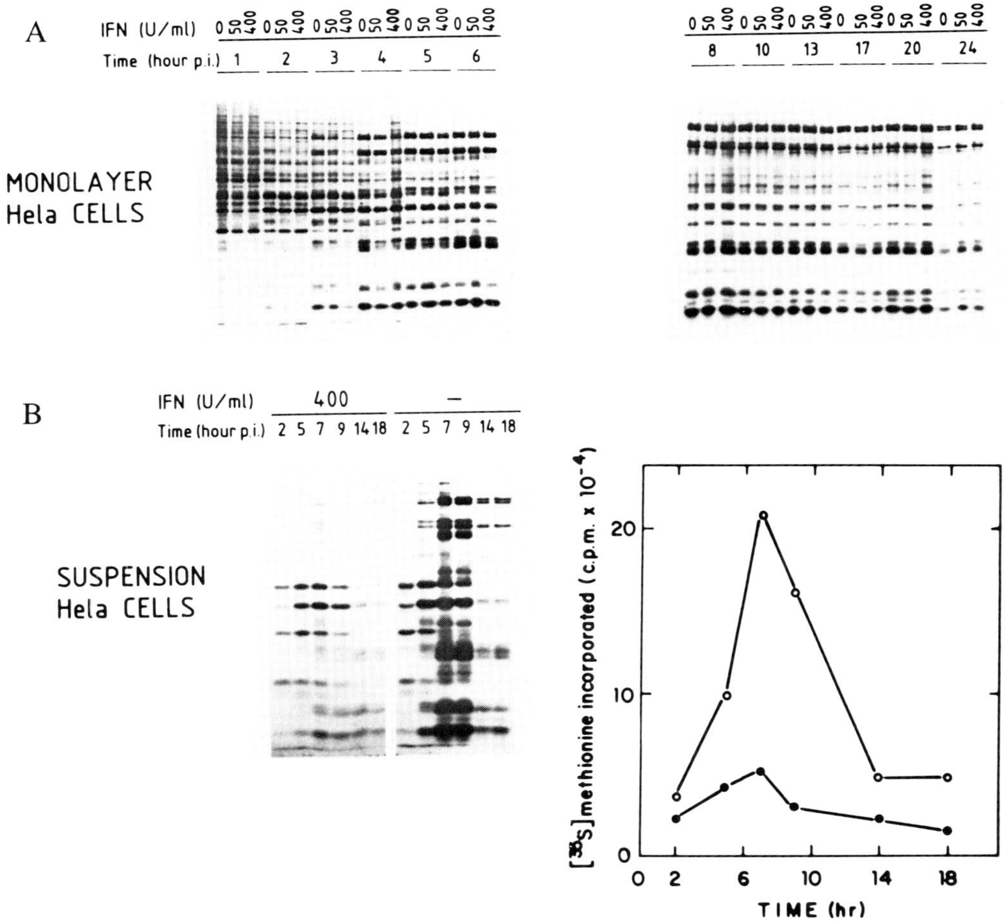

FIGURE 5. Effect of interferon on protein synthesis in monolayers (A) or suspension cultures (B) of vaccinia virus-infected human HeLa cells. Cells grown in 30 mm diameter petri dishes (A) or in suspension (HeLa S_3 cells) were treated (●) or not (○) with 400 IU/mℓ human lymphoblastoid IFN 18 hr before infection were labeled for 1 (panel A) or 2 hr (panel B) with (^{35}S)-methionine and analyzed by polyacrylamide gel electrophoresis.

produced in treated cells penetrate into cells and uncoat to a lesser extent than those grown in untreated cells. They are, however, still able to cause shut-off of host translation, and give rise to normal viral proteins.[173] Using another approach, Carrasco and Esteban[175] studied the effect of IFN on changes in cell membrane permeability following vaccinia virus infection of mouse L cell monolayers. Infection with this virus leads to an increase in the permeability of the plasma membrane, allowing the passage of small molecules such as the translation inhibitor hygromycin B (550 M_r) and even macromolecules such as the protein toxin-sarcin (16,800 M_r) into the cytoplasm.

In summary, IFN treatment is not able to prevent vaccinia virus-infected cells from dying. Two general situations are found depending on the conditions of cell culture. In suspension cultures, IFN-treated vaccinia virus-infected mouse L cells[152,153,156] or human HeLa S_3 cells[176] develop a drastic and rapid CPE and die even before untreated virus-infected cells (Figure 5). This cellular suicide is more pronouced at the high multiplicities of infection/IFN dose ratios, and does not allow the normal viral replication cycle to take place in infected cells. Thus, virus multiplication and spreading is blocked by a process that shows an obvious beneficial role when a cell population or tissue is considered.

IFN treatment of cells growing in monolayers does not greatly affect early steps of the vaccinia virus cycle. Most viral proteins are synthesized at levels that are comparable to untreated cells (Figure 5). In agreement with this finding, both treated and untreated cells exhibit a similar and parallel CPE that causes them to die simultaneously, although repeated IFN treatment p.i. can delay this process.[177]

3. Herpesviruses

This group comprises large viruses (180 to 200 nm in diameter) that have dsDNA as their genome. This DNA is located, together with a number of proteins, in an internal core that is surrounded by an icosahedral protein capsid, a tegument consisting of several globular proteins and a membrane (known as the envelope) composed of cell-derived lipids and viral glycoproteins.[118,178] Several types of herpesviruses exist that share a variety of natural hosts. Some of the best known are herpes simplex type 1 and 2 (HSV-1, HSV-2). They productively infect many cells of different origins giving rise to the progeny virus and cell death. Another characteristic of many herpesviruses is the production of latent infections in host organism for prolonged periods with periodic recurrences. The herpesvirus infectious cycle begins with the fusion of the virion envelope and the cellular plasma membrane, with the viral nucleocapsid being released into the cytoplasm. The nucleocapsid migrates to the cell nucleus, where the viral DNA is uncoated and transcribed. RNA molecules are processed (polyadenylation, capping) and transported to the cytoplasm. Once viral polypeptides are synthesized, many of them return to the nucleus where the assembly of virus progeny will take place. Virus transcription and translation occur sequentially.[179,180] Initially, a group of predominantly nonstructural polypeptides (α) are synthesized. Their synthesis is necessary for the appearance of the second group of proteins (β), including viral enzymes involved in viral DNA synthesis and in the inhibition of cellular macromolecular synthesis. Their synthesis stops their own transcription and triggers that of a third group (γ) of proteins. This last set of polypeptides is virion structural components.

The replication of viral DNA starts at about 3 hr p.i. and continues for 5 to 7 hr. It is dependent on initiation of β proteins. More than 50 viral polypeptides have been detected in herpes virus-infected cells with M_r ranging from 280 to 15 kdalton.[180,181] The core and the capsid are assembled in the nucleus of infected cells. At the end of viral infection, the nucleocapsid locates itself on the internal lamella of the nuclear membrane which has been devoided of cellular proteins and takes this membrane with it. Finally, it buds through regions of the plasma membrane that contain clusters of previously incorporated viral glycoproteins.

Though IFN efficiency against herpesvirus infection in humans is well-documented,[11,182,183] these viruses were earlier considered to be relatively insensitive to IFN in cell cultures.[184,185] Antiherpetic activity of IFN in patients was thought to be mediated via stimulation of the immune response. Indeed, this effect does take place and a potentiation of the natural killer (NK) activity against HSV-1-infected cells has been illustrated.[186-188] Besides this indirect effect, an inhibition of HSV-1 and HSV-2 multiplication has also been observed in cell culture systems.[189,190] Friedman and co-workers[191] found that IFN concentrations of 1000 U/mℓ were able to protect Ly mouse cell cultures against the CPE induced by high multiplicities (10 PFU/cell) of HSV-1. 24 hr p.i., the titers of HSV-1 were reduced by 3 log in IFN-treated cultures, diminishing progressively until they reached undetectable levels at 5 days p.i. without additional IFN treatments. Immunofluorescence analyses showed that HSV-1 antigens were absent from IFN-treated cells 24 hr p.i.

Treatment of cells with human IFN before cytomegalovirus infection leads to the inhibition of immediate-early protein synthesis and, perhaps as a consequence, to the suppression of early viral RNA synthesis.[192] In agreement with this finding, Panet and co-workers[193,194] reported that inhibition by IFN of HSV-1 β and γ proteins, perhaps as a consequence of an

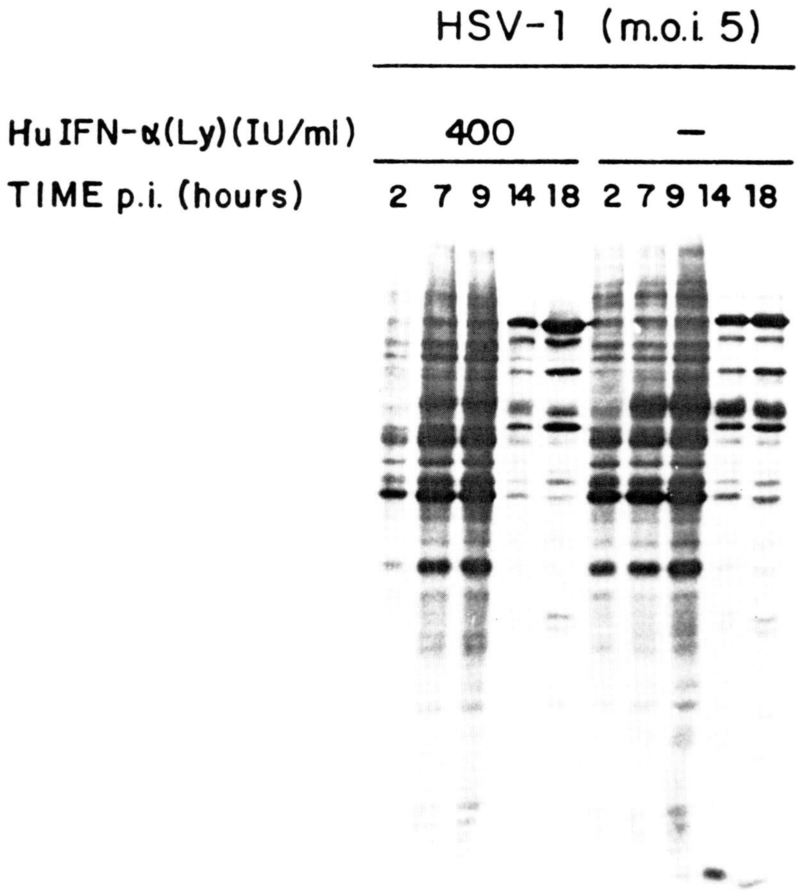

FIGURE 6. Protein synthesis in HSV-1-infected human HeLa cells. Cells were grown on 30 mm diameter petri dishes and treated or not with 400 IU/mℓ human lymphoblastoid IFN added 18 hr before HSV-1 infection (5 PFU/cell). Analysis of proteins was as described in Figure 5.

earlier blockade of immediate-early proteins. Moreover, the transcription of at least one β-mRNA was inhibited. This early effect of IFN does not seem to be mediated by the activation of the 2-5A synthetase/RNase pathway. Thus, there is no correlation between enzyme induction and HSV-1 inhibition. Furthermore, HSV-1 protein synthesis can be observed in doubly-infected cells treated with low IFN concentrations when EMC virus translation is effectively blocked.

In contrast with Friedman and colleagues,[191] we[127] found that human lymphoblastoid IFN did not protect HeLa cells when they were infected with herpes simplex viruses types 1 or 2. Only when low multiplicities of infection were used (below 0.5 PFU/cell) and with repeated IFN treatment, was a delay in the CPE observed. Comparative studies with other human IFN preparations (i.e., leukocytic, fibroblast, gamma, and α_2 recombinant IFNs) corroborated this observation. All preparations were only able to protect HeLa cell monolayers against HSV-1 infection at low multiplicities.[195] Analysis of the proteins synthesized in a single round cycle throughout infection revealed that shut-off of cellular protein synthesis and herpesvirus translation takes place in cells treated with all of the above-mentioned IFN preparations at levels almost comparable to those of control-untreated cells (Figure 6). Only a general, but not drastic, reduction in viral proteins late in infection was observed. Although

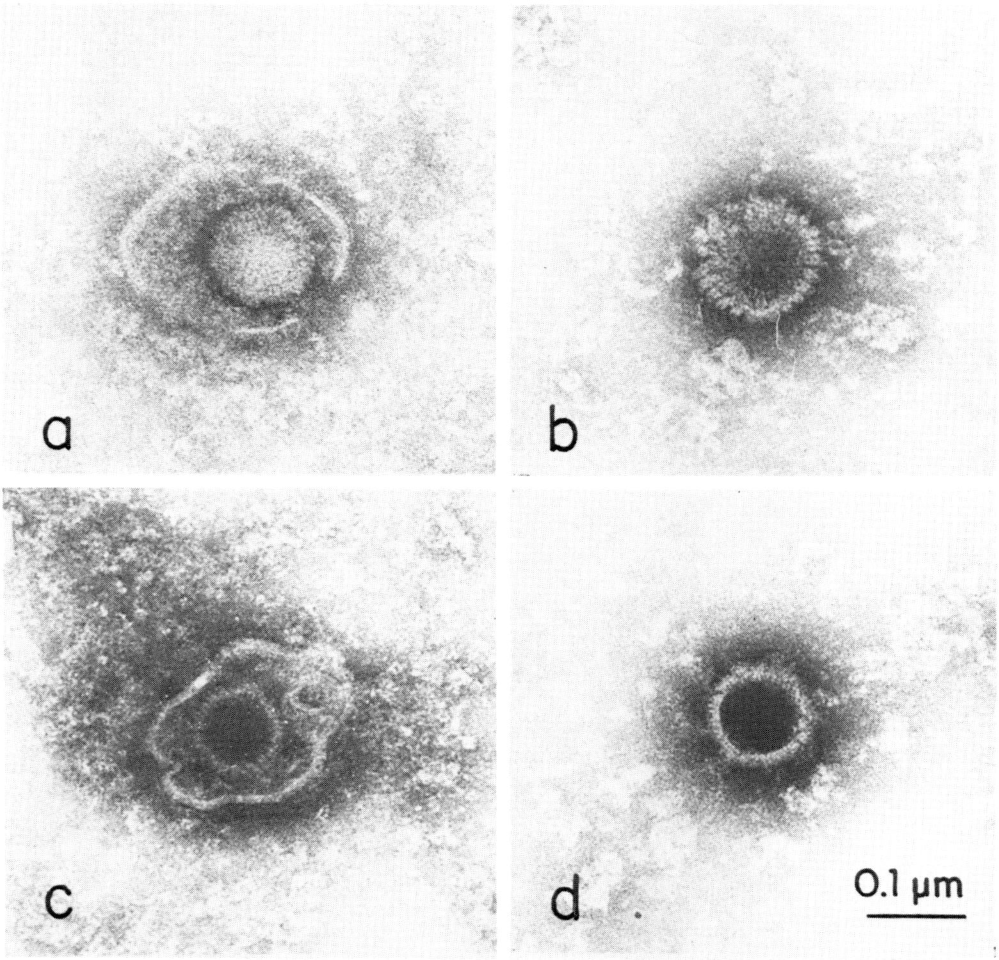

FIGURE 7. Morphological types of HSV-1 particles formed in control and IFN-treated human HeLa cells. Virions were extracted from infected cells 18 hr p.i. and negatively stained with 2% phosphotungstic acid. Percentages of each type were similar in both untreated or treated cells: full, enveloped (type a) 8%; full, naked (type b) 4%; empty, enveloped (type c) 76%; and empty, naked (type d) 17/12%.

all IFN preparations allowed the synthesis of the bulk of late viral proteins, some differences in the expression of specific viral proteins were seen. Slight differences were also found among different types of IFN. For example, the synthesis of a minor viral β-(glyco)protein probably involved in NK cell recognition and/or destruction of HSV-1-infected HeLa cells was stimulated by IFN.[188]

An effective inhibition of the production of infectious particles by lower IFN doses than those reported by other authors was observed. Gross alterations were not observed either in the pattern of phosphorylation and glycosylation of viral proteins, or in the synthesis of viral DNA, that might have accounted for that inhibition.[196] A comparable number of new virion particles is formed in control- and IFN-treated cells (Figure 7). However, analysis of the virus progeny reveals differences in the amount of some proteins.[197] This finding suggests an alteration in the assembly of new virions in IFN-treated cells that could render them defective. On the other hand, the replication of the viral genome was not affected, as might be expected, because the synthesis of late proteins occurs at high levels. In the same cells,

HSV-2 translation is more sensitive to IFN. This is in agreement with the higher inhibition in the production of infective virions.[196]

Virions formed in IFN-treated HSV-1-infected cells can attach to and enter cells, but their replication is hampered since virus proteins are not detected.[196,197] These results are, in some respects, similar to those obtained by Esteban[173] in vaccinia virus infections of mouse cell monolayers as discussed in the previous section. Obviously, neither RNase L nor protein kinase must be acting in these systems because viral mRNA is present, and the components of the protein-synthesizing machinery remain active. This can be explained by the absence of dsRNAs during the HSV-1 replication or by a HSV-1-induced blockade of the activity of these enzymes. Thus, Lewis et al.[198] and Mengheri et al.[199] described that the sensitivity to IFN of a thymidine kinase (tk)-deficient strain of mouse L929 cells could be restored by introducing the tk gene of herpes simplex virus into these cells. Subsequently, they demonstrated that the capacity to restore IFN sensitivity depends on the introduction of a given segment of herpesvirus DNA that contains sequences other than those coding for the tk enzyme. Therefore, it remains to be established whether the tk gene or closely related sequences are responsible for the activation of IFN sensitivity.[200]

In human Chang cells, both leukocyte and fibroblast IFNs did not prevent the shut-off of host protein synthesis induced by HSV-1.[201] A more profound inhibition of viral protein synthesis than in the case of HeLa cells takes place. In contrast with that system, HSV-2 mRNA translation and virus yield show a lower sensitivity to IFN. Doses of IFN that inhibit HSV-1 are much higher than those necessary to inhibit encephalomyocarditis (EMC) virus. Despite similar levels of 2-5A accumulated with each of these viruses, less ribosomal RNA degradation is observed in HSV-1-infected Chang cells. Two 2-5A derivative compounds capable under appropriate conditions of inhibiting the activation of RNase L by 2-5A have been found in these cells. This could explain the low levels of activation of RNase L, although a partial inactivation by HSV-1 infection itself cannot be excluded.[101]

In agreement with our results obtained in HeLa cells, Chatterjee and co-workers[202] found that both cloned $\alpha 2$ and β human IFNs effectively reduced HSV-1 and HSV-2 virus production in human foreskin fibroblasts. Doses as low as 10 U/mℓ IFN-β inhibited HSV-2 replication by 95%. In these experiments, cells were treated with IFN from 18 hr before infection until 24 hr p.i. A drawback of this study could be the reduced virus titers obtained in the absence of IFN treatment, 2 to 5 $\times$ 10^3-fold lower than those reported by other groups.[196,201]

In that system, IFN blocked a late step in the replication cycle of herpes simplex virus since, as occurs in HeLa cells, similar levels of viral DNA were synthesized in both IFN-treated and untreated cells. This finding excludes an effect at a stage prior to viral DNA synthesis. In contrast to our studies, the inhibition of infection correlates with the amount of total extracellular virus particles, as assessed by electron microscopy analysis, indicating that a late stage in viral morphogenesis was affected by IFN.[203]

Similar results have been obtained by Bielefeldt-Ohmann et al.[204] in the bovine system. Recombinant bovine α_1 IFN reduces both the number of cultured alveolar macrophages productively infected by bovine HSV-1 virus type 1 and virus yield, causing a delay in detectable virus production without affecting the synthesis of viral polypeptides. Domke et al.[205] have also studied the effect of IFN on HSV-1 infection of cells of the immune system. They reported a high sensitivity of HSV-1 replication in mouse and human macrophages to different IFN types (α, β, γ). IFNs decrease the transcription of HSV-1 DNA polymerase, a β-protein, suggesting an effect at an early step during viral replication cycle in this case. Treatment of macrophage cultures with IFN results in the induction of the 2-5A synthetase and RNase L.

Eppstein et al.[206] have shown a synergistic inhibition of HSV-2 replication by human IFNs (α, β, γ) in combination with 9-(1,3-Dihydroxy-2-propoxy)methyl guanine (DHPG).

An additive to synergistic effect of human IFN-β and acyclovir in reducing HSV-1 and HSV-2 productions has been also reported.[190]

In conclusion, herpesviruses appear to be sensitive to IFN in varying degrees, dependent on the virus type (or even strain), and also on the host cell used. The mechanism of this inhibition is not fully determined, although the dsRNA-activated enzymatic pathways do not seem to play an important role. Rather, it is likely that IFN blocks a late event in the replication cycle, such as virus assembly or budding.

4. Adenoviruses

Adenoviruses are complex dsDNA-containing viruses. Their dsDNA (20 to 25 $\times$ 10^6 M_r with a coding capacity for about 20 to 25 proteins) has a protein covalently attached to the 5′ ends and is surrounded by an icosahedral capsid formed by 252 capsomers, of which 240 are called hexons and the other 12, located in the vertex, are designated as pentons. Each penton is composed of a penton base and a projection known as the fiber. Virions are complex, consisting of at least 15 different polypeptide species.[207] Adenovirus infection can produce two types of effects: transformation or productive infection.[208,209] During productive infection of permissive cells, early and late phases separated by viral DNA replication are distinguished. In the early phase, the enzymes required for replication and late transcription are synthesized. The late phase starts 8 to 10 hr p.i., with a drastic change in transcription and translation producing most of the viral structural proteins.[210,211] Once synthesized in the cytoplasm, viral proteins are transported to the nucleus where mature virions are formed.[210]

Only a few reports exist in the literature dealing with the effect of IFN on adenovirus infection. This is probably due to the low sensitivity to IFN observed with different adenovirus types both in human[136,211,213] and monkey cells.[138,139] Adenovirus T antigen synthesis is, in general, 100-fold less sensitive to IFN than SV40 T antigens during a lytic infection.[139]

Indeed, human lymphoblastoid IFN is unable to prevent the CPE induced by adenovirus type 5 in HeLa cells. Even if IFN is continuously present, before and after infection, IFN-treated cells die simultaneously with untreated cells.[177] The time course of translation in both kinds of cells is similar, and analysis of the proteins synthesized throughout infection reveals that adenovirus late proteins are unaffected by IFN-treatment.[176]

Recently, we have observed that adenovirus type 5 coinfection can reduce the inhibition of poliovirus by IFN in HeLa cells.[216] This finding suggests an adenovirus-induced antagonistic action on IFN activity as occurs with vaccinia virus.[167,169-171]

B. Effect of Interferon on RNA Viruses

1. Retroviruses

This group comprises spherical, enveloped viruses of about 80 to 100 nm in diameter that have an icosahedral capsid of ribonucleoprotein. The genome is composed of two identical chains of single stranded RNA.[217] A distinctive and unique characteristic of retroviruses is a virion RNA-dependent DNA polymerase, more commonly known as reverse transcriptase.[218,219] An essential requisite for virus multiplication is the formation of dsDNA (provirus) from virion RNA, which then integrates into the cellular genome. Thus, two ways of retrovirus transmission exist: horizontal — when the virus produced in a given host cell infects another cell by direct contact, and vertical — when the integrated provirus genome passes from the parents to the offspring through the germinal line. Retrovirus development does not cause cell death, but induces the cell to proliferate. Assembly of progeny virus occurs in association with the plasma membrane where several viral proteins are localized.[220] The new virions bud from the cell surface forming their envelope from the plasma membrane. Therefore, cells become chronically infected giving rise to a continuous production of viruses.

High doses of IFN can prevent cell transformation and virus production by affecting an early step, probably the synthesis and integration of proviral DNA of MSV/MLV,[221,222] or avian retroviruses.[223] In agreement with this, IFN inhibits proviral DNA supercoiling,[224] which is considered to be a requisite for subsequent integration.

Besides this effect, IFN exerts another inhibitory action on retroviruses, when the provirus is already integrated into the host genome. In some virus/cell systems, IFN treatment induces a drastic reduction in the production of extracellular virions.[225] IFN induces defects in virus assembly and release, resulting in accumulation of numerous virus particles on the cell surface. This effect could be due to IFN-induced changes in the properties of the cell surface, i.e., decrease in the fluidity of the plasma membrane,[102] since Herz and Sen[226] described a partial reversal of the antiretroviral effects of IFN by drugs that increase membrane fluidity.

A second possibility has been documented where production of new virions in IFN-treated cells is not inhibited, but most of them are noninfectious particles containing all virion
avian retroviruses.[223] In agreement with this, IFN inhibits proviral DNA supercoiling,[224] which is considered to be a requisite for subsequent integration.
components that are released normally.[227-229] The defect most probably lies in the way these particles were assembled.

Most commonly, IFN treatment results both in lower virus production and in reduced infectivity of the virus particles released. In some systems, other processes can be affected, such as cleavage of protein precursors[230,231] or RNA-protein interactions,[232] although the exact molecular basis of both effects is not yet known.

Recombinant human IFN α 2(αA) inhibits the replication and reverse transcriptase in vitro activity of human T leukemia virus type III (HTLV-III), the etiological agent of acquired immune deficiency syndrome (AIDS).[233] Inhibition is observed at doses as low as 4 U/mℓ and is maximal at 1024 U/mℓ.

Herz et al.[234] reported that human and β interferons inhibited RD-114 retrovirus production in a chronically infected cell line without affecting vesicular stomatitis virus and encephalomyocarditis virus, suggesting that the molecular mechanisms by which IFN blocks these viruses are different.

In summary, IFN blocks retrovirus multiplication affecting one or more still not well-defined late events of their cycle and, at least in one case, possibly the provirus integration (see Figure 8). Once again, the presence of normal amounts of viral components (RNA and proteins) in IFN-treated cells rules out the involvement of the 2-5A system and the protein kinase as the mechanism of such inhibition.

2. *Reoviruses*

Viruses of this family contain ten fragments of dsRNA in their genome. The virion is composed of two icosahedral capsids with ten different protein species.[235] Associated with the core is a virion transcriptase that synthesizes mRNA from the dsRNA genome. Thus, during the early phase of the reproductive cycle of reoviruses, transcription is resistant to translation inhibitors. The late phase starts when the dsRNA is replicated and is followed by an increase in the transcription of late mRNAs from immature progeny virus particles. The synthesis of cellular proteins is inhibited during this late phase, and at least 16 different viral proteins can be detected.[236,237]

The sensitivity of reovirus 2 multiplication to IFN in human cells and reovirus 3 in mouse cells was initially described by Oie and Loh[238] and Gauntt,[239] respectively. Gauntt[239] and Vassef et al.[240] reported a dose-dependent reduction in the accumulation of reovirus 3 RNA (single and double stranded) in IFN-treated mouse L929 cells. Similar results were obtained by Gupta et al.,[44] who detected the specific inhibition of viral protein synthesis by IFN treatment during the early phase of infection in such cells. This discriminatory effect was not reproducible in vitro, since both cellular and viral mRNA translation was inhibited in extracts from IFN-treated cells.

Wiebe and Joklik[241] observed the inhibition of translation of several viral mRNAs (36 to

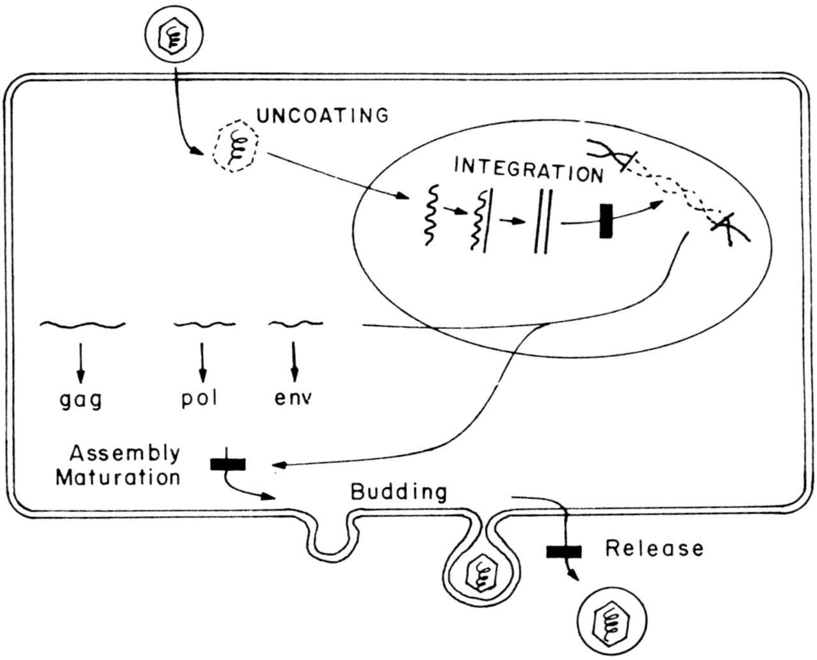

FIGURE 8. Scheme of the effects of interferon on retrovirus replication cycle.

72% with 75 U/mℓ IFN) under conditions of only slight reduction (15 to 25%) in their level of transcription. Messenger RNA coding for λ_1 polypeptide was particularly sensitive to IFN. The fact that three late viral functions, formation of dsRNAs progeny, transcription of late mRNAs, and formation of infectious virus progeny, were as sensitive as λ_1 mRNA translation, led to the conclusion that the blockade of viral mRNAs, particularly that of λ_1 polypeptide, was the cause of reovirus inhibition in mouse L929 cells. By contrast, Galster and Lengyel[242] have found similar compositions in dsRNA and protein of subviral particles formed in control- and IFN-treated cells. In human amniotic U cells, Samuel and Knutson[243] did not detect any inhibition of reovirus replication despite the induction of the kinase activity by different types of IFN.

Nilsen et al.[244] found nanomolar levels of 2-5A in suspension cultures of human HeLa cells treated with HuIFN-β 2 hr after reovirus infection at multiplicities higher than 5 PFU/cell. The 2-5A reached maximum levels at 6 hr p.i. and then decreased progressively. At that stage, no viral mRNAs of normal-size can be found with a concomitant degradation of ribosomal and cellular messenger RNAs. These authors postulate that the action of the 2′-5′A activated RNase is sufficient to account for the inhibition of reovirus replication observed in the early phase of infection. However, they did not give any explanation for the viral inhibition that takes place at low multiplicities of infection when 2-5A is not detected, or for its presence at times prior to virus uncoating. On the other hand, it has been postulated that during the reovirus replication cycle, there is no free viral dsRNA in the cellular cytoplasm that could be responsible for the activation of the IFN-induced 2-5A pathway.[235]

These same investigators later reported no inhibition of protein synthesis in reovirus-infected IFN-treated HeLa cells despite the elevation in 2-5A synthetase levels.[245] In an attempt to reconcile these contradictory results, they suggested that the degradation of cellular mRNAs would not result in a reduction in the level of translation if an increase in the synthesis of new RNA molecules was to compensate for its degradation. However, this suggestion was not correct (see below).

Surprisingly enough, activation of the kinase was not detected in such cells, but was observed in another HeLa cell strain under particular culture conditions.[246] Reovirus infection causes an increase in the phosphorylation of the α subunit of eIF-2 in IFN-treated cells when compared to uninfected cells. More recently, this group has reported that a large fraction of reoviral mRNA is not associated with polyribosomes in IFN-treated mouse L cells.[247] These viral mRNAs are, however, undegraded. Although only small amounts of eIF-2-α are phosphorylated in these cells, they propose that viral nonpolysomal mRNAs remain associated with this fraction of eIF-2-αP in inactive complexes.

The results reported by Baglioni's group[244-247] are not clear with respect to the possible role of both systems (2-5A synthetase and protein kinase) in the inhibition of reovirus replication in HeLa cells grown in suspension. Moreover, they are markedly in contrast with the results that we and other groups have observed (see below) of a reduction of only about 25 to 30% in the level of reovirus protein synthesis in this system using lymphoblastoid IFN. Gupta et al.[248] have also observed an increase in the phosphorylation of a 67 kdalton M_r polypeptide in IFN-treated mouse L cells after reovirus infection. No such increase was detected for the eIF-2. Moreover, the absence of data on the kinetics of protein synthesis makes it impossible to draw any firm conclusions.

In the course of a study of the effect of human lymphoblastoid IFN on infection by RNA-containing viruses of HeLa cells grown in monolayers, we[249] observed that the bulk of reovirus 3 proteins was synthesized in IFN-treated cells. In agreement with this finding, IFN pretreatment failed to prevent the shut-off of cellular protein synthesis and the induction of the CPE. Only when IFN treatment was continued p.i. at low multiplicity was a delay in cell death observed.[177]

Comparable levels of virus particles are formed in control- and IFN-treated cells. Since the production of infectious virions is reduced by IFN, some alteration in the content or assembly of the virus progeny must take place. Thus, at least in this particular system, IFN seems to block a late event in reovirus replication, as occurs for retroviruses and herpes viruses. In agreement with our results, reovirus translation was also resistant to human gamma IFN in human U cells.[250]

Recent results from our laboratory show that reovirus translation is also resistant to IFN in HeLa cell monolayers doubly-infected with other RNA viruses (polio, Semliki Forest virus, encephalomyocarditis, or vesicular stomatitis virus; see Figures 9 and 10).[151,252] In doubly infected HeLa cells, only reovirus translation is detected, suggesting that reovirus infection does not destroy the antiviral state induced by IFN and that IFN does not induce an unspecific inhibition of viral protein synthesis, due to massive RNA degradation. Degradation of cellular ribosomal RNA and inactivation of eIF-2 factor can also be ruled out as reovirus proteins are synthesized at control levels.

An increase in factor eIF-2 phosphorylation in reovirus-infected IFN-treated mouse L929 fibroblasts has been reported by Samuel and co-workers,[253] although at high doses (1000 U/mℓ) of a human recombinant hybrid leukocyte IFN (α A/D). The extent of specific eIF-2 phosphorylation was low, while about 5 to 10% of the eIF-2 was phosphorylated in vivo in untreated uninfected cells and 15 to 20% is observed with either IFN-treatment or reovirus infection; the level in reovirus-infected IFN-treated cells increased to 30%. Thus, even after stimulation, only a relatively low proportion of the cellular eIF-2 pool is phosphorylated. A more detailed study of the production of reovirus infectious particles and analysis of the proteins synthesized in such cells would be necessary to estimate the relevance of this observation. Also the analysis of eIF-2 phosphorylation at lower IFN concentrations, the use of homologous IFN, and analysis of other multiplicities of infection (only 50 PFUs/cell are used) would be of interest. These authors have recently reported an increase in phosphorylation of P1 and eIF-2 proteins in human amnion U cells treated with IFN-αA, but not with recombinant IFN-γ. As only IFN-αA inhibits viral protein synthesis,[253] these phosphorylations may be important in the regulation of viral blockade.

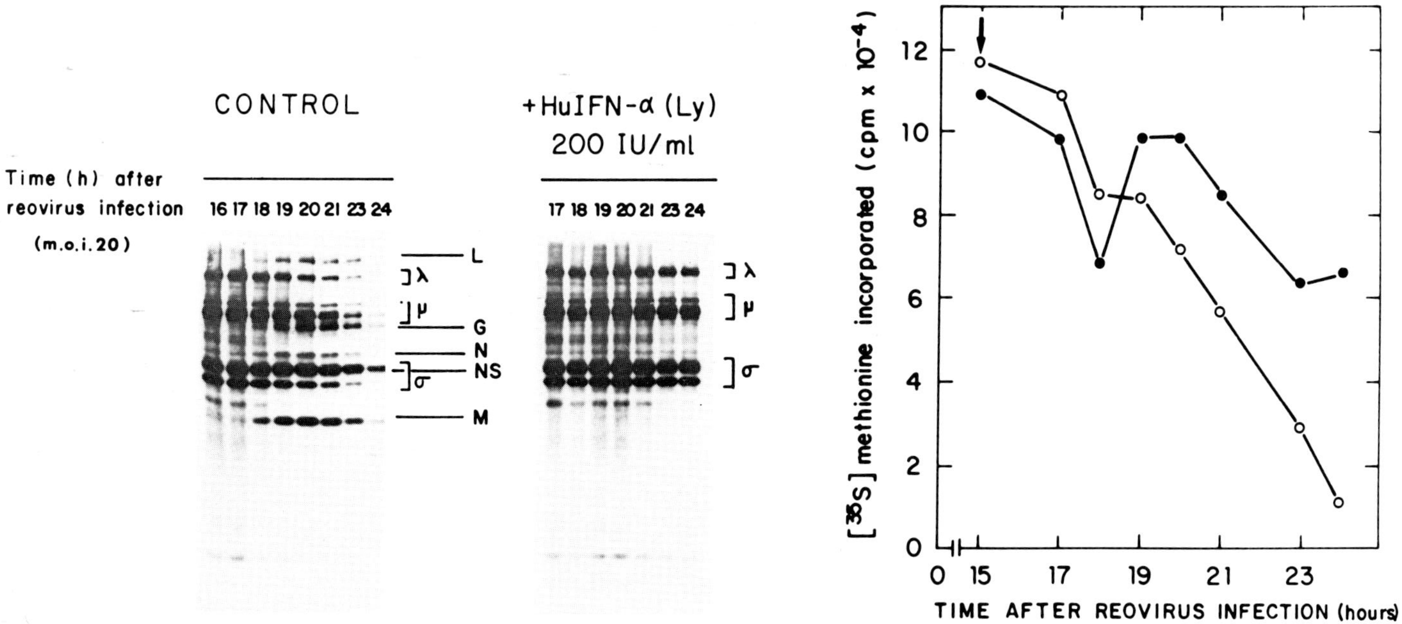

FIGURE 9. Effect of lymphoblastoid interferon on protein synthesis in human HeLa cells doubly-infected with reovirus and vesicular stomatitis virus (VSV). Control and IFN-treated cells were infected with reovirus (20 PFU/cell) and 16 hr later superinfected with VSV (100 PFU/cell). Reovirus (λ, μ, σ) and VSV (L, G, N, NS, M) proteins are indicated.

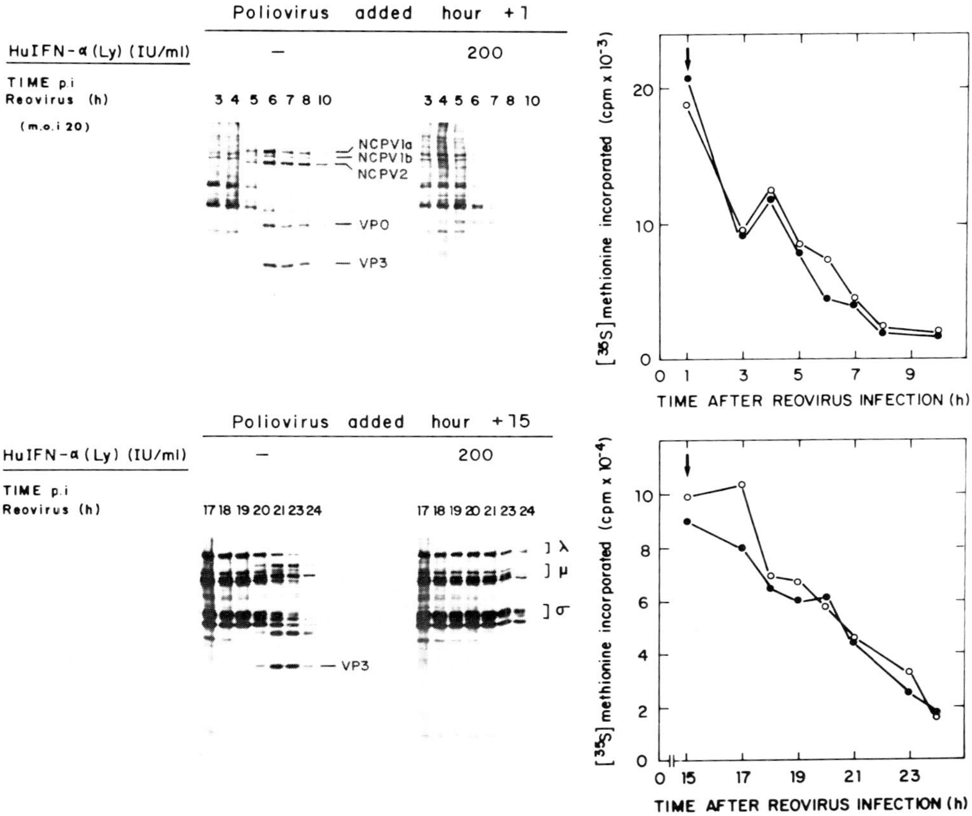

FIGURE 10. Protein synthesis in IFN-treated human HeLa cells infected with reovirus and poliovirus. Cells were treated (or not treated) with 200 IU/mℓ human lymphoblastoid IFN for 18 hr and infected with 20 PFU/cell reovirus (hr − 1). Two (hr + 1) or 16 (hr + 15) later cells were superinfected with 10 PFU/cell poliovirus. Some poliovirus (ncpv1a, NCPV1b, NCPV2, VPO, VP3) and reovirus (λ, μ, σ) proteins are indicated.

3. *Orthomyxoviruses*

The genome of viruses of this family is composed of several fragments of single-stranded RNA of negative polarity. Probably the most representative is influenza virus. Eight viral genes code for at least ten different proteins. The virus genome is transcribed in the cell nucleus into mRNAs. Viral protein synthesis is required for secondary transcription and amplification of the genome.

Although IFN was first described as a protecting agent against influenza virus,[15] only a few studies have been carried with viruses from this group. Influenza virus does not seem to be particularly sensitive to IFN.[254-257]

Primary transcription was suggested to be inhibited by IFN in chick fibroblasts.[258] However, Repik et al.[259] did not find any effect of IFN on this process. They proposed a later action of IFN in influenza infections of chick and mouse cells.

IFN-induced resistance against influenza viruses in mouse cells is dependent on the existence of the dominant allele of a host gene(Mx).[260] Mouse alpha and beta, but not gamma, IFN induce a unique 72 to 75 kdalton M_r protein (Mx) that is associated with an efficient antiviral state against influenza virus.[261,262] Gamma IFN, which does not induce the Mx protein in Mx^+ cells, is still able to protect them against vesicular stomatitis virus showing the specificity of the Mx protein activity for influenza virus.[262]

Therefore, different mechanisms are activated by the different IFN types acting against different viruses.[263] Mx protein accumulates in the cell nucleus of IFN-treated Mx^+ cells.[264] Neither virus attachment, penetration, nor uncoating are blocked by the mixture of α and β IFNs.[265,266] At the highest concentration, IFN causes only a 45% inhibition of viral primary transcription in cultures of both Mx^+ and Mx^- mouse macrophages. Viral transcripts remain intact and translatable in in vitro systems. However, IFNs markedly inhibit influenza protein synthesis only in Mx^+ macrophages.[266] Thus viral translation seems to be the site of action of the Mx protein-mediated anti-influenza activity of these IFNs.

A factor that complicates the study of human IFN action in this system is the production of abortive infections by influenza virus in human cell cultures. In monkey LLC-MK_2 cells, influenza A virus translation is not inhibited by doses of human lymphoblastoid IFN that effectively block other viruses.[249] The pattern and time course of protein synthesis is similar in both untreated and IFN-treated influenza virus-infected cells. Accordingly, CPE and cell destruction are not modified by IFN treatment. Neither RNase L activity nor eIF-2 inactivation is compatible with these results.

4. Rhabdoviruses

A typical virus of this family and one of the most studied as regards IFN sensitivity is vesicular stomatitis virus (VSV). The bullet-shaped virion is composed of an internal nucleoprotein containing the genomic single-stranded RNA associated with a protein (N). Two other proteins (NS and L) are present in the nucleocapsid and are involved in an RNA-dependent RNA polymerase activity.[267] The virion has an external lipid envelope containing a glycoprotein (G) that forms projections. The matrix protein (M) is located under the envelope. After fusion with the cellular plasma membrane, the genomic negative stranded RNA is transcribed in the cytoplasm to a precursor RNA that is cleaved to produce the five mRNAs that are then capped and polyadenylated. Each mRNA directs the synthesis of a single protein. The full-length positive RNA, copied from the genomic RNA molecule, serves as a template for the synthesis of progeny virion RNA. VSV infection causes a rapid shut-off of cellular translation and the development of a drastic CPE that leads to cell destruction in a few hr.

VSV replication is very sensitive to IFN action in different cell lines.[213,268-271] No viral proteins are synthesized in IFN-treated cells. The primary viral RNA transcription was found to be the process affected by IFN.[272,273] The likely target should be the blockade of the initiation of this process rather than the activation of nucleases. Results from other groups indicate a primary effect of IFN on the translation of viral proteins, since normal synthesis of complementary viral RNA[259] and messenger RNAs was detected.[274] In fact, in a comparative study, Thacore[275] observed that the particular step of the VSV replication cycle that is primarily affected by IFN varies from one cell line to another.[275] Moreover, Marcus and Sekellick[276] suggested a dual action whereby IFN could reduce the rate of viral mRNA synthesis (about fourfold at 50 U/mℓ) and inhibit its subsequent translation.

Although virus proteins are blocked, IFN treatment cannot prevent the development of the CPE and inhibition of host translation in rabbit RK cells infected with high multiplicities of VSV.[277] Katz et al.[278] also observed that IFN increased the susceptibility of mouse L929 cells to lysis by VSV, suggesting that this effect could be due to IFN-induced alterations in the cellular plasma membrane.

More recently, Simili et al.[279] detected normal-sized viral mRNAs in human HeLa cells treated with HuIF-β, that were, however, inefficient in translation. Later, this group described a defect in viral mRNA methylation that could have resulted in a reduced ribosomal binding capacity.[280] As dsRNA molecules have not been found in VSV-infected cells, the effect of IFN against this virus must be mediated by an activity independent of dsRNA-activated enzymes.[281]

Maheswari and co-workers[282-284] reported the production of noninfectious virions that were deficient in the viral glycoprotein G in IFN-treated cells. This inhibitory effect on protein G glycosylation has been ruled out by other studies;[249,286] also, specific inhibition of the incorporation of G and M proteins into virions has been claimed to be the basis of the antiviral effect of IFN against VSV in mouse cells.[287,288] This effect does not take place in simian BSC-1 cells,[288] and likewise does not agree with the uniform blockade of viral protein synthesis observed in both human HeLa[249] and amnion U cells.[250,289-291] In the latter system, human recombinant alpha and gamma IFN do not affect either virion entry, uncoating, or primary viral transcription. Viral mRNAs are not structurally or functionally altered, being fully translatable in in vitro systems. Thus, the general inhibition of VSV translation appears to be the major, if not the only, basis of alpha-IFN action.[250,289] For gamma IFN, however, other viral processes seem to be altered.[291] Surprisingly, Whitaker-Dowling et al.[292] have recently reported an inhibition by IFN of VSV entry in human, mouse, and chicken fibroblasts. However, this effect is not very potent since only 7% inhibition is observed with 10 U/mℓ and 65% when 1000 U/mℓ is used. This activity could not account for the inhibition of viral replication and should be considered as a first line of defense that is followed by an effective blockade of viral protein synthesis.

On the other hand, IFN cannot prevent the development of CPE induced by VSV in human HeLa cells. IFN-treated virus-infected cells die, although later than untreated cells. The degree of cell survival is directly proportional to the IFN dose/virus input ratio. This is true for all kinds of human IFN preparations assayed.[177,195] When the multiplicity of infection is sufficiently high, IFN is not able to prevent the shut-off of cellular protein synthesis, although viral translation is still inhibited.[249]

Different observations were made by Friedman et al.[191] in mouse Ly cells, where IFN pretreatment protects against VSV-induced CPE during the first few days p.i., with a significant reduction in virus titers in the culture medium. Both parameters, however, rise rapidly 5 days p.i., and cell monolayers are fully destroyed 4 days later. Continuous IFN treatment of cell cultures p.i. prevents this rise. Virus titers in culture fluids decrease to undetectable levels and cells appear to be definitively "cured" of VSV infection with no trace of VSV or its genome. However, these "cured" cells are sensitive to a new VSV infection and do not produce IFN spontaneously.

Controversial results have been obtained regarding the possible role of the protein kinase in the IFN-induced blockade of VSV. While in mouse embryonal carcinoma cells, IFN fails to induce the kinase activity and does not inhibit viral replication,[293] the opposite occurs in human fibroblast where VSV is blocked in the absence of protein kinase induction.[294] Finally, resistance to VSV infection and protein kinase induction can be induced by IFN in mouse fibroblasts.[295] Furthermore, Robbins et al.[296] showed lack of correlation between protein kinase activity induced by oxidized glutathione in mouse L cells and the antiviral state against VSV.

5. *Paramyxoviruses*

Viruses of this group are similar to rhabdoviruses, but their genome is bigger. Newcastle disease virus (NDV) is one of the most representative and most-studied member of this family with respect to its sensitivity to IFN. NDV infection leads to the inhibition of host protein synthesis that is followed by the synthesis of virus-coded proteins.[297] At least five viral proteins are detected after p.i.[249] NDV replication in chicken cells is quite resistant to homologous IFN in comparison with other viruses.[298-301] A different picture emerges in other systems (rabbit, bovine) where homologous IFN effectively blocks NDV replication.[254,302,303] In human HeLa cells treated with human lymphoblastoid IFN, NDV protein synthesis is inhibited, resulting in the consequent reduction in virus production.[247] IFN treatment cannot prevent the progressive decrease in the level of host translation in a situation analogous to that described for VSV. When moderate multiplicities of infection are employed (3 PFU/cell), IFN causes a delay in the appearance of CPE in HeLa cell monolayers. A small proportion of cells remain viable 10 days p.i. irrespective of whether IFN is maintained continuously in the culture medium or not.[179]

6. Picornaviruses

This family of animal viruses includes RNA-containing viruses such as poliovirus, encephalomyocarditis (EMC) virus, and mengovirus. Virion particles are composed of an icosahedral capsid containing 60 copies of four viral polypeptides (VP1, VP2, VP3, and VP4) that enclose the genome, a single-stranded RNA molecule. Unlike cellular mRNAs, this RNA does not possess a cap structure.[304]

Picornavirus development takes place in the cytoplasm of the infected cell. After decapsidation, the genomic RNA is translated to viral proteins, including a subunit of the viral replicase.[305] Once the whole replicase has been formed, the genome RNA serves as a template to synthesize negative-stranded RNA. This process occurs in association with membranes of the endoplasmic reticulum[306] and yields replicative forms (RF) of double-stranded RNA.[307] In addition, replicative intermediates (RI) formed by portions of dsRNA with ssRNA tails of variable length are synthesized during the transcription of negative-stranded RNAs.[308,309] Picornavirus infection of cultured cells leads to a specific and profound inhibition of host cell protein synthesis.[118] Viral RNA is translated as one polyprotein that is subsequently cleaved to give the mature virion proteins. Picornaviruses induce rapid and drastic CPE that lead to cell death in a short period of time.

The mechanism of action of IFN on picornaviruses was first studied by Levy and Carter,[310] who observed that IFN pretreatment of mouse L cells prevented the binding of mengovirus RNA to the small 40S ribosomal subunit, resulting in the inhibition of viral protein synthesis and, of course, expression of viral replicase. In this system, although no virus replication took place in IFN-treated cells, the blockade of host translation by viral infection followed a kinetics similar to control untreated cells and a clear cytotoxicity developed from about 4 to 5 hr p.i.[311,312] An analogous result was reported later by Kerr et al.,[313] who observed a strong inhibition of both cellular and viral protein synthesis in EMC virus-infected L cells treated with homologous IFN. However, a recovery of cellular translation is possible if IFN is present in the culture medium after mengovirus infection.[314] Falcoff and Sanceau[314] postulate that IFN treatment does not prevent the initial translation of the input virus RNA, allowing the production of small quantities of viral replicase. The activity of this enzyme, in turn, would initiate the RF complexes that could induce the 2-5A and protein kinase systems, leading to a blockade of viral and cellular translation. However, cells could recuperate their normal activity hours later when viral dsRNA was degraded, and hence the enzymatic systems inactivated. In extracts from these cells, the initiation of mRNA translation is more likely to be the step blocked, since reduced level of (^{35}S) met-tRNA-40S complexes have been observed.[315] Nevertheless, an enhanced ribonuclease and phosphoprotein kinase activity has been found in vitro in sedimentable fractions of IFN-treated mengovirus-infected cells known to contain viral replication complexes.[316] The fact that both cellular and viral mRNAs could be extracted and translated in vitro[313] indicated that the 2-5A system was not operating. This result contradicts the finding of significant amounts of 2-5A in mengovirus[315] and EMC virus-infected[318,319] cells. In addition, other results support the involvement of the 2-5A system:

1. Ribosomal RNA degradation takes place in IFN-treated EMC virus-infected mouse cells in an identical fashion to the activity of 2-5A-activated RNase L in vitro.[70]
2. 2-5A-inhibits protein synthesis when introduced in permeabilized cells,[320,321] and is able to block viral multiplication.[322]

3. Viral (EMC virus, VSV) replication is not inhibited by IFN in cell lines lacking RNase L.[319,323]
4. IFN treatment prevents the inactivation or inhibition of RNase L induced by EMC virus infection in several cell lines.[324] This result could explain the absence of the antiviral state in cells with high levels of 2-5A, and focuses IFN action on the prevention of the blockade of RNase L activation by the virus.

The situation regarding the involvement of 2-5A in the inhibition of picornavirus translation is not very clear. Differences have been reported in the 2-5A content of EMC virus-infected human HeLa cells. Thus, while Silverman et al.[325] found 5 and 8 n*M* 2-5A in control and IFN-treated cells, respectively, Baglioni and Nilsen[85] did not detect such nanomolar levels. These differences may be due to the different experimental conditions employed that could influence the synthesis of detectable quantities of 2-5A. In order to obtain evidence for a localized activation of RNase L by 2-5A, Baglioni and Nilsen analyzed EMC virus RNA in IFN-treated HeLa cells. However, as viral RNA synthesis is inhibited more than 1000-fold, insufficient amounts are formed to allow comparison of their integrity with those extracted from untreated cells.[85] Furthermore, when low doses of IFN are used to overcome this problem, no differences can be seen in RI from both kinds of cells. Therefore, one could speculate that if IFN inhibits the formation of EMC virus RNA, the fragments of RNA found may be the result of a defective transcription rather than an endonuclease activity. However, it has not been possible to demonstrate the localized activation of the RNase L in intact cells up to now.[84] This scheme could explain the reduced levels of 2-5A formed and the specific inhibition of EMC virus protein synthesis observed at low multiplicities of infection.[92] In contrast, a rapid blockade of total protein synthesis takes place when low IFN dose/MOI ratios are used. In parallel, cells develop a drastic CPE that leads them to death earlier than the untreated cells.[92] As UV-irradiation or heat treatment of virions prevents this shut-off phenomenon, some virus gene expression must occur in IFN-treated cells, resulting in synthesis of an undetectable virus product that may be responsible for the inhibition. A rapid destruction of the cellular plasma membrane does not seem to be the cause of this phenomenon.[92]

Therefore, two different situations are possible. EMC virus replication is inhibited both under low or high IFN dose/MOI ratios, although the fate of infected cells differs in each situation. While in the former situation, continuous treatment with IFN allows cell survival for a long period of time (a month), in the latter, cells die almost simultaneously with untreated cells.[177]

Yau et al.[326] also reported the specific inhibition of EMC translation in IFN-treated HeLa cells. Based on the similarity of effects on the rate of total cellular translation and virus production obtained with inhibitors such as cycloheximide and anisomycin, they claimed that IFN might be causing a nonspecific delay in one or more steps of the polypeptide chain elongation. Besides, EMC virus mRNA translation is more sensitive to elongation inhibitors than that of cellular mRNAs.[327] Later Ramabhadran and Thach[328] found a partial inhibition of cellular protein synthesis following EMC virus infection. After an initial decrease, it remained at 30 to 50% control for a prolonged period of time, when sufficient IFN doses to completely block viral translation were employed, indicating the selectivity of IFN action in this system. They did not, however, observe any reduction in the process of elongation that might be attributable to IFN.

In contrast to EMC virus, there are only a few reports on the effect of IFN on poliovirus infection in the literature.[324] In view of the results obtained with EMC virus, we decided to analyze the mechanism of IFN action in poliovirus-infected human HeLa cell monolayers. We already knew that IFN fails to protect infected cells from death even if IFN is continuously added to the culture medium.[127] Only when very low multiplicities of infection are used and a few cells of the monolayer become infected, a considerable protection of the cell monolayer is achieved, reflecting the ⩾90% reduction in poliovirus replication caused by doses of 50 or more U/mℓ.[330] Any human IFN preparation used (i.e., leukocytic, fibroblast, lymphoblastoid, gamma, or recombinant alpha 2) efficiently inhibits the synthesis of poliovirus proteins in such cells.[195,330] Nevertheless, the shut-off of host translation is not prevented, and at 3 hr p.i., the level of total protein synthesis is completely abated.

Although viral RNA synthesis is considerably inhibited, a certain amount of viral RNA is detected, suggesting that some viral replication takes place. Moreover, as UV-irradiation of input virions or treatment of infected cells with cycloheximide, but not with guanidine, blocks the shut-off of protein synthesis, it seems probable that viral protein synthesis from the input RNA in IFN-treated cells may be responsible for the blockade of cellular protein synthesis.[330] On the other hand, as in the case of EMC virus, no major changes in membrane permeability to monovalent cations occur at early stages of poliovirus infection in IFN-treated cells.

In order to determine whether this profound inhibition of cellular translation could be explained by an endonucleolytic activity, the integrity of RNAs in IFN-treated poliovirus-infected cells was investigated.[331] Analysis of ribosomal and messenger RNA extracted from these cells revealed that ribosomal RNA was not appreciably degraded at the point of maximal inhibition of protein synthesis. When the purified mRNAs were added to cell-free translation systems, the synthesis of cellular proteins was detected.[331] This result demonstrates that host mRNAs are undegraded and biologically active, despite the fact that they are not translated, and suggests that the blockade of poliovirus replication in IFN-treated HeLa cells may involve the inhibition of viral RNA production, since no viral proteins are synthesized in vitro. These findings disagree with the results of Vaquero et al.[317] who detected translatable mengovirus RNAs in IFN-treated mouse L cells.

The lack of correlation between the inhibition of translation and host mRNAs breakdown, together with the low level of viral RNA synthesis and the lack of evidence demonstrating a localized degradation of viral RNA in IFN-treated picornavirus-infected cells, argues against the involvement of the 2-5A pathway in this system. Moreover, the specific blockade of EMC virus translation in HeLa cells infected at low multiplicity and the high level of cellular protein synthesis are not compatible with a role for the protein kinase. Rather, it seems more likely that IFN affects an early stage of the picornavirus cycle prior to the formation of the dsRNA-containing replicative intermediates, possibly the synthesis of viral RNA.

7. Togaviruses

Representative species of this group are Semliki Forest virus (SFV) and Sindbis virus. Like picornaviruses, the virion contains a molecule of single-stranded RNA of positive polarity that is bound to a core protein (protein C).[332] Surrounding the nucleocapsid is a lipid envelope containing three viral glycoproteins. Their replicative cycle consists of two phases. During the early phase, the virion 42S mRNA is translated to several nonstructural proteins, including some components of the viral RNA polymerase. In the late phase, a 26S mRNA is formed. This 26S mRNA directs the synthesis of all the virus structural proteins. During this late phase of infection, host protein synthesis is inhibited.[118]

Both SFV and Sindbis virus are the most sensitive viruses to IFN.[218,331-335] However, only a few studies have been carried out on the molecular basis of this inhibition. Martin and Sonnabend[336] demonstrated that IFN blocked the synthesis of SFV RNA polymerase in chicken cells. In agreement with this finding, Friedman reported a strong inhibition of the SFV-induced proteins in the same cell line under conditions in which viral RNA synthesis was unaffected.[337] Also, IFN prevents the appearance of SFV proteins in HeLa cells.[249] Whether the translation of the 26S mRNA is the step initially affected by IFN, or if this is

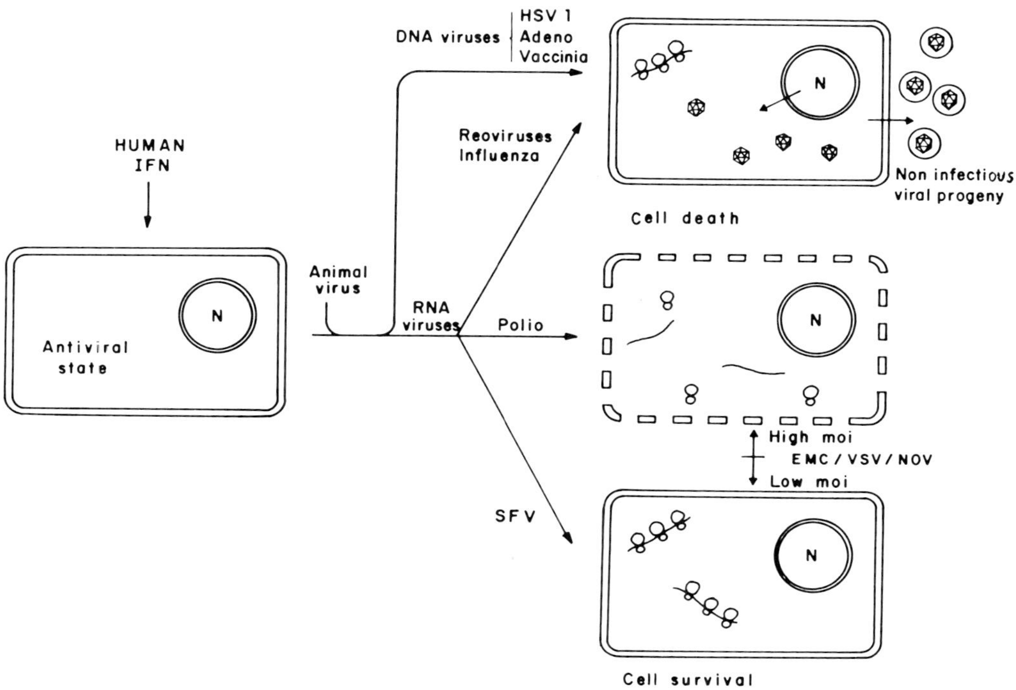

FIGURE 11. Action of interferon on virus-induced cytophatic effects.

the result of a previous blockade in RNA synthesis, remains to be determined. In this system, SFV replication is inhibited to a higher extent than several other viruses studied. Cells can survive SFV infection for 3 to 4 weeks if IFN is continuously present. After this period, cell cultures can be maintained and propagated even without further IFN treatment, and are thus cured from SFV infection[177] (Figure 11).

As occurs in the case of EMC infections at low multiplicities, IFN prevents the shut-off of host protein synthesis in SFV-infected HeLa cells. This result rules out the possibility that IFN-induced inactivation of eIF-2 or RNAse L activity is involved in the antiviral action of IFN.

IV. CONCLUSIONS

The main conclusion derived from all of the findings reviewed in this contribution is the variety of effects that IFN treatment has on virus-infected cells. IFN can inhibit the multiplication of many animal viruses in cells from different species to a variable extent. The stage of virus replication blocked depends on the type of IFN used, the species, type and strain of cells, and the viruses under study. More importantly, the particular conditions of each experiment, i.e., IFN dose, multiplicity of infection, culture conditions, etc. influence the effects of IFN. Thus, it is not surprising that no single mechanism is involved in that inhibition. Addition of IFN to homologous cells results in the induction of several proteins. Undoubtedly, the study of these proteins (characterization, expression, and activity) will help to elucidate the mechanisms leading to virus impairment. Both the availability of homogeneous IFN molecules and the cloning of mRNAs coding for IFN-induced proteins will be of great help in these studies. Up to date, a large number of experiments carried out in cell cultures has been designed or interpreted in light of the effects of IFN in cell-free extracts. However, as occurs with other biological systems, the in vitro results cannot be directly extrapolated to the in vivo situation because the inhibition of translation by the protein kinase or the 2-5A system cannot account for the wide array of effects observed in intact cells.

From our current knowledge of IFN action, a few patterns can be clearly established. IFN does not usually protect infected cells from death. Viruses can enter and initiate their replication cycle in IFN-treated cells until a certain step that differs from one virus to another is reached. In some cases, the step blocked is early in infection (transcription, translation) and in others is late (maturation, assembly, release).

During the past few years the enormous attention given to IFN, due to its possible clinical usefulness, has mainly focused on the study of its production, purification, and clinical efficacy. It is obvious that a better knowledge of the mechanism of action of IFN against each animal virus will not only benefit the scientific interest of this system but will also be of practical importance in obtaining the maximum benefit from its application in medicine.

ACKNOWLEDGMENTS

We acknowledge FIS and CAICYT for financial support. Ms. M. Teresa Aguado is acknowledged for kindly typing this manuscript.

REFERENCES

1. **Rubinstein, M., Levy, W. P., Moschera, J. A., Lai, C. Y., Hershberg, R. D., Bartlett, R. T., and Pestka, S.,** Human leukocyte interferon: isolation and characterization of several molecular forms, *Arch. Biochem. Biophys.*, 210, 307, 1981.
2. **Yip, Y. K., Barrowclough, B. S., Urban, C., and Vilcek, Y.,** Purification of two subspecies of human (immune) interferon, *Proc. Natl. Acad. Sci. U.S.A.*, 79, 1820, 1982.
3. **Weck, P. K., Apperson, S., May, L., and Stebbing, N.,** Comparison of the antiviral activities of various cloned human interferon subtypes in mammalian cell cultures, *J. Gen. Virol.*, 57, 233, 1981.
4. **Streuli, M., Hall, A., Boll, W., Stewart, W. E., II, Nagata, S., and Weissmann, C.,** Target cell specificity of two species of human interferons-alpha produced in *Escherichia coli* and of hybrid molecules derived from them, *Proc. Natl. Acad. Sci. U.S.A.*, 78, 2848, 1981.
5. **Taylor-Papadimitrou, J.,** Effects of interferons on cell growth and function, in *Interferon 2*, Gresser, I., Ed., Academic Press, London, 1980, 13.
6. **Taylor-Papadimitrou, J.,** Effects of interferons on cell growth and function, in *Interferon 1: General and Applied Aspects*, Billiau, A., Ed., Elsevier, Amsterdam, 1984, 139.
7. **Gresser, I. and Tovey, M. G.,** Antitumor effects of interferon, *Biochim. Biophys. Acta*, 516, 231, 1978.
8. **De Maeyer, E.,** Interferons and the immune system: a review (limited to α and β interferons), in *The Biology of Interferon System*, De Maeyer, E., Galasso, G., and Schellekens, H., Eds., Elsevier/North Holland, Amsterdam, 1981, 203.
9. **De Maeyer, E.,** Interferons and the immune system, in *Interferon 1: General and Applied Aspects*, Billiau, A., Ed., Elsevier, Amsterdam, 1984, 167.
10. **Kirchner, H.,** Interferons, a group of multiple lymphokines, *Springer Semin. Immunopathol.*, 7, 347, 1984.
11. **Scott, G. M.,** Interferon: perspectives for clinical problems, in *Problems of Antiviral Therapy*, Stuart-Harris, C. H. and Oxford, J., Eds., Academic Press, London, 1983, 209.
12. **Sikora, K.,** *Interferon and Cancer*, Plenum Press, New York, 1983.
13. **Borden, E. C.,** *Interferons and Cancer: How the Promise is Being Kept*, Gresser, I., Ed., Academic Press, London, 1983, 45.
14. **Herberman, R. B.,** Interferon and cytotoxic effector cells, in *Interferon 2: Interferons and the Immune System*, Vilcek, I. and De Maeyer, E., Eds., Elsevier, Amsterdam, 1984, 61.
15. **Isaacs, A. and Lindenmann, J.,** Virus interference. I. The interferon, *Proc. R. Soc. London, Ser. B*, 147, 258, 1957.
16. **Radke, K. L., Colby, C., Kates, J. R., Krider, H. M., and Prescott, D. M.,** Establishment and maintenance of the interferon-induced antiviral state: studies in enucleated cells, *J. Virol.*, 13, 623, 1974.

17. **Young, C. H. S., Pringle, C. R., and Follet, E. A. C.,** Effect of interferon in enucleated cells, *J. Virol.*, 15, 428, 1975.
18. **Taylor, J.,** Inhibition of interferon action by actinomycin, *Biochem. Biophys. Res. Commun.*, 14, 447, 1964.
19. **Lockart, R. Z., Jr.,** The necessity for cellular RNA and protein synthesis for viral inhibition from interferon, *Biochem. Biophys. Res. Commun.*, 15, 513, 1964.
20. **Friedman, R. M. and Sonnabend, J. A.,** Inhibition of interferon action by puromycin, *J. Immunol.*, 95, 696, 1965.
21. **Friedman, R. L., Manly, S. P., McMahon, M., Kerr, I. M., and Stark, G. R.,** Transcriptional and posttranscriptional regulation of interferon-induced gene expression in human cells, *Cell*, 38, 745, 1984.
22. **Knight, E., Jr. and Korant, B. D.,** Fibroblast interferon induces synthesis of four proteins in human fibroblast cells, *Proc. Natl. Acad. Sci. U.S.A.*, 76, 1824, 1979.
23. **Weil, J., Epstein, C. J., Epstein, L. B., Sedmak, J. J., Sabran, J. L., and Grossberg, S. E.,** A unique set of polypeptides is induced by γ interferon in addition to those induced in common with α and β interferons, *Nature (London)*, 301, 437, 1983.
24. **Larner, A. C., Jonak, G., Cheng, Y.-S. E., Korant, B., Knight, E., and Darnell, J. E., Jr.,** Transcriptional induction of two genes in human cells by β interferon, *Proc. Natl. Acad. Sci. U.S.A.*, 81, 6733, 1984.
25. **Branca, A. A. and Baglioni, C.,** Evidence that types I and II interferons have different receptors, *Nature (London)*, 294, 768, 1981.
26. **Anderson, P., Yip, Y. K., and Vilcek, J.,** Specific binding of ^{125}I-human interferon-γ to high affinity receptors on human fibroblasts, *J. Biol. Chem.*, 257, 11301, 1982.
27. **Tovey, M. G.,** Interferon and cyclic nucleotides, in *Interferon 4*, Gresser, I., Ed., Academic Press, London, 1982, 23.
28. **Sarkar, F. H. and Gupta, S. L.,** Interferon receptor interaction. Internalization of interferon α_2 and modulation of its receptor on human cells, *Eur. J. Biochem.*, 140, 461, 1984.
29. **Zoon, K. C. and Arnheiter, H.,** Studies of the interferon receptors, *Pharm. Ther.*, 24, 259, 1984.
30. **Vengris, V. E., Stollar, B. D., and Pitha, P. M.,** Interferon externalization by producing cell before induction of antiviral state, *Virology*, 65, 410, 1975.
31. **Huez, G., Silhol, M., and Lebleu, B.,** Microinjected interferon does not promote an antiviral response in HeLa cells, *Biochem. Biophys. Res. Commun.*, 110, 155, 1983.
32. **Marcus, P. I. and Salb, J. M.,** Molecular basis of interferon action: inhibition of viral RNA translation, *Virology*, 30, 502, 1966.
33. **Carter, W. A. and Levy, H. B.,** Interaction of mammalian ribosomes with cellular and viral RNAs, *Arch. Biochem. Biophys.*, 120, 563, 1967.
34. **Carter, W. A. and Levy, H. B.,** The recognition of viral RNA by mammalian ribosomes an effect of interferon, *Biochem. Biophys. Acta*, 155, 437, 1968.
35. **Kerr, I. M., Sonnabend, J. A., and Martin, E. M.,** Protein-synthetic activity of ribosomes from interferon-treated cells, *J. Virol.*, 5, 132, 1970.
36. **Content, J., Lebleu, B., Nudel, V., Zilberstein, A., Berissi, H., and Revel, M.,** Blocks in elongation and initiation of protein synthesis induced by interferon treatment in mouse L cells, *Eur. J. Biochem.*, 54, 1, 1975.
37. **Kerr, I. M.,** Protein synthesis in cell-free systems: an effect of interferon, *J. Virol.*, 7, 448, 1971.
38. **Falcoff, E., Falcoff, R., Lebleu, B., and Revel, M.,** Interferon treatment inhibits Mengo RNA and haemoglobin mRNA translation in cell-free extracts of L cells, *Nature London New Biol.*, 240, 145, 1972.
39. **Gupta, S. L., Sopori, M. L., and Lengyel, P.,** Inhibition of protein synthesis directed by added viral and cellular messenger RNAs in extracts of interferon-treated Ehrlich acites tumor cells. Location and dominance of the inhibitor(s), *Biochem. Biophys. Res. Commun.*, 54, 777, 1973.
40. **Samuel, C. E. and Joklik, W. K.,** A protein synthesizing system from interferon-treated cells that discriminates between cellular and viral messenger RNAs, *Virology*, 58, 476, 1974.
41. **Zilberstein, A., Dudock, B., Berissi, H., and Revel, M.,** Control of messenger RNA translation by minor species of leucyl-transfer RNA in extracts from interferon-treated cells, *J. Mol. Biol.*, 108, 43, 1976.
42. **Friedman, R. M., Metz, D. H., Esteban, R. M., Tovell, D. R., Ball, L. A., and Kerr, I. M.,** Mechanism of interferon action: inhibition of viral messenger RNA translation in L cell extracts, *J. Virol.*, 10, 1184, 1972.
43. **Falcoff, E., Falcoff, R., Lebleu, B., and Revel, M.,** Correlation between the antiviral effect of interferon treatment and the inhibition of in vitro mRNA translation in noninfected L cells, *J. Virol.*, 12, 421, 1973.
44. **Gupta, S. L., Grazidei, W. D., III, Weideli, M., Sopori, M. L., and Lengyel, P.,** Selective inhibition of viral protein accumulation in interferon-treated cells; nondiscriminate inhibition of the translation of added viral and cellular messenger RNAs in their extracts, *Virology*, 57, 49, 1974.

45. **Hiller, G., Winkler, I., Vienhauser, G., Jungwirth, C., Bodo, G., Dube, S., and Ostertag, W.,** Failure of globin mRNA to stimulate globin synthesis in cell-free extracts of interferon-treated globin-synthesizing mouse erythro-leukemic cells, *Virology,* 69, 360, 1976.
46. **Kerr, I. M., Brown, R. E., and Ball, L. A.,** Increased sensitivity of cell-free protein synthesis to double-stranded RNA after interferon treatment, *Nature (London),* 250, 57, 1974.
47. **Roberts, W. K., Hovanessian, A., Brown, R. E., Clemens, M. I., and Kerr, I. M.,** Interferon-mediated protein kinase and low-molecular-weight inhibitor of protein synthesis, *Nature (London),* 264, 477, 1976.
48. **Zilberstein, A., Kimchi, A., Schmidt, A., and Revel, M.,** Isolation of two interferon-induced translational inhibitors: a protein kinase and an oligo-isoladenylate synthetase, *Proc. Natl. Acad. Sci. U.S.A.,* 75, 4734, 1978.
49. **Lebleu, B., Sen, G. C., Shaila, S., Cabrer, B., and Lengyel, P.,** Interferon, double-stranded RNA and protein phosphorylation, *Proc. Natl. Acad. Sci. U.S.A.,* 73, 3107, 1976.
50. **Sen, G. C., Lebleu, B., Brown, G. E., Kawakita, M., Stattery, E., and Lengyel, P.,** Interferon, double-stranded RNA and mRNA degradation, *Nature (London),* 264, 370, 1976.
51. **Sen, G. C., Taira, H., and Lengyel, P.,** Interferon, double-stranded RNA, and protein phosphorylation: characteristics of a double-stranded RNA-activated protein kinase system partially purified from interferon-treated Ehrlich ascites tumor cells, *J. Biol. Chem.,* 253, 5915, 1978.
52. **Hovanessian, A. G., Galabru, J., Svab, J., Laurent, A., Krust, B., and Robert, N.,** Double-Stranded RNA Dependent Protein Kinase System Consist of Two Proteins Induced by Interferon: p68 Kinase and eIF2 Kinase, presented at Int. Symp. on Interferon System, Rome, May 8 to 11, 1985, 40.
53. **Sen, G. C., Gupta, S. L., Brown, G. E., Lebleu, B., Rebello, M. A., and Lengyel, P.,** Interferon treatment of Ehrlich ascites tumor cells: effects on exogenous mRNA translation and tRNA inactivation in the cell extract, *J. Virol.,* 17, 191, 1976.
54. **Kerr, J. M., Brown, R. E., Clemens, M. J., and Gilbert, C. S.,** Interferon mediated inhibition of cell-free protein synthesis in response to double-stranded RNA, *Eur. J. Biochem.,* 69, 551, 1976.
55. **Ball, L. A. and White, C. N.,** Oligonucleotide inhibitor of protein synthesis made in extracts of interferon-treated chick embryo cells: comparison with the mouse low molecular weight inhibitor, *Proc. Natl. Acad. Sci. U.S.A.,* 75, 1167, 1978.
56. **Kerr, I. M. and Brown, R. E.,** pppA2′p5′A2′p5A: an inhibitor of protein synthesis synthesized with an enzyme fraction from interferon-treated cells, *Proc. Natl. Acad. Sci. U.S.A.,* 75, 256, 1978.
57. **Baltimore, D.,** Structure of the poliovirus replicative intermediate RNA, *J. Mol. Biol.,* 32, 359, 1968.
58. **Nilsen, T. W., Maroney, P. A., and Baglioni, C.,** Double-stranded RNA causes synthesis of 2′,5′-oligo (A) and degradation of messenger RNA in interferon-treated cells, *J. Biol. Chem.,* 256, 7806, 1981.
59. **Colby, C. and Duesberg, P. H.,** Double-stranded RNA in vaccinia virus infected cells, *Nature (London),* 222, 940, 1969.
60. **Lucas, J. J. and Ginsberg, H. S.,** Identification of double-stranded virus-specific ribonucleic acid in KB cell infected with type 2 adenovirus, *Biochem. Biophys. Res. Commun.,* 49, 39, 1972.
61. **Ball, L. A. and White, C. N.,** Induction, purification and properties of 2′5′-oligo-adenylate synthetase, in *Regulation of Macromolecular Synthesis,* Koch, G. and Richter, D., Eds., Academic Press, New York, 1979, 303.
62. **Ferbus, D., Justesen, J., Besacon, F., and Thang, M. N.,** The 2′5′ oligoadenylate synthetase has a multifunctional 2′5′ nucleotidyl-transferase activity, *Biochem. Biophys. Res. Commun.,* 100, 847, 1981.
63. **Samantha, H., Dougherty, J. P., and Lengyel, P.,** Synthesis of (2′-5′) $(A)_n$ from ATP: characteristics of the reaction catalyzed by (2′-5′)$(A)_n$ synthetase purified from mouse Ehrlich ascites tumor cells treated with interferon, *J. Biol. Chem.,* 225, 9807, 1980.
64. **Minks, M. A., West, D. K., Benvin, S., and Baglioni, C.,** Structural requirements of dsRNA for activation of 2-5A polymerase and protein kinase of interferon-treated HeLa cells, *J. Biol. Chem.,* 254, 10180, 1979.
65. **Nilsen, T. W. and Baglioni, C.,** Mechanism for discrimination between viral and host mRNA in interferon-treated cells, *Proc. Natl. Acad. Sci. U.S.A.,* 76, 2600, 1979.
66. **Stark, F. R., Dower, W. J., Schimke, R. T., Brown, R. E., and Kerr, I. M.,** 2-5A synthetase: assay distribution and variation with growth or hormone status, *Nature (London),* 278, 471, 1979.
67. **Cayley, P. I., White, R. F., Antiniw, J. F., Walesby, H. J., and Kerr, I. M.,** Distribution of the ppp(A2′p)$_n$A-binding protein and interferon-related enzymes in animals, plants and lower organisms, *Biochem. Biophys. Res. Commun.,* 108, 1243, 1982.
68. **Shulman, L. and Revel, M.,** Interferon-dependent induction of mRNA activity for (2′5′)oligo-isoadenylate synthetase, *Nature (London),* 288, 98, 1980.
69. **Baglioni, C. and Maroney, P. A.,** Mechanisms of action of human interferons. Induction of 2′-5′-oligo (A) polymerase, *J. Biol. Chem.,* 255, 8390, 1980.
70. **Krishnan, I. and Baglioni, C.,** Increased levels of (2′-5′) oligo(A) polymerase activity in human lymphoblastoid cells treated with glucocorticoids, *Proc. Natl. Acad. Sci. U.S.A.,* 77, 6506, 1980.
71. **Wreschner, D. H., McCauley, J. W., Skehel, J. J., and Kerr, I. M.,** Interferon action, sequence specificity of the ppp(A2′p)$_n$-dependent ribonuclease, *Nature (London),* 289, 414, 1981.

72. **Floyd-Smith, G., Slattery, E., and Lengyel, P.,** Interferon action: RNA cleavage pattern of a 2′-5′ oligoadenylate-dependent endonuclease, *Science*, 122, 1030, 1981.
73. **Ratner, L., Sen, G. C., Brown, G. E., Lebleu, B., Kawakita, M., Cabrer, B., Slattery, E., and Lengyel, P.,** Interferon, dsRNA and RNA degradation. Characteristics of an endonuclease activity, *Eur. J. Biochem.*, 79, 565, 1977.
74. **Baglioni, C., Minks, M. A., and Maroney, P. A.,** Interferon action may be mediated by activation of a nuclease by pppA2′p5′ A2′p5′A, *Nature (London)*, 273, 684, 1978.
75. **Kimchi, A., Shulman, L., Schmidt, A., Chernajovsky, Y., Fradin, A., and Revel, M.,** Kinetics of the induction of 3 translation-regulatory enzymes by interferon, *Proc. Natl. Acad. Sci. U.S.A.*, 76, 3208, 1979.
76. **Verhaegen-Lewalle, M. and Content, J.,** 2′-Phosphodiesterase activity in human cell lines treated or untreated with human interferon, *Eur. J. Biochem.*, 126, 639, 1982.
77. **Field, A. R., Tytell, A. A., Lampson, G. P., and Hilleman, M. R.,** Inducers of interferon and host resistance. II. Multistranded synthetic polynucleotide complexes, *Proc. Natl. Acad. Sci. U.S.A.*, 58, 1004, 1967.
78. **Stewart, W. E.,** II, *The Interferon System*, 2nd ed., Springer-Verlag, New York, 1981.
79. **Ho, M.,** Induction and inducers of interferon, in *Interferon 1: General and Applied Aspects*, Billiau, A., Ed., Elsevier, Amsterdam, 1984, 79.
80. **Lengyel, P.,** Mechanisms of interferon action: the $(2'\text{-}5')(A)_n$ synthetase-RNase L pathway, in *Interferon 3*, Gresser, I., Ed., Academic Press, London, 1981, 65.
81. **Jelinek, K. and Darnell, J. E.,** Double-stranded regions in heterogenous nuclear RNA from HeLa cells, *Proc. Natl. Acad. Sci. U.S.A.*, 69, 2537, 1972.
82. **Calvet, J. P. and Pederson, T.,** Heterogeneous nuclear RNA double-stranded regions probed in living HeLa cells by crosslinking with the psoralen derivative aminomethyltrioxsalen, *Proc. Natl. Acad. Sci. U.S.A.*, 76, 755, 1979.
83. **Revel, M., Kimchi, A., Shulman, L., Fradin, A., Shuster, R., Yakobson, E., Chernajovsky, Y., Schmidt, A., Shure, A., and Bendori, R.,** Role of interferon-induced enzymes in the antiviral and antimitogenic effects of interferon, *Ann. N.Y. Acad. Sci.*, 350, 459, 1980.
84. **Nilsen, T. W. and Baglioni, C.,** Mechanism for discrimination between viral and host mRNA in interferon-treated cells, *Proc. Natl. Acad. Sci. U.S.A.*, 76, 2600, 1979.
85. **Baglioni, C. and Nilsen, T. W.,** Mechanisms of antiviral action of interferon, in *Interferon 5*, Gresser, I., Ed., Academic Press, London, 1983, 23.
86. **Safer, B. and Jagus, R.,** The regulation of eIF2 function during protein synthesis initiation, *Biochimie*, 63, 709, 1981.
87. **Safer, B.,** 2B or not 2B:regulation of the catalytic utilization of eIF-2, *Cell*, 33, 7, 1983.
88. **Siekierka, J., Manne, V., and Ochoa, S.,** Mechanism of translational control by partial phosphorylation of the subunit of eukaryotic initiation factor 2, *Proc. Natl. Acad. Sci. U.S.A.*, 81, 352, 1984.
89. **Jagus, R., Anderson, W. F., and Safer, B.,** The regulation of initiation of mammalian protein synthesis, *Prog. Nucl. Acids. Res. Mol. Biol.*, 25, 127, 1981.
90. **Kaempfer, R., Israeli, R., Rosen, J., Knoller, S., Zilberstein, A., Schmidt, A., and Revel, M.,** Reversal of the interferon-induced block of protein synthesis by purified preparations of eukaryotic initiation factor 2, *Virology*, 99, 170, 1979.
91. **Samuel, C. E., Duncan, R., Knutson, G. S., and Hershey, J. W. B.,** Mechanism of interferon action. Increase phosphorylation of protein synthesis initiation factor eIF-2 in interferon-treated, reovirus-infected mouse L929 fibroblasts in vitro and in vivo, *J. Biol. Chem.*, 259, 13451, 1984.
92. **Muñoz, A. and Carrasco, L.,** Protein synthesis and membrane integrity in interferon-treated HeLa cells infected with encephalomyocarditis virus, *J. Gen. Virol.*, 56, 153, 1981.
93. **Masters, P. A. and Samuel, C. E.,** Mechanism of interferon action: inhibition of vesicular stomatitis virus replication in human amnion U cells by cloned human leukocyte interferon, *J. Biol. Chem.*, 258, 12026, 1983.
94. **Hovanessian, A. G. and Kerr, I. M.,** The oligoadenylate (pppA2′-5′A2′-5′A) synthetase and protein kinase(s) from interferon-treated cells, *Eur. J. Biochem.*, 93, 515, 1979.
95. **De Benedetti, A. and Baglioni, C.,** Inhibition of mRNA binding to ribosomes by located activation of dsRNA-dependent protein kinase, *Nature (London)*, 311, 79, 1984.
96. **Gupta, S. L. Sopori, M. L., and Lengyel, P.,** Release of the inhibition of messenger RNA translation in extracts of interferon-treated Ehrlich ascites tumor cells by added transfer RNA, *Biochem. Biophys. Res. Commun.*, 57, 763, 1974.
97. **Content, J., Lebleu, B., Zilberstein, A., Berissi, H., and Revel, M.,** Mechanism of the interferon-induced block of mRNA translation in mouse L cells: reversal of the block by transfer RNA, *FEBS Lett.*, 41, 125, 1974.
98. **Weissenbach, J., Dirheimer, G., Falcoff, R., Sanceau, J., and Falcoff, E.,** Yeast tRNA leu (Anticodon U-A-G) translates all six leucine codons in extracts from interferon-treated cells, *FEBS Lett.*, 82, 71, 1977.

99. **Falcoff, R., Lebleu, B., Sanceau, J., Weissenbach, J., Dirheimer, G., Ebel, J. P., and Falcoff, E.,** The inhibition of translation of synthetic polyribonucleotides and Mengo RNA translation in extracts from interferon-treated L cells and its reversion by yeast tRNAs, *Biochem. Biophys. Res. Commun.*, 68, 1323, 1976.
100. **Zilberstein, A., Dudock, B., Berissi, H., and Revel, M.,** Control of messenger RNA translation by minor species of leucyl-transfer RNA in extracts from interferon-treated L cells, *J. Mol. Biol.*, 108, 43, 1976.
101. **Mayr, U., Bermayer, H. P., Weidinger, G., Jungwirth, C., Gross, H. J., and Bodo, G.,** Release of interferon-induced translation inhibition by tRNA in cell-free extracts from mouse erythroleukemia cells, *Eur. J. Biochem.*, 72, 541, 1977.
102. **Sen, G. C.,** Biochemical pathways in interferon action, *Pharm. Ther.*, 24, 235, 1984.
103. **Colby, C., Penhoet, E. E., and Samuel, C. E.,** A comparison of transfer RNA from untreated and interferon-treated murine cells, *Virology*, 74, 262, 1976.
104. **Schmidt, A., Chernajovski, Y., Shulman, L., Federman, P., Berissi, H., and Revel, M.,** Interferon-induced phosphodiesterase degrading (2′-5′)A and the CCA-terminus transfer RNA, *Proc. Natl. Acad. Sci. U.S.A.*, 76, 4788, 1979.
105. **Revel, M.,** Molecular mechanisms involved in the antiviral effects of interferon, in *Interferon 1*, Gresser, I., Ed., Academic Press, London, 1979, 101.
106. **Sen, G. C., Lebleu, B., Brown, G. E., Rebello, M. A., Furuichi, Y., Morgan, M., Shatkin, A. J., and Lengyel, P.,** Inhibition of reovirus messenger RNA methylation in extracts of interferon-treated Ehrlich ascites tumor cells, *Biochem. Biophys. Res. Commun.*, 65, 427, 1975.
107. **Sen, G. C., Shaila, S., Lebleu, B., Brown, G. E., Desrosiers, R. C., and Lengyel, P.,** Impairment of reovirus mRNA methylation in extracts of interferon-treated Ehrlich ascites tumor cells: further characteristics of the phenomenon, *J. Virol.*, 21, 69, 1977.
108. **Shaila, S., Lebleu, B., Brown, G. E., Sen, G. C., and Lengyel, P.,** Characteristics of extracts from interferon-treated HeLa cells: presence of a protein kinase and endoribonuclease activated by double-stranded RNA and of an inhibitor of mRNA methylation, *J. Gen. Virol.*, 37, 535, 1977.
109. **Samuel, C. E., Farris, D. A., and Epstein, D. A.,** Mechanism of interferon action, *Virology*, 83, 56, 1977.
110. **Ball, L. A. and White, C. N.,** Effect of interferon pretreatment on coupled transcription and translation in cell-free extracts of primary chick embryo cells, *Virology*, 84, 496, 1978.
111. **Desrosiers, R. C. and Lengyel, P.,** Impairment of reovirus mRNA 'cap' methylation in interferon-treated mouse L929 cells, *Biochim. Biophys. Acta*, 562, 471, 1979.
112. **Kroath, H., Gross, H. J., Jungwirth, C., and Bodo, G.,** RNA methylation in vaccinia virus-infected chick embryo fibroblasts treated with homologous interferon, *Nucl. Acids. Res.*, 5, 2441, 1979.
113. **Kahana, C., Yakobson, E., Revel, M., and Groner, Y.,** Increased methylation of RNA in SV40-infected interferon-treated cells, *Virology*, 112, 109, 1981.
114. **Sharma, O. K. and Goswani, B. B.,** Inhibition of vaccinia mRNA methylation by 2′5′-linked oligo (adenylic acid) triphosphate, *Proc. Natl. Acad. Sci. U.S.A.*, 78, 2221, 1981.
115. **Lebleu, B. and Content, J.,** Mechanisms of interferon action: biochemical and genetic approaches, in *Interferon 4*, Gresser, I., Ed., Academic Press, London, 1982, 47.
116. **Sen, G. C.,** Mechanism of interferon action: progress toward its understanding, *Prog. Nucl. Acids. Res. Mol. Biol.*, 27, 105, 1982.
117. **Salzman, N. P. and Khoury, G.,** Reproduction of papovaviruses, in *Comprehensive Virology*, Vol. 3, Fraenkel-Conrat, H. and Wagner, R. R., Eds., Plenum Press, New York, 1974, 66.
118. **Carrasco, L. and Smith, A. E.,** Molecular biology of animal virus infection, *Pharm. Ther.*, 9, 311, 1980.
119. **Martin, R. G. and Chou, J. Y.,** Simian virus 40 function required for the establishment and maintenance of malignant transformation, *J. Virol.*, 15, 599, 1975.
120. **Sleigh, M. J., Topp, W. C., Hanich, R., and Sambrook, J. F.,** Mutants of SV40 with an altered small t protein are reduced in their ability to transform cells, *Cell*, 14, 79, 1978.
121. **Cole, C. N., Landers, T., Goff, S. P., Mantevil-Brutlag, S., and Berg, P.,** Physical and genetic characterization of mutants of simian virus 40 constructed in vitro, *J. Virol.*, 24, 277, 1977.
122. **Tjian, R., Fey, G., and Graessman, A.,** Biological activity of purified simian virus 40 T antigen proteins, *Proc. Natl. Acad. Sci. U.S.A.*, 75, 1279, 1978.
123. **Mueller, C., Graessman, A., and Graessman, M.,** Mapping of early SV40-specific functions by microinjection of different early viral DNA fragments, *Cell*, 15, 579, 1978.
124. **Mozes, L. W. and Defendi, V.,** Effect of interferon on the early functions of simian virus-40, *Virology*, 85, 315, 1978.
125. **Yamamoto, Y., Yamaguchi, N., and Oda, K.,** Mechanism of interferon-induced inhibition of early simian virus 40 (SV 40) functions, *Virology*, 68, 58, 1975.
126. **Brennan, M. B. and Stark, G. R.,** Interferon pretreatment inhibits simian virus 40 infections by blocking the onset of early transcription, *Cell*, 33, 811, 1983.

127. **Oxman, M. N. and Levin, M. J.,** Interferon transcription of early virus-specific RNA in cells infected with Simian virus 40, *Proc. Natl. Acad. Sci. U.S.A.*, 68, 299, 1971.
128. **Yakobson, E., Prives, C., Hartman, J. R., Winocour, E., and Revel, M.,** Inhibition of viral protein synthesis in monkey cells treated with interferon late in simian virus 40 lytic cycle, *Cell*, 12, 73, 1977.
129. **Kingsman, S. M., Smith, M. D., and Samuel, C. E.,** Mechanism of interferon action: simian virus 40-specific early polypeptydes synthesized in untreated and interferon-treated monkey kidney cells, *Proc. Natl. Acad. Sci. U.S.A.*, 77, 2419, 1980.
130. **Metz, D. H., Levin, M. J., and Oxman, M. N.,** Mechanism of interferon action: further evidence for transcription as the primary site of action in simian virus 40 infection, *J. Gen. Virol.*, 32, 227, 1976.
131. **Graessman, A., Graessman, M., Hoffman, H., Niebel, J., Brandner, G., and Mueller, N.,** Inhibition by interferon of SV40 tumor antigen formation in cells injected with SV40 cRNA transcribed *in vivo*, *FEBS Lett.*, 39, 249, 1974.
132. **Kingsman, S. M. and Samuel, C. E.,** Mechanism of interferon action, *Virology*, 101, 458, 1980.
133. **Yakobson, E., Revel, M., and Winocour, E.,** Inhibition of simian virus 40 replication by interferon treatment late in the lytic cycle, *Virology*, 80, 225, 1977.
134. **Daher, K. A. and Samuel, C. E.,** Mechanism of interferon action. Stability and translation of simian virus-40 early mRNA coding for T-antigen is comparable in untreated and interferon-treated monkey cells, *Biochem. Biophys. Res. Commun.*, 101, 697, 1981.
135. **Daher, K. A. and Samuel, C. E.,** Mechanism of interferon action. Differential effect of interferon on the synthesis of simian virus 40 and reovirus polypeptides in monkey kidney cells, *Virology*, 117, 379, 1982.
136. **Oxman, M. N., Baron, S., Black, P. H., Takemoto, K. K., Habel, K., and Rowe, W. P.,** The effect of interferon on SV40 T antigen production in SV40-transformed cells, *Virology*, 32, 122, 1967.
137. **Oxman, M. N., Rowe, W. P., and Black, P. H.,** Studies of adenovirus-SV40 hybrid viruses. VI. Differential effects of interferon on SV40 and adenovirus T antigen formation in cells infected with SV40 hybrid viruses, *Proc. Natl. Acad. Sci. U.S.A.*, 57, 941, 1967.
138. **Oxman, M. N., Levin, M. J., and Lewis, A. M., Jr.,** Control of simian virus 40 gene expression in adenovirus-simian virus 40 hybrid viruses, *J. Virol.*, 13, 322, 1974.
139. **Mozes, L. W. and Defendi, V.,** The differential effect of interferon on T antigen production in simian virus 40-infected or transformed cells, *Virology*, 93, 558, 1979.
140. **Rigby, P. W. J. and Berg, P.,** Does simian virus 40 DNA integrate into cellular DNA during productive infection?, *J. Virol.*, 28, 475, 1978.
141. **Cheeseman, S. H., Black, P. H., Rubin, R. H., Cantell, K., and Hirsch, M. J.,** Interferon and BK papovavirus-clinical and laboratory studies, *J. Infect. Dis.*, 141, 157, 1980.
142. **Moss, B.,** Reproduction of poxviruses, in *Comprehensive Virology*, Vol. 3, Fraenkel-Conrat, H. and Wagner, R. R., Eds., Plenum Press, New York, 1974, 405.
143. **Bablanian, R.,** Structural and functional alterations in cultured cells infected with cytocidal viruses, *Prog. Med. Virol.*, 17, 40, 1975.
144. **Bablanian, R., Esteban, M., Baxt, B., and Sonnabend, J. A.,** Studies on the mechanisms of vaccinia virus cytophatic effects. I. Inhibition of protein synthesis in infected cells is associated with virus-induced RNA synthesis, *J. Gen. Virol.*, 39, 391, 1978.
145. **Person, A. and Beaud, G.,** Inhibition of host protein synthesis in vaccina virus infected cells in the presence of cordycepin (3′-Deoxyadenosine), *J. Virol.*, 25, 11, 1978.
146. **Mc. Auslan, B. R. and Kates, J. R.,** Poxvirus-induced acid deoxyribonuclease: regulation of synthesis: control of activity *in vivo;* purification and properties of the enzyme, *Virology*, 33, 709, 1967.
147. **Ohno, S. and Nozima, T.,** Inhibitory effect of interferon on the induction of thymidine kinase in vaccinia virus-infected chick embryo fibroblast, *Acta Virol.*, 8, 479, 1964.
148. **Ghosh, S. N. and Gifford, G. E.,** Effect of interferon on the dynamics of ^{3}H-thymidine incorporation and thymidine kinase induction in chick fibroblasts cultures infected with vaccinia virus, *Virology*, 27, 186, 1965.
149. **Levine, S., Magee, W. W., Hamilton, R. D., and Miller, O. V.,** Effect of interferon on early enzyme and viral DNA synthesis in vaccine virus infection, *Virology*, 32, 33, 1967.
150. **Bodo, G. and Jungwirth, C.,** Effect of interferon on deoxyribonuclease induction in chick fibroblast cultures infected with cow poxvirus, *J. Virol.*, 1, 466, 1976.
151. **Barban, S. and Baron, S.,** Differential inhibitory effects of interferon on deoxythymidine kinase induction of vaccinia-infected cell cultures, *Proc. Soc. Exp. Biol. Med.*, 127, 160, 1968.
152. **Joklik, W. K. and Merigan, T. C.,** Concerning the mechanism of action of interferon, *Proc. Natl. Acad. Sci. U.S.A.*, 56, 558, 1966.
153. **Metz, D. H. and Esteban, M.,** Interferon inhibits viral protein synthesis in L-cells infected with vaccinia virus, *Nature (London)*, 238, 385, 1972.
154. **Horak, I., Jungwirth, C., and Bodo, G.,** Poxvirus specific cytopathic effect in interferon-treated L cells, *Virology*, 45, 456, 1971.

155. **Stewart, W. E., II, DeClerq, E., and De Somer, P.,** Specificity of interferon-induced enhancement of toxicity for double-stranded ribonucleic acids, *J. Gen. Virol.*, 18, 237, 1973.
156. **Metz, D. H., Esteban, M., and Danielescu, G.,** The effect of interferon on the formation of virus polyribosomes in L-cells infected with vaccinia virus, *J. Gen. Virol.*, 27, 197, 1975.
157. **Osterhoff, J., Jäger, M., Jungwirth, C., and Bodo, G.,** Inhibition of poxvirus specific functions induced in chick-embryo fibroblasts by treatment with homologous interferon, *Eur. J. Biochem.*, 69, 535, 1976.
158. **Bialy, H. S. and Colby, C.,** Inhibition of early vaccinia virus ribonucleic acid synthesis in interferon treated chicken embryo fibroblasts, *J. Virol.*, 9, 286, 1972.
159. **Klessel, N., Wolf, W., Miller, G., Jungwirth, C., and Bodo, G.,** Synthesis of poly (A) sequences in poxvirus-infected cells treated with interferon or poly (rI)-poly(rC), *Virology*, 57, 570, 1974.
160. **Esteban, M., Benavente, J., and Paez, E.,** Effect of interferon on integrity of vaccinia virus and ribosomal RNA in infected cells, *Virology*, 134, 40, 1984.
161. **Woodson, B.,** Vaccinia mRNA synthesis under conditions that prevent uncoating, *Biochem. Biophys. Res. Commun.*, 27, 169, 1967.
162. **Kates, J. R. and Mc. Auslan, B. R.,** Poxvirus DNA dependent RNA polymerase, *Proc. Natl. Acad. Sci. U.S.A.*, 58, 134, 1967.
163. **Benavente, J., Páez, E., and Esteban, M.,** Indiscriminate degradation of RNAs in interferon-treated, vaccinia virus-infected mouse L cells, *J. Virol.*, 51, 866, 1984.
164. **Páez, E. and Esteban, M.,** Resistance of vaccinia virus to interferon is related to an interference phenomenon between the virus and the interferon system, *Virology*, 134, 12, 1984.
165. **Rice, A. P., Roberts, W. K. and Kerr, I. M.,** 2-5A accumulates to high levels in interferon-treated, vaccinia virus-infected cells in the absence of any inhibition of virus replication, *J. Virol.*, 50, 220, 1984.
166. **Muñoz, A. and Carrasco, L.,** unpublished data, 1985.
167. **Páez, E. and Esteban, M.,** Nature and mode of action of vaccinia virus products that block activation of the interferon-mediated ppp(A2′p)$_n$A-synthetase, *Virology*, 134, 29, 1984.
168. **Rice, A. P. and Kerr, I. M.,** Interferon-mediated, double-stranded RNA-Dependent protein kinase is inhibited in extracts from vaccinia virus-infected cells, *J. Virol.*, 50, 229, 1984.
169. **Whitaker-Dowling, P. and Youngner, J. S.,** Vaccinia rescue of VSV from interferon-induced resistance: reversal of translation block and inhibition of protein kinase activity, *Virology*, 131, 128, 1983.
170. **Youngner, J. S. and Whitaker-Dowling, P.,** Vaccinia virus overcomes the antiviral effect of interferon and inhibits the interferon-induced protein kinase, in *The Biology of the Interferon System*, Kirchner, H. and Schellekens, H., Eds., Elsevier, Amsterdam, 1984, 347.
171. **Youngner, J. S., Whitaker-Dowling, P., and Jagus, R.,** Vaccinia-Mediated Inhibition of Interferon Action and Effects on Protein Synthesis, presented at Int. Symp. on Interferon System, Rome, May 8 to 11, 1985, 8.
172. **Esteban, M.,** Analysis of replicating vaccinia DNA in interferon-treated, virus-infected cells, *J. Interferon Res.*, 4, 179, 1984.
173. **Esteban, M.,** Defective vaccinia virus particles in interferon-treated infected cells, *Virology*, 133, 220, 1984.
174. **Rösel, L. and Jungwirth, C.,** Isolation of early viral proteins from poxvirus-infected chick embryo fibroblast by DNA-cellulose chromatography and inhibition of their synthesis, *Eur. J. Biochem.*, 132, 361, 1983.
175. **Carrasco, L. and Esteban, M.,** Modification of membrane permeability in vaccinia virus-infected cells, *Virology*, 117, 62, 1982.
176. **Muñoz, A.,** Mecanismo de Acción del Interferón Humano en Células Infectadas por Virus Animales, Ph. D. thesis, Universidad Autónoma, Madrid, 1983.
177. **Muñoz, A. and Carrasco, L.,** Do cells treated with human interferon survive virus infection?, *FEMS Microbiol. Lett.*, 20, 317, 1983.
178. **Spear, P. G. and Roizman, B.,** Herpes simplex viruses, in *DNA Tumor Viruses*, Tooze, J., Ed., Cold Spring Harbor Laboratory, Cold Spring Harbor, N.Y., 1981, 615.
179. **Honess, R. W. and Roizman, B.,** Regulation of herpesvirus macromolecular synthesis. I. Cascade regulation of the synthesis of three groups of viral proteins, *J. Virol.*, 14, 8, 1974.
180. **Roizman, B., Kozak, M., Honess, R. W., and Hayward, G. S.,** Regulation of herpesvirus macromolecular synthesis: evidence for multilevel regulation of herpes simplex 1 RNA and protein synthesis, *Cold Spring Harbor Symp. Quant. Biol.*, 37, 687, 1974.
181. **Strand, B. C. and Aurelian, L.,** Proteins of herpesvirus type 2. III. Isolation and immunologic characterization of a large molecular weight viral protein, *Virology*, 87, 401, 1978.
182. **Dunnick, J. K. and Galasso, G. J.,** Clinical trials with exogenous interferon: summary of a meeting, *J. Infect. Dis.*, 139, 109, 1979.
183. **Merigan, T. C.,** Present appraisal of and future hopes for the clinical utilization of human interferon, in *Interferon 3*, Gresser, I., Ed., Academic Press, New York, 1981, 135.
184. **Lockart, R. Z., Jr.,** Interference and interferon with respect to herpesviruses, in *The Herpesviruses*, Kaplan, A., Ed., Academic Press, New York, 1975, 262.

185. **Rasmussen, L. E. and Farley, L. B.,** Inhibition of herpesvirus hominis by human interferon, *Infect. Immunol.,* 12, 104, 1975.
186. **Fitzgerald, P. A., Von Wussow, P., and Lopez, C.,** Role of interferon in natural kill of HSV-1-infected fibroblasts, *J. Immunol.,* 129, 819, 1982.
187. **Muñoz, A., Carrasco, L., and Fresno, M.,** Enhancement of susceptibility of HSV-1-infected cells to natural killer lysis by interferon, *J. Immunol.,* 131, 783, 1983.
188. **Alarcón, B., Muñoz, A., and Fresno, M.,** Reconocimiento de las glicoproteínas del virus herpes simplex tipo 1 por las células citotóxicas naturales NK, *Inmunología,* 3, 131, 1984.
189. **Overall, J. C., Jr., Yeh, T-J., and Kern, E. R.,** Sensitivity of herpes simplex virus types 1 and 2 to three preparations of human interferon, *J. Infect. Dis.,* 142, 943, 1980.
190. **Stanwick, T. L., Schimazi, R. F., Campbell, D. E., and Nahmias, A. J.,** Combined antiviral effect of interferon and acyclovir on herpes simplex virus types 1 and 2, *Antimicrob. Agents Chemother.,* 19, 672, 1981.
191. **Friedman, R. M., Costa, J. C., Ramseur, J. M., Meyers, M. W., Jay, F. T., and Chang, E. H.,** Persistence of the viral genome in interferon-treated cells infected with oncogenic or nononcogenic viruses, *J. Infect. Dis.,* 133, A43, 1976.
192. **Stinski, M. F., Thompsen, D. R., and Rodriguez, J. E.,** Synthesis of human cytomegalovirus-specified RNA and protein in interferon-treated cells at early times after infection, *J. Gen. Virol.,* 60, 261, 1982.
193. **Panet, A. and Falk, H.,** Inhibition by interferon of herpes simplex virus thymidine kinase and DNA polymerase in infected and biochemically transformed cells, *J. Gen. Virol.,* 64, 1999, 1983.
194. **Panet, A., Gloger, I., and Falk, H.,** Mechanisms of herpes simplex virus inhibition by interferon, in *The Biology of the Interferon System,* Kirchner, H. and Schellekens, H., Eds., Elsevier, Amsterdam, 1985, 325.
195. **Muñoz, A. and Carrasco, L.,** Comparison of the antiviral action of different human interferons against DNA and RNA viruses, *FEMS Microbiol. Lett.,* 21, 105, 1984.
196. **Muñoz, A. and Carrasco, L.,** Formation of non-infective herpesvirus particles in cultured cells treated with human interferon, *J. Gen. Virol.,* 65, 1069, 1984.
197. **Muñoz, A. and Carrasco, L.,** Acción del interferón linfoblastoide humano frente a la infección por virus herpes simplex tipo 1, *Interferón Biotechnología,* 1, 17, 1984.
198. **Lewis, J. A., Mengheri, E., and Esteban, M.,** Induction of an antiviral response by interferon requires thymidine kinase, *Proc. Natl. Acad. Sci. U.S.A.,* 80, 26, 1983.
199. **Mengheri, E., Esteban, M., and Lewis, J. A.,** Thymidine kinase genes and the induction of antiviral response by interferon, *FEBS Lett.,* 157, 301, 1983.
200. **Lewis, J. A. and Esteban, M.,** Induction of an antiviral response and 2′,5′-oligoA synthetase by interferon in several thymidine kinase-deficient cell lines, *Virology,* 133, 464, 1984.
201. **Cayley, P. J., Davies, J. A., McCullagh, K. G., and Kerr, I. M.,** Activation of the ppp(A2′p)$_n$A system in interferon-treated, herpes simplex virus-infected cells and evidence for novel inhibitors of the ppp(A2′p)$_n$A-dependent RNase, *Eur. J. Biochem.,* 143, 165, 1984.
202. **Chatterjee, S., Lakeman, A. D., Whitley, R. J., and Hunter, E.,** Effect of cloned human interferons on the replication of and cell fusion induced by herpes simplex virus, *Virus Res.,* 1, 81, 1984.
203. **Chatterjee, S., Hunter, E., and Whitley, R.,** Effect of Human Interferons on the Morphogenesis and Infectivity of Herpes Simplex Virus, Proc. 5th Int. Congr. Virology, Sendai, 1984, 159.
204. **Bielefeldt-Ohmann, H., Gilchrist, J. E., and Babiuk, L. A.,** Effect of recombinant DNA-produced bovine interferon alpha (BoIFN-$_1$) on the interaction between bovine alveolar macrophages and bovine herpesvirus type 1, *J. Gen. Virol.,* 65, 1487, 1984.
205. **Domke, I., Straub, P., Kirchner, H., Panet, A., and Jacobsen, H.,** Inhibition of Replication of Herpes Simplex Virus in Mouse Macrophages by Interferon, presented at Int. Symp. Interferon System, Rome, May 8 to 11, 1985, 139.
206. **Eppstein, D. A., Marsh, Y. V., Fraser-Smith, E. B., and Matthews, T. R.,** Potent synergistic inhibition of herpes simplex virus-2 by combination of α, β, or γ interferons with DHPG [9-((1,3-Dihydroxy-2-propoxy)methyl)-guanine]: differences *in vitro* vs. *in vivo,* in *The Biology of the Interferon System,* Kirchner, H. and Schellekens, H., Eds., Elsevier, Amsterdam, 1985, 361.
207. **Philipson, L.,** Structure and assembly of adenoviruses, in *Molecular Biology of Adenoviruses 1, Current Topics in Microbiology and Immunology,* Vol. 109, Doerfler, W., Ed., Springer-Verlag, Berlin, 1983, 1.
208. **Philipson, L. and Lindberg, V.,** Reproduction of adenoviruses, in *Comprehensive Virology,* Fraenkel-Conrat, H. and Wagner, R. R., Eds., Plenum Press, New York, 1974, 143.
209. **Tooze, J.,** *DNA Tumor Viruses,* 2nd ed., Cold Spring Harbor Laboratory, Cold Spring Harbor, N.Y., 1981.
210. **Ginsberg, H. S., Bello, L. J., and Levine, A. J.,** Control of biosynthesis of host macromolecules in cells infected with adenovirus, in *The Molecular Biology of Viruses,* Colter, J. S. and Paranchych, W., Eds., Academic Press, New York, 1967, 547.

211. **Anderson, C. W., Baum, S. G., and Gesteland, R. F.,** Processing of adenovirus 2-induced proteins, *J. Virol.,* 12, 241, 1973.
212. **Sundqvist, B., Everitt, E., Philipson, L., and Höglund, S.,** Assembly of adenoviruses, *J. Virol.,* 11, 449, 1973.
213. **Gallagher, J. G. and Khoobyarian, N.,** Adenovirus susceptibility to interferon: sensitivity of types 2, 7 and 12 to human interferon, *Proc. Soc. Exp. Biol. Med.,* 130, 137, 1969.
214. **Gallagher, J. G. and Khoobyarian, N.,** Sensitivity of adenoviruses types 1, 3, 4, 5, 8, 11 and 18 to human interferon, *Proc. Soc. Exp. Biol. Med.,* 136, 920, 1971.
215. **Gallagher, J. G. and Khoobyarian, N.,** Adenoviruses susceptibility to human interferon during one-step replication, *Infect. Immunol.,* 5, 905, 1972.
216. **Feduchi, E. and Carrasco, L.,** unpublished data, 1985.
217. **King, A. M. Q.,** High molecular weight RNA from Rous sarcoma virus and Moloney murine leukemia virus contain two subunits, *J. Biol. Chem.,* 251, 141, 1976.
218. **Temin, H. M.,** On the origin of RNA tumor viruses, *Annu. Rev. Genet.,* 8, 155, 1974.
219. **Verma, I. M.,** The reverse transcriptase, *Biochem. Biophys. Acta,* 473, 1, 1977.
220. **Bader, J. P.,** Reproduction of RNA tumor viruses, in *Comprehensive Virology,* Vol. 4, Fraenkel-Conrat, H. and Wagner, R. R., Eds., Plenum Press, New York, 1975, 253.
221. **Morris, A. G. and Burke, D. C.,** An interferon-sensitive early step in the establishment of infection of murine cells by murine sarcoma/leukemia virus, *J. Gen. Virol.,* 43, 173, 1979.
222. **Avery, R. I., Norton, J. D., Jones, J. S., Burker, D. C., and Morris, A. G.,** Interferon inhibits transformation by murine sarcoma viruses before integration of provirus, *Nature (London),* 288, 93, 1980.
223. **Strube, W., Strube, M., Kroath, H., Jungwirth, C., Bodo, G., and Graf, T.,** Interferon inhibits establishment of fibroblast infection with avian retroviruses, *J. Interferon Res.,* 2, 37, 1982.
224. **Huleihel, M. and Aboud, M.,** Inhibition of retrovirus DNA supercoiling in interferon-treated cells, *J. Virol.,* 48, 120, 1983.
225. **Canivet, M., Jovanny, C., Fourcade, A., Lasneret, I., Rhodes-Feuillette, A., and Peries, J.,** Effect of human interferon on type D retroviruses multiplication in chronically infected cell lines, *J. Interferon Res.,* 3, 53, 1983.
226. **Herz, R. E. and Sen, G. C.,** Effect of procaine and oxyphenylbutazone on interferon-mediated inhibition of murine leukemia virus production, *Antiviral Res.,* 3, 189, 1983.
227. **Friedman, R. M., Chang, E. H., Ramseur, J. M., and Meyers, M. W.,** Interferon-directed inhibition of chronic murine leukemia virus production in cell cultures: lack of effect on intracellular viral markers, *J. Virol.,* 16, 569, 1975.
228. **Pitha, P. A., Rowe, W. P., and Oxman, M. N.,** Effect of interferon on exogenous, endogenous and chronic murine leukemia, *Virus Infect.,* 70, 324, 1976.
229. **Billiau, A., Heremans, H., Allen, P. T.,** De Maeyer-Guinard, J., and De Somer, P., Trapping of oncornavirus particles at the surface of interferon treated cells, *Virology,* 73, 537, 1976.
230. **Chang, E. I. and Friedman, R. M.,** A large glycoprotein of Moloney leukemia virus derived from interferon-treated cells, *Biochem. Biophys. Res. Commun.,* 77, 392, 1977.
231. **Pitha, P. M., Fernie, B., Moldaretti, F., Haltman, T., and Wivel, N. A.,** Effect of interferon on mouse leukemia virus (MuLV). V. Abnormal protein in virions of Rauscher MuLV produced in the presence of interferon, *J. Gen. Virol.,* 46, 97, 1980.
232. **Bandyopadhyay, A. D., Chang, E. H., Levy, C. C., and Friedman, R. M.,** Structural abnormalities in murine leukemia viruses produced by interferon-treated cells, *Biochem. Biophys. Res. Commun.,* 87, 983, 1979.
233. **Ho, D. D., Harshorn, K. L., Rota, T. R., Andrews, C. A., Kaplan, J. C., Schooley, R. T., and Hirsch, M. S.,** Recombinant human interferon alpha-A suppresses HTLV-III replication, *in vitro, Lancet,* 1, 602, 1985.
234. **Herz, R. E., Rubin, B. Y., and Sen, G. C.,** Human interferon-α and -β-mediated inhibition of retrovirus production in the absence of an inhibitory effect on vesicular stomatitis virus and encephalomyocarditis virus replication in RD-114 cells, *Virology,* 125, 246, 1983.
235. **Joklik, W. K.,** Structure and function of the reovirus genome, *Microbiol. Rev.,* 45, 483, 1981.
236. **Both, G. W., Lavi, S., and Shatkin, A. J.,** Synthesis of all the gene products of the reovirus genome, *in vivo* and *in vitro, Cell,* 4, 173, 1975.
237. **Cross, R. K. and Fields, B. N.,** Reovirus-specific polypeptides: analysis using discontinuous gel electrophoresis, *J. Virol.,* 19, 162, 1976.
238. **Oie, H. K. and Loh, P. C.,** Reovirus type 2: production of and sensitivity to interferon in human amnion cells, *Proc. Soc. Exp. Biol. Med.,* 127, 1210, 1968.
239. **Gauntt, C. J.,** Effect of interferon on synthesis of ssRNA in reovirus type-3-infected L cell cultures, *Biochem. Biophys. Res. Commun.,* 47, 1228, 1972.
240. **Vassef, A., Beaud, G., Paucker, K., and Lengyel, P.,** Interferon assay based on the inhibition of double-stranded reovirus RNA accumulation in mouse L-cells, *J. Gen. Virol.,* 19, 81, 1973.

241. **Wiebe, M. F. and Joklik, W. K.,** The mechanism of inhibition of reovirus replication by interferon, *Virology*, 66, 229, 1975.
242. **Galster, R. L. and Lengyel, P.,** Formation and characteristics of reovirus subviral particles in interferon-treated mouse L-cells, *Nucl. Acids. Res.*, 3, 581, 1976.
243. **Samuel, C. E. and Knutson, G. S.,** Mechanism of interferon action: cloned human leukocyte interferons induce protein kinase and inhibit vesicular stomatitis virus but not reovirus replication in human amnion cells, *Virology*, 114, 302, 1981.
244. **Nilsen, T. W., Maroney, P. A., and Baglioni, C.,** Synthesis of (2′-5′)oligo-adenylate and activation of an endoribonuclease in interferon-treated HeLa cells infected with reovirus, *J. Virol.*, 42, 1039, 1982.
245. **Nilsen, T. W., Maroney, P. A., and Baglioni, C.,** Maintenance of protein synthesis in spite of mRNA breakdown in interferon-treated HeLa cells infected with reovirus, *Mol. Cell. Biol.*, 3, 64, 1983.
246. **Nilsen, T. W., Maroney, P. A., and Baglioni, C.,** Inhibition of protein synthesis in reovirus-infected HeLa cells with elevated levels of interferon-induced kinase activity, *J. Biol. Chem.*, 257, 14593, 1982.
247. **Baglioni, C. and De Benedetti, A.,** The Mechanism of Action of the eIF-2 Protein Kinase Induced by IFN: Inhibition of Viral Messenger RNA Translation in IFN-Treated L-Cells Infected with Reovirus, presented at Int. Symp. Interferon System, Rome, May 8 to 11, 1985, 6.
248. **Gupta, S. L., Holmes, S. L., and Mehra, L. L.,** Interferon action against reovirus: activation of interferon-induced protein kinase in mouse L929 cells upon reovirus infection, *Virology*, 120, 495, 1982.
249. **Muñoz, A. and Carrasco, L.,** Action of human lymphoblastoid interferon on HeLa cells infected with RNA-containing animal viruses, *J. Gen. Virol.*, 65, 377, 1984.
250. **Samuel, C. E., Ulker, N., Knutson, G. S., Zhang, X., and Masters, P. S.,** Molecular mechanisms of the antiviral action of human alpha and gamma interferons, in *The Biology of the Interferon System 1984*, Kirchner, H. and Schellekens, H., Eds., Elsevier, Amsterdam, 1985, 131.
251. **Muñoz, A., Alonso, M. A., Feduchi, E., and Carrasco, L.,** Effect of Interferon on Human HeLa Cells Doubly Infected with Different Animal Viruses, Int. Symp. Interferon System, Rome, May 8 to 11, 1985.
252. **Muñoz, A., Alonzo, M. A., and Carrasco, L.,** unpublished data, 1985.
253. **Samuel, C. E., Berry, M. J., Bischoff, J. R., Knutson, G. S., Munemitsu, S. M., Ulker, N., and Zhang, X.,** Phosphorylation of Protein P1 and Initiation Factor eIF-2 *In Vitro* and *In Vivo* in Untreated and Interferon-Treated Cells, Int. Symp. Interferon System, Rome, May 8 to 11, 1985, 19.
254. **Oh, J. O. and Gill, E. J.,** Role of interferon-like viral inhibitor in endotoxin-induced corneal resistance to newcastle disease virus, *J. Bacteriol.*, 91, 251, 1966.
255. **Burke, D. C. and Isaacs, A.,** Interferon: relation to heterologous interference and lack of antigenicity, *Acta Virol.*, 4, 215, 1960.
256. **Riley, B. P., Toy, S. T., and Gifford, G. E.,** Relative effects of interferon on virus plaque formation, *Proc. Soc. Exp. Biol. Med.*, 122, 1142, 1966.
257. **Merigan, T. C., Reed, S. E., Hall, T. S., and Tyrell, D. A. J.,** Inhibition of respiratory virus infection by locally applied interferon, *Lancet*, 1, 563, 1973.
258. **Bean, W. J. and Simpson, R. W.,** Primary transcription of the influenza virus genome in permissive cells, *Virology*, 56, 646, 1973.
259. **Repik, P., Flamand, A., and Bishop, D. H. L.,** Effect of interferon upon the primary and secondary transcription of vesicular stomatitis and influenza viruses, *J. Virol.*, 14, 1169, 1974.
260. **Haller, O. H., Arnheiter, J., Lindenmnann, J., and Gresser, I.,** Host gene influences sensitivity to interferon action selectively for influenza virus, *Nature (London)*, 282, 660, 1980.
261. **Horisberger, M. A., Staeheli, P., and Haller, O.,** Interferon induces a unique protein in mouse cells bearing a gene for resistance to influenza virus, *Proc. Natl. Acad. Sci. U.S.A.*, 80, 1910, 1983.
262. **Staeheli, P., Horisberger, M. A., and Haller, O.,** Mx-dependent resistance to influenza viruses is induced by mouse interferon alpha and beta but not gamma, *Virology*, 132, 456, 1984.
263. **Arnheiter, H. and Haller, O.,** Mx gene control of interferon action: different kinetics of the antiviral state against influenza virus and vesicular stomatitis virus, *J. Virol.*, 47, 626, 1983.
264. **Dreiding, P., Staeheli, P., and Haller, O.,** Interferon-induced protein Mx accumulates in nuclei of mouse cells expressing resistance to influenza viruses, *Virology*, 140, 192, 1985.
265. **Horisberger, M. A., Haller, O., and Arnheiter, H.,** Interferon-dependent genetic resistance to influenza virus in mice: virus replication in macrophages is inhibited at an early step, *J. Gen. Virol.*, 50, 205, 1980.
266. **Meyer, T. and Horisberger, M. A.,** Combined action of mouse α and β interferons in influenza virus-infected macrophages carrying the resistance gene Mx, *J. Virol.*, 49, 709, 1984.
267. **Naito, S. and Ishihama, A.,** Function and structure of RNA polymerase from vesicular stomatitis virus, *J. Biol. Chem.*, 251, 4307, 1976.
268. **Steward, W. E., II, and Lockart, R. Z., Jr.,** Relative antiviral resistance induced in homologous and heterologous cells by cross-reacting interferons, *J. Virol.*, 6, 795, 1970.
269. **Rabson, A. S., Tyrell, S. A., and Levy, H.,** Inhibition of human cytomegalovirus *in vitro* by double-stranded polyribocytidylic-inosinic acid (Poly rI: rC), *Proc. Soc. Exp. Biol. Med.*, 131, 495, 1969.

270. **Kern, E. R., Overall, J. C., and Glasgow, L. A.,** Herpesvirus hominis infection in newborn mice: treatment with interferon inducer poly I.C., *Antimicrob. Agents Chemother.*, 7, 793, 1975.
271. **Youngner, J. S., Thacore, H. R., and Kelly, M. E.,** Sensitivity of ribonucleic acid and deoxyribonucleic acid viruses to different species of interferon in cell cultures, *J. Virol.*, 10, 171, 1972.
272. **Marcus, P. I., Englehardt, D. L., Hunt, J. M., and Sekellick, M. J.,** Interferon action: inhibition of vesicular stomatitis virus RNA synthesis induced by virion-bound polymerase, *Science*, 174, 593, 1971.
273. **Manders, E. K., Tiller, J. G., and Huang, A. S.,** Interferon-mediated inhibition of virion-directed transcription, *Virology*, 49, 573, 1972.
274. **Baxt, B., Sonnabend, J. A., and Bablanian, R.,** Effect of interferon on vasicular stomatitis virus transcription and translation, *J. Gen. Virol.*, 35, 325, 1977.
275. **Thacore, H. R.,** Effect of interferon on transcription and translation of vesicular stomatitis virus in human and simian cells cultures, *J. Gen. Virol.*, 41, 421, 1978.
276. **Marcus, P. I. and Sekellick, M. J.,** Interferon action. III. The rate of primary transcription of vesicular stomatitis virus is inhibited by interferon action, *J. Gen. Virol.*, 38, 391, 1978.
277. **Yamakazi, S. and Wagner, R. R.,** Action of interferon: kinetics and differential effect on viral functions, *J. Virol.*, 6, 421, 1970.
278. **Katz, L. J., Lee, S. H. S., and Rozee, K. R.,** Interferon-mediated increased lysis of L-cells by vesicular stomatitis virus, *Can. J. Microbiol.*, 20, 1077, 1974.
279. **Simili, M., De Ferra, F., and Baglioni, C.,** Inhibition of RNA and protein synthesis in interferon-treated HeLa cells infected with vesicular stomatitis virus, *J. Gen. Virol.*, 47, 373, 1980.
280. **De Ferra, F. and Baglioni, C.,** Viral messenger RNA methylated in the 5′-terminal guanosine in interferon-treated HeLa cells infected with vesicular stomatitis virus, *Virology*, 112, 426, 1981.
281. **Nilsen, T. W., Wood, D., and Baglioni, C.,** Cross-linking of viral RNA by 4′-aminomethyl-4,5′, 8-trimethylpsoralen in HeLa cells infected with encephalomyocarditis virus and the tsG114 mutant of vesicular stomatitis virus, *Virology*, 109, 82, 1981.
282. **Maheswari, R. K. and Friedman, R. M.,** Effect of interferon treatment on vesicular stomatitis virus (VSV): release of unusual particles with low multiplicity, *Virology*, 101, 399, 1980.
283. **Maheswari, R. K., Banerjee, D. K., Waetcher, C. J., Olden, K., and Friedman, R. M.,** Interferon treatment inhibits glyocosylation of a viral protein, *Nature (London)*, 287, 454, 1980.
284. **Maheswari, R. K., Jay, F. T., and Friedman, R. M.,** Selective inhibition of glycoprotein and membrane protein of vesicular stomatitis virus from interferon-treated cells, *Science*, 207, 540, 1980.
285. **Maheswari, R. K., Dempsey, A. E., Mohanty, S. B., and Friedman, R. M.,** Interferon-treated cells release vesicular stomatitis virus particles lacking glycoprotein spikes: correlation with biochemical data, *Proc. Natl. Acad. Sci. U.S.A.*, 77, 2284, 1980.
286. **Olden, K., Bernard, B. A., Turner, W., and White, S. N.,** Effect of interferon on protein glycosylation and comparison with tunicamycin, *Nature (London)*, 300, 290, 1982.
287. **Jay, F. T., Dawood, M. R., and Friedman, R. M.,** Interferon induces the production of membrane protein deficient and infectivity-defective vesicular stomatitis virions through interference in the virion assembly process, *J. Gen. Virol.*, 64, 707, 1983.
288. **Drebot, M. A., Lee, S. H. S., and Rozee, K. R.,** Differential effect of interferon on glycoprotein and membrane protein of vesicular stomatitis virus released from murine and simian cells, *J. Interferon Res.*, 4, 167, 1984.
289. **Masters, P. S. and Samuel, C. E.,** Mechanism of interferon action. Inhibition of vesicular stomatitis virus in human amnion U cells by cloned human leukocyte interferon, *Biochem. Biophys. Res. Commun.*, 119, 326, 1984.
290. **Ulker, N. and Samuel, C. E.,** Mechanism of interferon action: inhibition of vesicular stomatitis virus replication in human amnion U cells by cloned human γ-interferon. I. Effect on early and late stages of the viral multiplication cycle, *J. Biol. Chem.*, 260, 4319, 1985.
291. **Ulker, N. and Samuel, C. E.,** Mechanism of interferon action: inhibition of vesicular stomatitis virus replication in human amnion U cells by cloned human γ-interferon. II. Effect on viral macromolecular synthesis, *J. Biol. Chem.*, 260, 4324, 1985.
292. **Whitaker-Dowling, P. A., Wilcox, D. K., Widnell, C. C., and Youngner, J. S.,** Interferon mediated inhibition of virus penetration, *Proc. Natl. Acad. Sci. U.S.A.*, 80, 1083, 1983.
293. **Wood, J. N. and Hovanessian, A. G.,** Interferon enhances 2-5A synthetase in embryonal carcinoma cells, *Nature (London)*, 282, 74, 1979.
294. **Holmes, S. L. and Gupta, S. L.,** Interferon action in human fibroblasts: induction of 2′5′-oligodenylate synthetase in the absence of detectable protein kinase activity, *Arch. Virol.*, 72, 137, 1982.
295. **Hovanessian, A. G., Meurs, E., and Montagnier, L.,** Lack of systematic correlation between the interferon mediated antiviral state and the levels of 2-5A synthetase and protein kinase in three different types of murine cells, *J. Interferon Res.*, 1, 179, 1981.

296. **Robbins, C. H., Kramer, G., Saneto, R., Hardesty, B., and Johnson, H. M.,** Dissociation of protein kinase activity and the induction of the antiviral state in a cell line responsive to the antiviral effects of interferon, *Biochem. Biophys. Res. Commun.*, 103, 103, 1981.
297. **Hightower, L. E. and Bratt, M. A.,** Protein synthesis in Newcastle disease virus-infected chicken embryo cells, *J. Virol.*, 13, 788, 1974.
298. **Isaacs, A., Porterfield, J. S., and Baron, S.,** The influence of oxygenation on virus growth. II. Effect on the antiviral action of interferon, *Virology*, 14, 450, 1961.
299. **Grossberg, S. E. and Scherer, W.,** Inapparent viral infection of cells *in vitro*. II. An interferon produced in chicken embryonic cell cultures inoculated with Japanese encephalitis virus, *Am. J. Pathol.*, 45, 519, 1964.
300. **Portnoy, J. and Merigan, T. C.,** The effect of interferon and interferon inducers on avian influenza, *J. Infect. Dis.*, 124, 545, 1971.
301. **Merigan, T. C.,** Purified interferons: physical properties and species specificity, *Science*, 145, 811, 1964.
302. **Schachter, N., Galin, M., Weissenbacher, M., Baron, S., and Billiau, A.,** Comparison of antiviral action of interferon, interferon inducers and IDU against herpes simplex and other viruses, *Ann. Opthalmol.*, 2, 795, 1970.
303. **Ahl, R. and Rump, A.,** Assay of bovine interferons in cultures of the porcine cell line IB-RS-2, *Infect. Immunol.*, 14, 603, 1976.
304. **Shatkin, A. J.,** Capping of eukaryotic mRNAs, *Cell*, 9, 645, 1976.
305. **Shatkin, A. J.,** Animal RNA viruses: genome structure and function, *Annu. Rev. Biochem.*, 43, 643, 1974.
306. **Traub, A., Diskin, B., Rosenberg, H., and Kalmer, E.,** Isolation and properties of the replicase of encephalomyocarditis virus, *J. Virol.*, 18, 375, 1976.
307. **Bishop, J. M. and Koch, G.,** Purification and characterization of poliovirus-induced infectious double-stranded RNA, *J. Biol. Chem.*, 242, 1736, 1967.
308. **Bishop, J. M., Koch, G., Evans, B., and Merriman, M.,** Poliovirus replicative intermediate: structural basis of infectivity, *J. Mol. Biol.*, 46, 235, 1967.
309. **Savage, T., Granboulan, N., and Girard, M.,** Architecture of the poliovirus replicative intermediate RNA, *Biochimie*, 53, 533, 1971.
310. **Levy, H. B. and Carter, W. A.,** Molecular basis of the action of interferon, *J. Mol. Biol.*, 31, 561, 1968.
311. **Levy, H. B.,** Studies on the mechanism of interferon action. II. The effect of interferon on some early events in mengovirus infection in L cells, *Virology*, 22, 575, 1964.
312. **Gauntt, C. J. and Lockart, R. Z., Jr.,** Destruction of L-cell by mengovirus: use of interferon to study the mechanism, *J. Virol.*, 2, 567, 1968.
313. **Kerr, I. M., Friedman, R. M., Esteban, M., Brown, R. E., Ball, L. A., Metz, D. H., Risby, D., Tovell, D. R., and Sonnabend, J. A.,** The control of protein synthesis in interferon-treated infected cells, *Adv. Biosci.*, 11, 109, 1973.
314. **Falcoff, R. and Sanceau, J.,** Mengovirus-induced inhibition of cell protein synthesis and subsequent recovery of cellular activity in interferon pretreated cells, *Virology*, 98, 433, 1979.
315. **Vaquero, C., Sanceau, J., and Falcoff, R.,** Protein synthesis in extracts from interferon-treated Mengo virus infected cells, *Biochem. Biophys. Res. Commun.*, 107, 974, 1982.
316. **Aujean, O., Sanceau, J., Falcoff, E., and Falcoff, R.,** Location of enhanced ribonuclease activity and of a phosphoprotein kinase in interferon-treated Mengovirus-infected cells, *Virology*, 92, 583, 1979.
317. **Vaquero, C., Aujean-Rigaud, D., Sanceau, J., and Falcoff, R.,** Effect of interferon on transient shut-off of cellular RNA and protein synthesis induced by Mengo virus infection, *Antiviral Res.*, 1, 123, 1981.
318. **Williams, B. R. G., Golgher, R. R., Brown, R. E., Gilbert, C. S., and Kerr, I. M.,** Natural occurrence of 2-5A in interferon-treated EMC virus-infected L cells, *Nature (London)*, 282, 582, 1979.
319. **Knight, M., Cayley, P. J., Silverman, R. H., Wreschner, D. H., Gilbert, C. S., Brown, R. E., and Kerr, I. M.,** Radioimmune, radiobinding and HPLC analysis of 2-5A and related oligonucleotides from intact cells, *Nature (London)*, 288, 189, 1980.
320. **Williams, B. R. G. and Kerr, I. M.,** Inhibition of protein synthesis by 2′-5′-linked adenine oligonucleotides in intact cells, *Nature (London)*, 88, 1978.
321. **Hovanessian, A. G., Wood, J., Meurs, E., and Montagnier, L.,** Increased nuclease activity in intact cells treated with pppA2′p5′A2′ p5′A, *Proc. Natl. Acad. Sci. U.S.A.*, 76, 3261, 1979.
322. **Hovanessian, A. G. and Wood, J. N.,** Anticellular and antiviral effects of pppA(2′p5′A)n, *Virology*, 101, 81, 1980.
323. **Czarniecky, C. W., Sreevalsan, T., Friedman, R. M., and Panet, A.,** Dissociation of interferon effects on murine leukemia virus and encephalomyocarditis virus replication in mouse cells, *J. Virol.*, 37, 827, 1981.
324. **Cayley, P. I., Knight, M., and Kerr, I. M.,** Virus-mediated inhibition of the ppp(A2′p)nA system and its prevention by interferon, *Biochem. Biophys. Res. Commun.*, 104, 376, 1982.

325. **Silverman, R. H., Cayley, P. J., Knight, M., Gilbert, C. S., and Kerr, I. M.,** Control of the ppp(A2′p)nA system in HeLa cells. Effects of interferon and virus infection, *Eur. J. Biochem.*, 124, 131, 1982.
326. **Yau, P. M. P., Godefroy-Colburn, T., Birge, C. H., Ramabhadran, T. V., and Thach, R. E.,** Specificity of interferon action in protein synthesis, *J. Virol.*, 27, 648, 1978.
327. **Jen, G., Birge, C. H., and Thach, R. E.,** Comparison of initiation rates of encephalomyocarditis virus and host protein synthesis in infected cells, *J. Virol.*, 27, 640, 1978.
328. **Ramabhadran, T. V. and Thach, R. E.,** Translational elongation rate changes in encephalomyocarditis virus-infected and interferon-treated cells, *J. Virol.*, 39, 573, 1981.
329. **Buckler, C. E., Wong, K. T., and Baron, S.,** Induction of the interferon system by various inducers, *Proc. Soc. Exp. Biol. Med.*, 127, 1258, 1968.
330. **Muñoz, A. and Carrasco, L.,** Effect of interferon treatment on blockade of protein synthesis induced by poliovirus infection, *Eur. J. Biochem.*, 137, 623, 1983.
331. **Muñoz, A., Harvey, R., and Carrasco, L.,** Cellular RNA is not degraded in interferon-treated HeLa cells after poliovirus infection, *FEBS Lett.*, 160, 87, 1983.
332. **Strauss, J. H., Jr., Burge, B. W., Pfefferkom, E. R., and Darnell, J. E., Jr.,** Identification of the membrane protein and 'core' protein of Sindbis virus, *Proc. Natl. Acad. Sci. U.S.A.*, 59, 533, 1968.
333. **Glasgow, L. A., Hanshaw, J. B., Merigan, T. C., and Petralli, J. K.,** Interferon and cytomegalovirus *in vivo* and *in vitro, Proc. Soc. Exp. Biol. Med.*, 125, 843, 1967.
334. **Stewart, W. E., II, Allen, R., and Sulkin, S. E.,** Persistent infection in bats and bat cell cultures with Japanese encephalitis virus, *Bacteriol. Proc.*, V, 283, 1969.
335. **Ho, M. and Enders, J. F.,** Further studies on an inhibitor of viral activity appearing in infected cell cultures and its role in chronic viral infection, *Virology*, 9, 446, 1959.
336. **Martin, E. M. and Sonnabend, J. A.,** Ribonucleic acid synthesis of double-stranded arbovirus ribonucleic acid, *J. Virol.*, 1, 97, 1967.
337. **Friedman, R. M.,** Interferon inhibition of arbovirus protein synthesis, *J. Virol.*, 2, 1081, 1968.

INDEX

D

E

F

G

H

I

K

L

M

N

O

P

R

S

T

V

W

Y